PMI-ACP® Exam Prep

A Course in a Book® for Passing the
PMI Agile Certified Practitioner (PMI-ACP)® Exam

By Mike Griffiths, PMI-ACP, PMP

RMC PUBLICATIONS™

PMI-ACP® Exam Prep

Printed in the United States of America

Fourth Printing

ISBN: 978-1-932735-98-7

Library of Congress Control Number: 2015948973

Figures 1.13, 2.16, 2.23, 2.24, 2.25, 2.26, 3.12, 3.13, 4.9, 4.14, 5.19, 5.20, 5.24, 5.26, Umbrella Man, Shovel Man, Vision Man, and Cart Man were adapted from images originally published by Leading Answers, Inc., original image copyright © 2012 Leading Answers, Inc. Images reproduced and adapted with permission from Leading Answers, Inc., www.leadinganswers.com.

Agile FASTrack is a trademark of RMC Project Management, Inc.

PMI, PMI-ACP, PMP, PMBOK, and SeminarsWorld are registered marks of the Project Management Institute, Inc.

This publication uses the following terms trademarked by the Project Management Institute, Inc.: Project Management Institute (PMI)®, PMI Agile Certified Practitioner (PMI-ACP)®, Project Management Professional (PMP)®, and *A Guide to the Project Management Body of Knowledge (PMBOK® Guide)*.

PRINCE2 is a registered trademark of AXELOS Limited.

Phone: 952.846.4484
Fax: 952.846.4844
Email: info@rmcls.com
Web: www.rmcls.com

Dedication

To Jake and Samantha.

Jake, your hard work and perseverance are a true inspiration. Whenever I feel disheartened about a task I think of how hard you try, how far you have come, and where you might go. Never stop trying, never stop working; you can do great things.

Sam, thank you for supporting me. You insulate the world from my random ideas and crazy plans. Thanks for being my best friend all these years.

Table of Contents

Table of Contents

© 2015 RMC Publications, Inc • 952.846.4484 • info@rmcls.com • www.rmcls.com

Table of Contents

© 2015 RMC Publications, Inc • 952.846.4484 • info@rmcls.com • www.rmcls.com

Author's Acknowledgments

I would like to recognize David Anderson for encouraging me to start blogging about agile project management in 2005 when we were working on the APLN board together. Thanks, it has opened many doors for me.

I am grateful, too, to Doug DeCarlo, who got me started writing articles for ProjectManagement.com and put me in touch with RMC and the discussions for this book.

The first edition of this book would not have happened without the help, support, and encouragement of my first editor, Laurie Diethelm. She intuitively understood agile and servant leadership, all the things I was trying to explain in the book—and she modeled them perfectly. She transformed my disjointed ramblings into a consistent voice and steered my rants toward useful exam preparation advice. However, the biggest thing she did was to provide consistent encouragement and a rational voice when competing demands for my time threatened the timeline.

Rose Brandt did a great job fact-checking the first edition of the book and then acting as the editor of the second edition. I am grateful for her patience and professionalism. We fixed up and improved many aspects of the second edition, in most part due to Rose's hard work.

Jason Craft and Whitney Thulin transformed my scribbles into comprehensible diagrams, and this book's consistent appearance is in large part due to their hard work.

Thanks are due to the reviewers of the first edition, too, who provided valuable feedback on sometimes incomplete chapters, making adaptation and improvement possible.

Finally, thanks to the indexers, typesetters, and publishers whose roles and work often go unrecognized, but without whom, quite literally, there would be no book.

To these people and those I have not directly mentioned here, but who still helped, thank you. My name may be on the cover, but your work and influence are in these pages, and I am grateful.

—Mike Griffiths

Publisher's Acknowledgments

RMC Learning Solutions (formerly RMC Project Management) would like to thank the following people for their invaluable contributions to this book:

Editing	Production/Layout Graphics
Rose Brandt	Whitney Thulin
Deborah Kaufman	Jason Craft

Reviewers (First Edition)

Dr. Alistair Cockburn
Co-author of the Manifesto for Agile Software Development, co-founder of the International Consortium for Agile, PMI-ACP Steering Committee Member

Dennis Stevens
PMI-ACP Steering Committee Member, PMI Agile Community of Practice Council Leader
Enterprise Agility Coach, Leading Agile

Michele Sliger, PMI-ACP, PMP
PMI-ACP Steering Committee Member, co-author of The Software Project Manager's Bridge to Agility
Sliger Consulting

Theofanis C. Giotis, PMI-ACP, MSc, Ph.D. C., PMP, MCT, PRINCE2
12PM Consulting

Ursula Kusay, PMI-ACP, PMP
CSC Deutschland Akademie GmbH

Anton Josef Müller, PMP, CBAP
CSC Deutschland Akademie GmbH

Andreas Buzzi, PMI-ACP, PMP, CSM, and OpenGroup certified Architect & IT Specialist

Carlos Sánchez-Sicilia, PMP, CBAP, CGEIT, CRISC

Jeffrey S. Nielsen, PMI-ACP, CSM, PgMP, PMP, PMI-RMP, MSCIS

Margo Kirwin, PMI-ACP, PMI-PBA, PMP, CPLP

Additional Contributors (First Edition)

Eric Rudolf, CSM, CSPO
Barbara A. Carkenord, PMI-ACP, PMP, PMI-PBA, CBAP

About the Author

 Mike Griffiths is a world-renowned project manager, trainer, consultant, and writer, holding multiple project management and agile-related certifications. Mike was on the original PMI-ACP Steering Committee, which defined the agile-related knowledge, skills, tools, and techniques to be tested on the PMI-ACP exam. He helped create *A Guide to the Project Management Body of Knowledge (PMBOK® Guide)—Fifth Edition*, as well as the *Software Extension to the PMBOK® Guide*.

Mike frequently presents at PMI® Global Congress conferences, and teaches a two-day Agile Project Management course for the Project Management Institute (PMI) as part of their traveling SeminarsWorld® program. He also helped start the PMI Agile Community of Practice, which is now the largest PMI community in existence.

Mike helped create the agile method DSDM (Dynamic Systems Development Method) and has been using agile methods including FDD, Scrum, and XP for the last 20 years. Mike also served on the Board of Directors for the Agile Alliance and the Agile Project Leadership Network (APLN). In addition to extensive agile methods experience, Mike is very active in traditional project management, holding both the PMP® and PRINCE2® certifications.

About RMC Learning Solutions

RMC Learning Solutions (formerly RMC Project Management) develops and trains project managers, business analysts, and agile practitioners by providing the hard skills, soft skills, and business knowledge necessary for them to succeed in their careers. Since 1991, RMC has helped more than 750,000 individuals—as well as executives across the Fortune 500 and in government agencies—accomplish the following:

» Develop practical knowledge immediately applicable to their careers.
» Deliver organizational results using proven techniques and best practices.
» Attract, develop, and retain top talent to work on key initiatives.
» Consistently bring projects in on time and under budget.
» Increase the adoption of training through coaching and mentoring.

Today, RMC delivers a wide range of project management, business analysis, and agile training in multiple learning formats including traditional classroom training, live online training, self-directed e-learning, books, and software. The company also continues to reinforce its training via ongoing coaching, mentoring, and organizational transformation services.

What Makes RMC Unique

In addition to the above, RMC Learning Solutions separates itself from what has become a very crowded training marketplace by offering the following value-added advantages:

» **Proven techniques.** RMC provides individuals and teams with the opportunity to learn and use highly effective project management, business analysis, and agile practices that can be applied immediately on their current projects. We listen to students and clients, and integrate their ideas and perspectives to enrich each and every learning experience.

» **Outcome-based learning.** All RMC instruction and content focuses on a continuous cycle of learning. We plan your learning, set goals, and teach—then measure progress toward goals and adjust accordingly. RMC incorporates real-time feedback into all of its programs wherever possible, and is always monitoring for additional learning opportunities.

» **Professional instructors.** RMC actively recruits high-performing instructors who are champions of—and experts in—their respective disciplines. All RMC instructors have a deep conceptual understanding of their practice area, and use a variety of strategies designed to help students understand AND apply what they are learning in the real world.

» **An international reach.** Through our extensive international distributor and training provider network, RMC is able to offer its training in 18 languages and more than 60 regions across the globe. This ensures all RMC students receive the same quality of training and delivery, and the same training experience—no matter where they may be located.

» **Multiple ways to learn.** In addition to traditional instructor-led training, RMC offers a variety of learning modes to meet both individual and corporate training needs. From live online classes and self-directed e-learning courses to books, software, and templates, RMC offers students the opportunity to learn in multiple formats and at multiple price points.

We Need You to Help Us Stop Copyright Infringement

As the publisher of some of the best-selling exam prep books on the market, RMC is also, unfortunately, one of the most illegally copied. It is true that many people use our materials legally and with our permission to teach exam preparation. However, from time to time, we are made aware of others who copy our exam questions, tips, and other content illegally and use them for their own financial gain.

If you recognize any of RMC's proprietary content being used in other exam prep materials or courses, please notify us at copyright@rmcls.com immediately. We will do the investigation. Please also contact us at the e-mail address above for clarification on how to use our materials in your class or study group without violating any laws.

Contact Us

We love to hear your feedback. Is there anything in this book that you wish was expanded? Is there anything we focus on too much, or is there anything not covered that you think should be here? We would love to hear from you. Send us an e-mail at agileprep@rmcls.com.

Free Updates and Extras

Purchasing this book gives you access to valuable supplemental study tools for the PMI-ACP exam at shop.rmcls.com/agileprep. Be sure to take advantage of those resources, which include a practice quiz that can help you determine if you are ready to take the exam. Also, be sure to check back from time to time, since we will continue to add more resources to that site. (You will need your copy of this book the first time you access that website.)

INTRODUCTION

About the PMI-ACP® Exam

Welcome to the second edition of *PMI-ACP® Exam Prep*. Since the first edition of this book, the PMI-ACP exam has gone from being a foray into new agile territory for PMI, to their fastest-growing credential ever. This credential has been growing even faster than the PMP credential, and it is now set to be a significant qualification in the domain of project delivery—not just for leaders and managers, but also for agile practitioners and team members.

The PMI Agile Certified Practitioner is now an established agile certification sought by professionals and hiring managers alike. In a marketplace that was filled with a confusing mix of easy-to-obtain credentials, the PMI-ACP has emerged as a meaningful measure of agile experience and knowledge. PMI has achieved this by requiring education, relevant experience, and training—as well as by applying the rigor of ISO/IEC 17024 requirements to ensure good question design and rigorous examination management. So there are no shortcuts, easy options, or free passes to earning this certification. If you are just starting on your path toward obtaining the PMI-ACP, this might sound daunting—but it's also what makes the credential so valuable. Everyone knows you have to earn it, and that it means you have verified skills. Fortunately, the book you are holding is the best resource you can find to help you pass the exam and obtain your PMI-ACP certification.

Since the PMI-ACP exam was first offered in 2011, the agile techniques that it tests have been validated by surveying thousands of practitioners who use agile methods. PMI wanted to make sure the techniques included in the exam content outline were in fact common practice, and used in proportions that matched the weightings of the exam questions. So in 2014 they conducted a role delineation study.

This survey ended up strongly validating the exam content. However, it also highlighted the importance of having an agile mindset (being agile) over merely using agile practices (doing agile). To reflect this, PMI created a new domain called "Agile Principles and Mindset." The material covered in this domain isn't new—the exam coverage has not expanded. Instead, the topics that were previously covered in an "Agile Framework" category outside of the exam domains have now been folded into their own domain.

Since the first publication of this book I have been fortunate to receive hundreds of letters of thanks and feedback. The book is consistently rated as the best exam preparation resource available for the PMI-ACP exam, and for many people this book is the only resource they use to prepare for, and pass, the exam. This community of successful readers has provided many thoughtful suggestions for improvements to the book. Their suggestions for more coverage of lean and Kanban approaches and a host of other ideas have been analyzed, researched, and applied to this new version of the book.

In short, as the PMI-ACP exam has gained in popularity it has been revisited and improved—and so too has this book. In the true agile spirit, through use, review, and adaptation we have produced a better increment of the product.

Welcome to the first step on your journey. The PMI-ACP is an important, valued credential, and you have what you need to prepare—so let's get started!

About the Exam

In this introduction, we'll discuss the scope of the PMI-ACP exam at a high level. The exam consists of 120 questions, and you will have three hours in which to complete it. Just like on a project, when taking the exam you need to know what is in scope, what is out of scope, and what you need to do to be successful. Let's start by addressing the qualification requirements, the organization of the exam content, and the exam's key assumptions.

Qualifying to Take the Exam

Passing the PMI-ACP exam is just one component of achieving your PMI-ACP certification. The other components are education, general project experience, agile project experience, and training. In order to qualify to take the exam, you need to have all of the following:

Education	General Project Experience	Agile Project Experience	Training in Agile Practices
High school diploma, associate's degree, or equivalent	2,000 hours (about 12 months) of project team experience within the last five years	1,500 hours (about 8 months) of agile project team or agile methodology experience within the last three years	21 hours

PMI may make changes to aspects of the exam, including the qualification requirements, the application process, the passing score, and the breakdown of questions in each domain. For the latest information, please visit www.pmi.org and read your authorization notice carefully. Any differences between what is listed here and what is communicated by PMI should be resolved in favor of PMI's information.

With 1,500 hours of agile experience, you may have already been exposed to many of the concepts that will be tested on the exam. However, there are several different agile methods (e.g., Scrum, XP, Kanban, etc.), and each one focuses on different concepts and uses different terminology. This means that the terms and ideas you encounter on the exam might be different from what you are familiar with from your projects. Also, the exam questions will draw from a range of agile methodologies—so knowing one methodology really well isn't enough to pass the exam. As you go through this book, make a note of any terms or concepts that differ from your real-world experience so you will be prepared when you see them on the exam.

Exam Content

After reviewing your qualifications for taking the exam, the next step is to find out what you need to know to pass the exam. PMI provides an exam content outline (the "PMI Agile Certified Practitioner (PMI-ACP)® Examination Content Outline") on its website, www.pmi.org.[1] This document outlines what you need to know for the exam, organizing it into seven domains, a set of Tools and Techniques, and a set of Knowledge and Skills. Let's explore these components of the exam content in more detail.

The Seven Domains

The exam content outline organizes the material that will be tested by breaking it down into seven thematically related groups, called domains. Here are the key points to understand about these domains:

» The domains are provided simply to organize the vast world of agile and show how PMI intends the exam material to be understood and taught. The exam won't test the domains themselves or how the topics are grouped by domain. So don't spend time trying to remember what is covered in which domain. The domains are just a study tool; they won't be tested on the exam.

» You can use the domains to identify the strengths and gaps in your agile knowledge. If your agile knowledge is uneven, you might be an expert in certain areas, but still fail the exam. To pass, you need to have a good basic understanding of all seven domains.

» You can use the domains to help you allocate your study time, since the questions on the exam are weighted by domain. (This is the only information PMI provides about how different topics are weighted on the exam.) For example, domain II has the most weight (20 percent) and domain VII has the least weight (9 percent). Now, this doesn't mean you can skip studying domain VII and still ace the exam with a score of 91 percent. The exam weighting isn't that straightforward. But it does mean that you should spend more time making sure you thoroughly understand the topics in domain II, since they will be required to accurately answer many more exam questions than the topics in domain VII.

Here are the name and weighting of each domain as defined in the exam content outline:

Weighting of Domains for PMI-ACP Exam

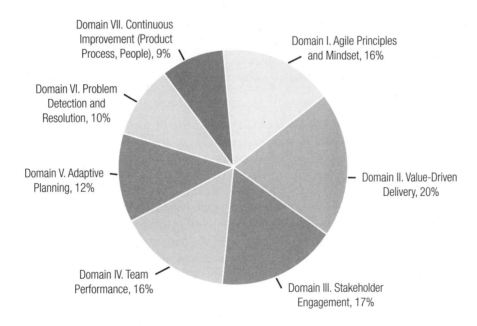

Domain VII. Continuous Improvement (Product Process, People), 9%

Domain I. Agile Principles and Mindset, 16%

Domain VI. Problem Detection and Resolution, 10%

Domain V. Adaptive Planning, 12%

Domain II. Value-Driven Delivery, 20%

Domain IV. Team Performance, 16%

Domain III. Stakeholder Engagement, 17%

Domain Overviews

Here's an overview of what is covered in each domain:

I.   **Agile Principles and Mindset:** This domain focuses on the agile mindset, its fundamental values and principles, the agile methodologies, and agile leadership.

II.   **Value-Driven Delivery:** This domain deals with maximizing business value, including prioritization, incremental delivery, testing, and validation.

III.   **Stakeholder Engagement:** This domain focuses on working with the project stakeholders, including establishing a shared vision, collaboration, communication, and interpersonal skills.

IV.   **Team Performance:** This domain focuses on building high-performing teams, including how teams form and develop mastery, team empowerment, collaborative team spaces, and performance tracking.

V.   **Adaptive Planning:** This domain deals with sizing, estimating, and planning, including adaptive planning, progressive elaboration, value-based analysis and decomposition, and release and iteration planning.

VI.   **Problem Detection and Resolution:** This domain deals with the agile practices used to prevent, identify, and resolve threats and issues, including catching problems early, tracking defects, managing risk, and engaging the team in solving problems.

VII.   **Continuous Improvement (Product, Process, People):** This final domain focuses on continuous improvement in the areas of product, process, and people, including process analysis and tailoring, product feedback methods, reviews, and retrospectives.

Chapters 1 through 7 of this book are aligned with these seven domains, with each chapter covering the domain of the equivalent number. At the start of each of these chapters, we provide a summary of the key information for that domain, including:

» The weighting of that domain for the exam (including the number of questions)
» The key topics in the domain that are most important to know for the exam
» A quick summary of the tasks that fall within that domain, as defined in the exam content outline

As described in the "How This Book Is Organized" section of this chapter, you can use the domain summaries to help allocate your study time and check your understanding of the exam concepts. When you finish reading each chapter, return to the chapter summary and make sure you understand all the key topics and tasks, based on the material discussed in the chapter.

The Agile Toolkit (T&Ts and K&Ss)

In addition to the seven domains, the exam content outline also provides two lists of topics that will be tested on the exam—tools and techniques (T&Ts) and knowledge and skills (K&Ss).

» **Tools and Techniques (T&Ts)**—These are specific practices you should be able to do or use, such as story maps, active listening, or daily stand-ups. The exam will test your ability to apply these practices, often with brief scenario questions that ask "what should you do" or "what happens next?"

© 2015 RMC Publications, Inc • 952.846.4484 • info@rmcls.com • www.rmcls.com

» **Knowledge and Skills (K&Ss)**—These are broader areas you should know or understand, such as incremental delivery, problem solving, or prioritization. Every question on the exam will be implicitly based on one or more of the knowledge and skills on this list. You can think of these topics as the invisible groundwork or understructure required to answer the exam questions. Also, like the T&Ts, you may encounter some scenario questions that ask about these topics directly.

Although the exam content outline doesn't associate the T&Ts and K&Ss with specific domains, in this book we have assigned each of these topics to the appropriate domain so that you don't have to guess which domain they are in. (We'll explain this breakdown shortly.)

Note: The exam content outline refers to only the T&Ts as the "agile toolkit," using the term "toolkit" interchangeably with "tools and techniques." However, this term won't be tested on the exam—it is simply used to organize the material. So in this book we'll use the term "toolkit" to refer all the T&Ts and the K&Ss, not just the T&Ts, since we'll generally be referring to these concepts as one group, and "toolkit" is easier to read than "tools and techniques and knowledge and skills."

The seven domains and the toolkit may seem overwhelming at first—but remember, if you have the required agile training and project experience, it's likely that you have already been exposed to many of the concepts the exam will be testing.

Agile Toolkit by Domain

To do well on the exam, you need to understand all the toolkit topics. To make sure you don't miss any of them, they are all listed in the chapter summaries and flagged with special "T&T" or "K&S" icons in the text. However, you don't want to only focus on the toolkit, since that won't cover everything you need to know for the exam. Many key agile concepts—such as distributed teams and self-organization—aren't in the toolkit; instead, they might be integrated into the tasks for each domain. (Note that any of the material discussed in this book could show up in an exam question, with the exception of the "background information" boxes described below.)

Nevertheless, the agile toolkit is a key study tool, and the table on the next two pages shows our interpretation of how the toolkit items fall into the different domains; this is the organization that we will follow in this book. However, the exam content outline not only doesn't specify which topics fall in which domain, it even associates some agile concepts with multiple domains. (That's understandable, since many of these concepts are important for multiple domains.) So our breakdown is necessarily somewhat arbitrary. However, you can use it as a study guide to make sure you understand the topics that are relevant for each domain. You may also find this breakdown to be valuable in helping you create a mental framework to group the concepts, so that it is easier to understand how they are related to each other.

Bear in mind that although you might encounter any of these topics on the exam, they are not equally important. As you read the chapters, you'll see that some of these concepts are much more essential for agile practice than others—and the exam will emphasize the material that's most important for agile practice, not the tricky details. For example, you're likely to see more exam questions on release and iteration planning than on the details of project accounting.

In the same vein, the number of toolkit items for a given domain isn't a reflection of that domain's importance for the exam. (For that, refer to the domain weightings given above.)

Domain	Tools and Techniques		Knowledge and Skills
I. Agile Principles and Mindset *(Chapter 1)*	Servant leadership		Agile frameworks and terminology Agile methods and approaches Agile values and principles Leadership
II. Value-Driven Delivery *(Chapter 2)*	Agile tooling Compliance Continuous integration Cumulative flow diagrams Customer-valued prioritization EVM for agile projects Frequent verification and validation Kanban board Kano analysis Minimal marketable feature (MMF)	Minimal viable product (MVP) MoSCoW Relative prioritization/ranking ROI/NPV/IRR Task board Testing, including exploratory and usability WIP limits Work in progress	Agile contracting Agile project accounting principles Incremental delivery Managing with agile KPIs Prioritization Regulatory compliance
III. Stakeholder Engagement *(Chapter 3)*	Active listening Agile modeling Brainstorming Chartering Collaboration Collaboration games Conflict resolution Definition of done Emotional intelligence Information radiator	Negotiation Personas Social media-based communication Two-way communications (trustworthy, conversation-driven) Wireframes Workshops	Agile project chartering Assessing and incorporating community and stakeholder values Communication management Facilitation methods Knowledge sharing/written communication Participatory decision models (convergent, shared collaboration) Stakeholder management

Domain	Tools and Techniques		Knowledge and Skills
IV. Team Performance *(Chapter 4)*	Adaptive leadership Burndown/burnup charts Osmotic communication for colocated and/or distributed teams	Team space Velocity	Building agile teams Developmental mastery models (Tuckman, Dreyfus, Shu-Ha-Ri) Global, cultural, and team diversity Physical and virtual co-location Team motivation Training, coaching, and mentoring
V. Adaptive Planning *(Chapter 5)*	Affinity estimating Architectural spike Backlog grooming/ refinement Daily stand-ups Ideal time Iteration and release planning Product roadmap Progressive elaboration	Relative sizing/story points/T-shirt sizing Requirements reviews Risk-based spike Story maps Timeboxing User stories/backlog Wideband Delphi/ planning poker	Agile discovery Agile sizing and estimation Value-based analysis and decomposition
VI. Problem Detection and Resolution *(Chapter 6)*	Control limits Cycle time Defect rate Lead time Risk-adjusted backlog	Risk burndown graphs Throughput/ productivity Variance and trend analysis	Problem solving
VII. Continuous Improvement *(Chapter 7)*	Approved iterations Feedback methods Fishbone diagram analysis Five Whys Kaizen Learning cycle	Pre-mortem (rule setting, failure analysis) Process tailoring Product feedback loop Retrospectives, intraspectives Reviews Value stream mapping	Agile hybrid models Continuous improvement PMI's Code of Ethics and Professional Conduct Principles of systems thinking (complex, adaptive, chaos) Process analysis Self-assessment tools and techniques

Exam tip

Remember that this breakdown won't be tested on the exam. You won't be asked to list the seven domains. You won't be asked whether "kaizen" falls in domain V or VII. You won't be asked whether "team motivation" is a T&T or a K&S. In fact, the domains and the toolkit won't even be mentioned on the exam—they are simply the background structure that PMI uses to organize the exam content. This information is only important because it shows what topics are covered on the exam, and how those topics are weighted.

Exam Assumptions

The following key assumptions can be extremely helpful when answering questions on the PMI-ACP exam, especially tricky situational questions.

1. Despite the amount of project experience and training required to qualify for the exam, the questions will focus on basic agile projects. Complex topics such as scaling agile practices or using an agile methodology outside of standard small project implementations are not covered on the exam. Keep this in mind for the context of the exam questions. In essence, the exam will test you on plain-vanilla agile.

2. The exam attempts to be methodology-agnostic. So unless a question specifies otherwise, when considering the answer options, think in terms of generic agile rather than a specific methodology.

3. Each agile methodology has a unique vocabulary, and the exam questions will use a mixture of terms from the most common methodologies—such as "sprint" or "ScrumMaster"—along with the generic agile terms for those concepts, such as "iteration" or "team leader." So to understand the exam questions, you'll need to be familiar with both Scrum and XP terms (as explained in chapter 1), as well as the many roughly equivalent terms for agile team roles (as explained in chapter 4).

Exam tip

To prepare you for the range of terms you'll encounter on the exam, we'll be using terms from different agile methodologies and generic agile interchangeably in this book. So you may see "sprint" and "iteration" or "ScrumMaster" and "team leader" used interchangeably, even in the same discussion. If your experience with agile is limited to one approach, this will help you learn to recognize the other terms that have the same meaning as the terms you are familiar with. Although it might seem odd that we use terms from different methodologies interchangeably, we are doing this intentionally, to make sure you won't have to struggle with any of the terms used in the exam questions.

This Book's Approach

Now that we've talked about the scope of the exam and its core assumptions, it can also be helpful to understand the approach and structure of this book.

How PMI Exams Are Created

To explain the approach I've taken in this book, I'll need to start by describing how PMI credential exams are created. PMI starts by forming a steering committee of experts who create an exam content outline for the new credential. This outline defines the main themes (domains) that will be examined, the tasks within each domain, and the toolkit that will be tested. From a project perspective, this is like defining a project by coming up with a list of what is in scope. As a member of the steering committee for the PMI-ACP exam, I participated in this initial definition process.

The next step is writing the questions—or using PMI's term, "item writing." PMI keeps the exam designers who are on the steering committee separate from the item writers. One reason for this is that the item writers have privileged insights into the exam, so they are not allowed to provide any exam preparation materials or training, which could create a conflict of interest.

So the exam designers simply pass the exam content outline they have developed to the item writers, along with a list of reference books that describe the topics to be examined. The item writers then use these reference books, as well as other sources, to create the exam questions.

Reference Books and Exam Topics

The twelve reference books we selected for the current version of the PMI-ACP exam are as follows:[2]

- » *Agile Estimating and Planning*, by Mike Cohn
- » *Agile Project Management: Creating Innovative Products*, 2nd ed., by Jim Highsmith
- » *Agile Retrospectives: Making Good Teams Great*, by Esther Derby and Diana Larsen
- » *Agile Software Development: The Cooperative Game*, 2nd ed., by Alistair Cockburn
- » *Coaching Agile Teams: A Companion for ScrumMasters, Agile Coaches, and Project Managers in Transition*, by Lyssa Adkins
- » *Effective Project Management: Traditional, Agile, Extreme*, by Robert K. Wysocki
- » *Exploring Scrum: The Fundamentals*, by Dan Rawsthorne with Doug Shimp
- » *Kanban in Action*, by Marcus Hammarberg and Joakim Sunden
- » *Kanban: Successful Evolutionary Change for Your Technology Business*, by David J. Anderson
- » *Lean-Agile Software Development: Achieving Enterprise Agility*, by Alan Shalloway, Guy Beaver, and James R. Trott
- » *The Software Project Manager's Bridge to Agility*, by Michele Sliger and Stacia Broderick
- » *User Stories Applied: For Agile Software Development*, by Mike Cohn

If you're wondering, "Oh no, do I need to read all those books!?" don't worry; the answer is no. The book you're holding now covers all the material the exam will test from these reference books, as well as the additional information you'll need to know from other sources. It gathers all that information together in one place, presented with a common voice and supported with relevant examples.

Of course, I'm not saying that you shouldn't read any of the reference books, if you have the time and the inclination. They are all useful, and any of them will greatly expand your understanding of agile. However, the only book you need to pass the exam is this one.

The steering committee chose these books because we thought they were the best overall resources for the exam topics and how people use agile methods today. However, these books contain much more information than you need to pass the PMI-ACP exam. As illustrated below, the information that will be tested on the exam comprises a small subsection of the material included in each book.

The Exam Topics Compared to the Resource Books

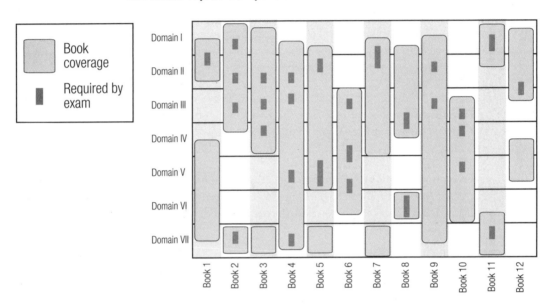

In this diagram, the vertical columns represent the reference books, and the horizontal rows represent the exam domains. The light green boxes represent all the material covered in a particular book, and the dark green rectangles represent the portions of that book that are tested on the PMI-ACP exam. Now, this diagram just illustrates the concept—it isn't trying to depict what exam material is actually covered in each book. But as you can see here, the reference books cover a lot of content that isn't tested on the exam. Although we considered those books to be the best resources for the exam topics, only a small subset of the information in each book is actually tested on the exam. Furthermore, the exam also tests many terms and concepts that aren't covered in any of the reference books.

So this book is the only resource that covers *all* the content required for the exam—whether or not it is in the resource books. It also includes background information and exercises that make that material more relevant, useful, and easy to remember. At over 400 pages, this book isn't a quick read—but it is considerably faster than reading the more than 4,000 pages contained in the resource books!

Watch Out for Free Online Quizzes!

I often get emails from people who read my book and score well on RMC's Agile FASTrack™ exam simulator, but then score poorly on some free online sample quizzes and become concerned. Unfortunately, many of the questions in free online practice quizzes are written from the reference books without checking the exam content outline.

For example, the question writer might open up *Kanban in Action*, see some material on Behavior-Driven Development (BDD), and then write a question about it. However, BDD isn't in the exam content outline—you don't need to know it for the exam—so it isn't covered in this book.

On the flip side, not everything you need to know for the exam is even in the reference books. There are many terms and concepts listed in the exam content outline that aren't even mentioned in those resources. In this book, however, I have explained every term included in the outline, even if it isn't covered in any of the reference books.

So you need to be very careful about taking free quizzes online—they could well mislead you, both by testing knowledge that isn't on the exam and also by *not* testing knowledge that *is* on the exam. Instead, I recommend you stick to study materials that you know are based on authoritative knowledge of the exam content, such as this book and RMC's Agile FASTrack™ and *Hot Topics*. (See "Other Study Resources for the PMI-ACP Exam" later in this chapter.)

How This Book Is Organized

To get the most out of this book, it's helpful to understand how it is organized and how to use the helpful features contained in the chapters. In this section we'll go over what you need to know to take full advantage of these tools.

Chapters and Domains

As noted earlier, this book is structured in the same way as the exam content outline—each chapter covers one of the PMI-ACP domains. To reflect the domain weightings, I've addressed each exam topic in the domain with which it is most closely associated. This makes it easier to see the totality of each domain and how its topics and tasks are interrelated. However, it also means that sometimes a key agile concept might be alluded to a few times in passing before it is fully explained later in the book. (For example, daily stand-ups aren't explained until chapter 5, although they are mentioned before that.)

So as you proceed through the book, if you encounter a term you aren't familiar with, I suggest that you just continue reading—that idea will most likely be explained in a later chapter. In any case, for exam preparation it's recommended that you go through this book more than once—and each time you review the material, you'll understand it better, gradually developing the holistic understanding of agile you'll need to pass the exam.

I suggest you cover one complete chapter at a time, but you don't have to read the chapters in sequence—and in your follow-up reviews you'll want to focus on the chapters that you're having the most difficulty with.

Domain Summaries

Each chapter starts out with a domain summary that provides a thumbnail reference guide for the domain. This summary includes the domain's weight for the exam, the key exam topics for that domain, and topline summaries of the tasks in that domain, as defined by the exam content outline. When you've finished reading the chapter, return to the domain summary to test your understanding of the key topics and tasks. See if you can recall what each concept means, how it's used, and why it is important for agile projects.

Once you fully understand all the summary items in each domain and how they work together, you should be well prepared for the exam.

Note: The key topics listed in the domain summaries include all the items in the agile toolkit (as well as associated terms and concepts you'll need to know for the exam). To help you locate the descriptions of the toolkit items in the chapter, the main discussion of each of these concepts will be flagged with one of the following icons:

T&T   K&S

Exam Tips and Background Information

In creating this book, I had to balance just covering the exam topics in a bare-bones format, or writing a primer for agile, including everything tested on the exam. I asked myself what kind of a guide I would like to read and talked to several candidates for the PMI-ACP exam to get their input. In the end, I decided to include a fair amount of context and examples, which I believe will best prepare most readers for the exam.

Since the first edition of this book, I've received a lot of positive feedback about this approach. However, there are also some readers who just want to focus on exam preparation. So to make it easier to identify which information is relevant to your study style, I've added two new features throughout the text—exam tips and background information, clearly labeled with special formatting.

The **Exam Tips** provide advice and tricks for the exam, addressing questions such as "How will this subject be tested?" "What will the questions about this topic be like?" "What's the most important point to remember about this concept?" These tips might give you additional insight into what you need to know about the exam, offer suggestions for studying for the exam, or explain how a concept might be tested on the exam. Be sure to read these tips carefully!

The **Background Information** sections provide context and further explanation of interesting points in the material that won't be tested on the exam. If you just want to focus on exam prep, you can skip these sections. Just bear in mind that if a concept is new to you, the background material might help you understand or remember it better.

Exercises

Each chapter includes several exercises to help you practice and apply the concepts being discussed. I strongly encourage you to actually do the exercises, rather than jumping right to the answers, even if the value of an exercise doesn't seem evident to you. If you make an effort to apply or work through an idea, it will "stick" in your mind much better.

The answers are provided immediately after each exercise. If you want to keep from seeing the answers, simply have a piece of paper handy to cover the answers until you have completed the exercise and are ready to review them.

Review Quizzes

The review quizzes at the end of each chapter offer a way to review the material covered in the chapter and check your knowledge. The quiz questions are designed to test your ability to identify, interpret, and apply agile concepts—and many of them are situational questions, just like the questions on the real exam.

© 2015 RMC Publications, Inc • 952.846.4484 • info@rmcls.com • www.rmcls.com

The best way to use the chapter review quizzes is to wait to take them until you have read through the entire book at least once. This will ensure that you are familiar with all the terms used in the question options and answer explanations. Also, taking the quizzes all at once (rather than as you finish each chapter), can help you evaluate whether you are ready to take the real exam.

The answers for each review quiz begin on the page after the quiz questions. If you can see the answers section while taking the quiz (which might happen on the last page), you may wish to cover it with a piece of paper until you are done taking the quiz.

Online Extras and RMC's Exam Prep System

Although this book can be used as a stand-alone study resource, it is also designed to be part of an exam prep system with RMC's Agile FASTrack™ exam simulator and *Hot Topics* flashcards. The next section will explain how to prepare for the exam, whether you have those other resources or are using the book alone. However, in either case, be sure to take advantage of the free extras provided at shop.rmcls.com/agileprep. We will be adding more resources to that site over time, including:

» What to do before you take the exam
» What to do while taking the exam
» Practice quiz (with questions from the review quizzes)
» Study tools for each domain
» Score sheet for Agile FASTrack™, chapter review quizzes, and practice quiz

The practice quiz on the extras website provides a sampling of the kinds of questions you will encounter on the real exam, drawn from the chapter review quizzes in this book. Although that quiz isn't long enough to check all your gaps, you can use it as a rough gauge to see if you are ready for the real exam. For example, if you ace the chapter review quizzes and the practice quiz, have a strong agile background, and are familiar with most of the concepts explained in this book, then you are probably ready to take the real exam.

However, if you struggle with the questions on the chapter review quizzes or the practice quiz, then you might benefit from further practice using the full-length exam simulator that's aligned with this book—Agile FASTrack™. In addition to providing a more detailed analysis of your gaps, that tool will allow you to practice taking a full-length exam and help you get familiar with a broader range of the kinds of questions you are likely to encounter on the exam.

Exam tip

Although the chapter review questions and the exam simulator are helpful study tools, you cannot prepare effectively for the PMI-ACP exam *only* by taking quizzes or exams. Cramming and being a "good quiz-taker" aren't enough, either. That's because the exam is actually testing the agile "mindset"—*being* rather than *doing* agile. To prepare for those kinds of questions, most of your study efforts should be based on this book, following the step-by-step study guide provided below, under "How to Study for the PMI-ACP Exam."

Other Study Resources for the PMI-ACP Exam

In addition to this book, RMC offers a complete PMI-ACP Exam Prep System, including:

Agile FASTrack™ Cloud Exam Simulator

RMC's Agile FASTrack™ Cloud exam simulator contains over 500 practice questions to help you prepare for the PMI-ACP exam. You can sort questions by domain, keyword, and PMI-ACP simulation. The software automatically scores and keeps records of your exams with its comprehensive grading and reporting capability. All questions are cross-referenced with this book, making it easy to go back and study your weak areas.

Hot Topics Flashcards *(in flipbook format)*

These flashcards feature the most important and difficult-to-recall terms and definitions you'll need to know for the PMI-ACP exam. They are an excellent study tool for people with busy schedules. You can use them at the office, on a plane, or in your car, adding instant mobility to your study routine.

PMI-ACP® Exam Prep—Online

This self-directed online course for the PMI-ACP exam covers everything you need to earn the PMI Agile Certified Practitioner credential with the least amount of study time. As a student of this course you will learn exclusive test-taking strategies and tips, and have access to real-world agile tools and techniques, interactive exercises, nine lessons worth of study content, and over 600 sample exam questions. If you are looking to earn the PMI-ACP credential *and* significantly boost your knowledge of agile methodologies, this online course offers everything you need, including:

» A 9-lesson interactive tutorial
» 600+ practice exam questions
» Dozens of interactive exercises
» Nearly 200 audio and video clips
» All 21 contact hours required for the PMI-ACP credential or 21 PDUs toward retaining a PMP certification

How to Study for the PMI-ACP Exam

To conclude this introductory section, let's examine how to use this book to prepare for the exam. To be successful, I recommend that you follow one of the two step-by-step study plans described below—Plan A or Plan B. Follow Plan A if you own RMC's complete PMI-ACP Exam Prep System (this book, Agile FASTrack™, and *Hot Topics*). Follow Plan B if you only have this book. After describing these two plans, I'll also explain how to use a study group to prepare for the exam.

Plan A: Using This Book with the PMI-ACP Exam Prep System

Note: This system includes the *PMI-ACP® Exam Prep* book, Agile FASTrack™, and *Hot Topics*.

Before you start, visit the free extras site for this book at shop.rmcls.com/agileprep. Download the score sheet referenced below and review the other study resources that are currently available. (Also check again later, since we will be continually adding new material.)

1. Read this book for the first time and complete all the exercises, focusing the most on the chapters where you have the most gaps in your agile knowledge or experience. Don't take the chapter review quizzes yet.

   - As you finish each chapter, go back and review the domain summary on the first page of the chapter. Make sure you understand each key topic and task and how it is used on agile projects. If there are any concepts you are unsure of, review that part of the chapter to improve your understanding. Use the *Hot Topics* flashcards to improve your recall and test your understanding of that chapter.

   - If possible, form a study group any time after you have read the book for the first time on your own. This can make your study time shorter and more effective. You will be able to ask questions, and the studying (and celebrating afterward) will be more fun. A study group should consist of only three or four people. (See the following discussion of "How to Use This Book in a Study Group.")

2. Once you are confident you know the material, test yourself with the chapter review quizzes and the first 10 questions from each domain in Agile FASTrack™. Use the results as a baseline against which to track your progress as you continue to study. This can help you determine how much more study time you need and which chapters to review more carefully.

3. Using the score sheet you downloaded, review each question you got wrong in Agile FASTrack™, writing down the specific reasons for each wrong answer. Assess why the correct choice is correct and why the other answers are wrong.

4. Continue to study this book, focusing on the areas in which you have gaps in your knowledge and skimming through the sections or chapters in which you did well. Use your list of why you got each question wrong to determine which material to study further.

5. Once you're confident you know the material, take a full PMI-ACP exam simulation in Agile FASTrack™. WARNING: Take no more than two full exam simulations in preparing for the exam. Otherwise, you diminish the value of Agile FASTrack™ and will start seeing some of the questions repeated.

6. Review the questions you got wrong on the Agile FASTrack™ simulation exam. As with step 3, use the score sheet you downloaded to identify in writing the specific reason you got each question wrong. Use your list of why you got each question wrong to determine which material to study further.

7. Once you have studied the material you missed on the last exam, take your second, and final, PMI-ACP simulation exam.

8. Use the score sheet to analyze the questions you got wrong, and study any topics you missed. Once you are ready for the exam, use the *Hot Topics* flashcards to retain the information you have studied until you take the real exam.

9. PASS THE EXAM!

Plan B: Using This Book as a Stand-Alone Study Resource

Before you start, visit the free extras site for this book at shop.rmcls.com/agileprep. Download the score sheet referenced below and review the other study resources that are currently available. (Also check again later, since we will be continually adding new material.)

1. Read this book for the first time and complete all the exercises, focusing the most on the chapters where you have the most gaps in your agile knowledge or experience. Don't take the chapter review quizzes yet.

   - As you finish each chapter, go back and review the domain summary on the first page of the chapter. Make sure you understand each key topic and task and how it is used on agile projects. If there are any concepts you are unsure of, review that part of the chapter to improve your understanding.

   - If possible, form a study group any time after you have read the book for the first time on your own. This can make your study time shorter and more effective. You will be able to ask questions, and the studying (and celebrating afterward) will be more fun. A study group should consist of only three or four people. (See the following discussion of "How to Use This Book in a Study Group.")

2. Once you are confident you know the material, test yourself with the chapter review quizzes in each domain, doing them all in one sitting. Use the results as a baseline against which to track your progress as you continue to study. This can help you determine how much more study time you need and which chapters to review more carefully.

3. Using the score sheet you downloaded, review each question you got wrong in the chapter reviews, writing down the specific reasons for each wrong answer. Assess why the correct choice is correct and why the other answers are wrong.

4. Continue to study this book, focusing on the areas in which you have gaps in your knowledge and skimming through the sections or chapters in which you did well. Use your list of why you got each question wrong to determine which material to study further. (You may wish to take the online practice quiz as well.)

5. Use the score sheet to analyze the questions you got wrong, and study any topics you missed. Make sure you have filled your gaps before taking the exam.

6. PASS THE EXAM!

How to Use This Book in a Study Group

To get started, pick someone to lead the discussion of each chapter (ideally someone who is still struggling with the chapter, because the presenter often learns and retains the most in the group). Each time you meet, go over questions about topics you do not understand and review key concepts on the exam using the *Hot Topics* flashcards, if you have them. Most groups meet for one hour per chapter. Either independently or with your study group, do further research on questions you do not understand or that you answered incorrectly.

Each member of the study group should have his or her own copy of the book. (Please note that it is a violation of international copyright laws to make copies of the material in this book or to create derivative works from this copyrighted book.)

> Note: Those leading a PMI-ACP exam preparation course using RMC's products may want to contact RMC for information on our Corporate Partnership program. Partners may be allowed to create slides or other materials using content from this book. Also, ask about other tools for study groups and independent instructors and how to receive quantity discounts on this book, Agile FASTrack™, or *Hot Topics*.

CHAPTER 1

Agile Principles and Mindset

Domain I Summary

This chapter discusses domain I in the exam content outline, which is 16 percent of the exam, or about 19 exam questions. This domain focuses on the agile mindset, its fundamental values and principles, the agile methodologies, and agile leadership.

Key Topics

» Agile frameworks and terminology
» Agile Manifesto
  – 4 values
  – 12 principles
» Agile methods and approaches
» Agile process overview

» Kanban
  – Five principles
  – Pull system
  – WIP limits
» Leadership practices and principles
  – Management vs. leadership
  – Servant leadership (4 duties)

» Lean
  – Core concepts
  – 7 wastes
» Scrum
  – Activities
  – Artifacts
  – Team roles
» XP
  – Core practices
  – Core values
  – Team roles

Tasks

1. Advocate for agile principles and values to create a shared mindset.
2. Ensure a common understanding of agile.
3. Support change through educating and influencing.
4. Practice transparency in order to enhance trust.
5. Create a safe environment for experimentation.
6. Experiment with new techniques and processes.
7. Share knowledge through collaboration.
8. Encourage emergent leadership via a safe environment.
9. Practice servant leadership.

In this chapter, we'll define some key concepts that provide a foundation for the material presented in the rest of this book. In essence, this chapter sets the stage for understanding agile for the PMI-ACP exam. It summarizes the agile mindset and methodologies, outlines the major agile methodologies, and explores the nature of agile leadership and how it differs from traditional management.

Even when an exam question deals with a specific tool or technique, an understanding of the broader agile framework will often be required to answer it correctly. So if you are faced with a question that appears to have two or more reasonable answers, think about the underlying values and principles of agile explained in this chapter to help you make the right selection.

Why Use Agile?

Let's start with a basic question. Why do we need another approach for running projects, anyway? Isn't a method like the one extensively documented in the *PMBOK® Guide* sufficient?

The answer is simple: different types of projects require different methods. In our everyday lives, we continually customize our approach depending on the situation, often in small and subtle ways. For example, we choose what information to communicate and how to present it based on our audience. We don't approach every issue we face in the exact same way, like a robot; instead, we adjust our methods to be most effective for each situation. This same concept applies to projects—and some projects, especially knowledge work projects in a fast-moving, complex environment, call for an agile approach.

Knowledge Work Projects Are Different

First let's discuss a little history that isn't tested on the exam but will help set the scene. Initially humans wandered the earth as hunter-gatherers. When people started planting crops and herding animals, it changed society and work. This was the Agricultural Revolution. As a result, people wandered less, and they lived and worked in one place.

The next big transformation came with the development of machines and factories, when people left their farms and villages to move into cities. This was the Industrial Revolution, which eventually led to the development of many classic project management tools and concepts, including Gantt charts, functional decomposition, and localized labor. In turn, these developments led to the creation of more advanced project management tools, such as the work breakdown structure (WBS).

The latest stage—which we are in now—is known as the Information Revolution. This revolution is focused on information and collaboration, rather than manufacturing. It places value on the ownership of knowledge and the ability to use that knowledge to create or improve goods and services.

The Information Revolution relies on knowledge workers. These are people with subject matter expertise who communicate their knowledge and take part in analysis or development efforts. Knowledge workers are not only found in the IT industry; they are also engineers, teachers, scientists, lawyers, doctors, writers, and many others employed today. In fact, knowledge workers have become the largest segment of the North American workforce.

So what makes knowledge work projects different from manufacturing projects? To find out, let's compare the key characteristics of industrial work and knowledge work:[1]

Characteristics of Industrial Work	Characteristics of Knowledge Work
Work is visible	Work is invisible
Work is stable	Work is changing
Emphasis is on running things	Emphasis is on changing things
More structure with fewer decisions	Less structure with more decisions
Focus on the right answers	Focus on the right questions
Define the task	Understand the task
Command and control	Give autonomy
Strict standards	Continuous innovation
Focus on quantity	Focus on quality
Measure performance to strict standards	Continuously learn and teach
Minimize cost of workers for a task	Treat workers as assets, not as costs

EXERCISE

Review the items in each column of the following table and place a check mark next to any that describe your job. When you are finished, look at the pattern of the check marks. Are they mostly on the left (industrial work) or mostly on the right (knowledge work)?

Industrial Work	Knowledge Work
☐ Work is visible	☑ Work is invisible
☐ Work is stable	☑ Work is changing
☐ Emphasis is on running things	☑ Emphasis is on changing things
☐ More structure with fewer decisions	☑ Less structure with more decisions
☐ Focus on the right answers	☑ Focus on the right questions
☐ Define the task	☑ Understand the task
☑ Command and control	☐ Give autonomy
☐ Strict standards	☑ Continuous innovation
☐ Focus on quantity	☑ Focus on quality
☑ Measure performance to strict standards	☑ Continuously learn and teach
☑ Minimize cost of workers for a task	☐ Treat workers as assets, not as costs

When knowledge work projects became more common, people found that the communication and collaboration involved in these projects made the work more uncertain and less definable than industrial work. As people tried to apply industrial work techniques to knowledge work projects, frustration—and project failures—increased. Agile methods were developed in response to this problem. Agile pioneers collected the most effective techniques for knowledge work and adapted them for use on projects, experimenting to see what worked best. This new initiative began in the software development field, but is now used in all kinds of knowledge work projects.

This development of agile methods took place over many years and was done by different people. As a result, agile has multiple methodologies that use different terminology. For example, Scrum calls its timeboxed development efforts "sprints," while XP calls them "iterations." This chapter will establish a framework for understanding the agile mindset and explain some fundamental concepts you need to understand to do well on the PMI-ACP exam.

Defined versus Empirical Processes

Another way of looking at the difference between knowledge work and industrial work is to examine the different kinds of processes they use. Industrial work typically uses a defined process, while knowledge work relies on an empirical process.

In a *defined* process, as the name implies, we can define the constituent steps in advance. If we know the optimum way to tie our shoelaces, then we can follow the same process each time. This is typically the most efficient way to proceed for a well-understood project in an unchanging environment, such as construction projects that use well-understood materials and building approaches. In fact, most industrial projects can be planned and managed by using a defined approach.

Other processes are not as well defined. When faced with a new or uncertain process, such as building an underwater home for the first time or using carbon nanotubes instead of steel, there will be many unknowns and uncertainties involved in the risks and solutions required for the new environment or materials.

When faced with such uncertainty, a process of trial and experiment is required to determine what works, surface issues, and incrementally build on small successes. The resulting process will be iterative and incremental, with frequent reviews and adaptation. The result is an *empirical* process. This approach is required for projects where the execution stage is characterized by uncertainty and risks—in other words, projects that would benefit from the agile approach.

K&S The Agile Mindset

Although there are a lot of tools, practices, and concepts that you'll need to know for the PMI-ACP exam, the underlying goal of the exam questions isn't really to test specific knowledge—instead, the exam wants to see if you have a comprehensive grasp of what it means to "be agile" (not just "do agile"). "Being agile" isn't simply a matter of using a certain set of tools or practices, or following a specific methodology. Agility really involves adopting a new mindset—way of thinking—that is based on agile values and principles.

The most important statement of these values and principles is the Agile Manifesto, which we'll examine in detail in the next section of this chapter. But first, let's look at some of the general characteristics of the agile mindset. We'll start with the "Declaration of Interdependence" (DOI) written in 2005 by the cofounders of the Agile Project Leadership Network (now the Agile Leadership Network). This document outlines six precepts:[2]

1. We **increase return on investment** by making continuous flow of value our focus.
2. We **deliver reliable results** by engaging customers in frequent interactions and shared ownership.
3. We **expect uncertainty** and manage for it through iterations, anticipation, and adaptation.

4. We **unleash creativity and innovation** by recognizing that individuals are the ultimate source of value, and creating an environment where they can make a difference.
5. We **boost performance** through group accountability for results and shared responsibility for team effectiveness.
6. We **improve effectiveness and reliability** through situationally specific strategies, processes, and practices.

Since the DOI is aimed at leaders—it focuses on the management side of agile projects—these six principles can serve as a kind of topline summary to introduce you to the agile mindset, although the DOI won't be tested on the exam.

Read through these six statements again and see if you understand them. Do they make sense to you? If this mindset sounds very different from the working environment you are used to, then you will need to carefully study this chapter in order to do well on the exam. The exam questions require you to have a firm grasp of "agile thinking"; they can't be answered simply by memorizing information.

Exam tip

In this section ("The Agile Mindset") we are presenting various perspectives on, or aspects of, the agile way of thinking. With the exception of the inverted agile triangle (discussed below) this information won't be directly tested on the exam. Here, our goal is simply to provide a basic answer to the question "What is agile?"

Moving on, here's another way of summing up the core principles of agile:

» Welcoming change
» Working in small value-added increments
» Using build and feedback loops
» Learning through discovery
» Value-driven development
» Failing fast with learning
» Continuous delivery
» Continuous improvement

As you think about this list, remember that agile isn't just a matter of adopting practices designed to achieve these outcomes. We actually have to take the agile mindset to heart and use it to guide our approach. Applying agile values and principles to how we use agile methods changes not only our approach, but also the effectiveness of the practices. This is the difference between "being agile" and "doing agile," as illustrated below.

Figure 1.1: Being Agile versus Doing Agile

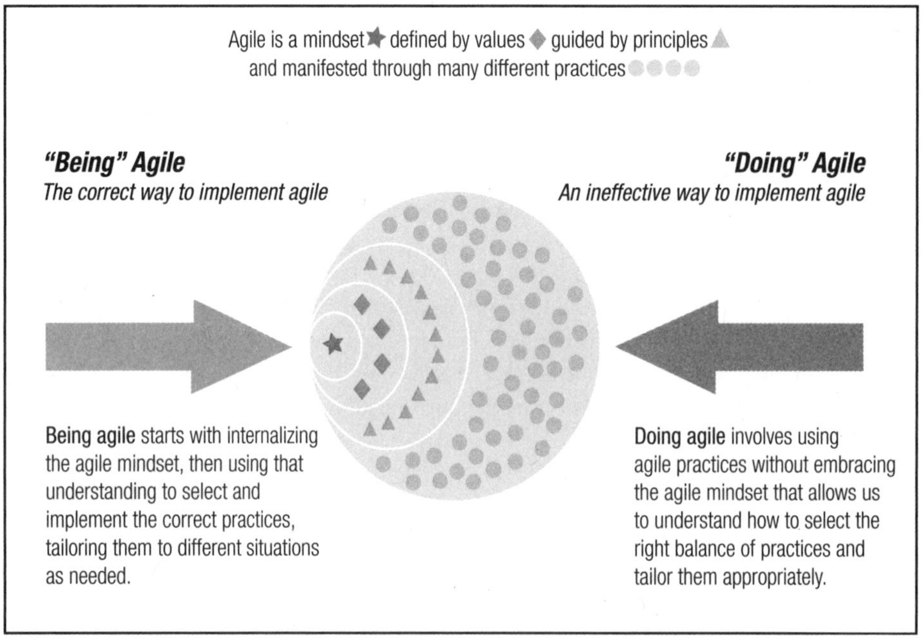

Agile is a mindset ★ defined by values ◆ guided by principles ▲
and manifested through many different practices ● ● ● ●

"Being" Agile
The correct way to implement agile

"Doing" Agile
An ineffective way to implement agile

Being agile starts with internalizing the agile mindset, then using that understanding to select and implement the correct practices, tailoring them to different situations as needed.

Doing agile involves using agile practices without embracing the agile mindset that allows us to understand how to select the right balance of practices and tailor them appropriately.

Image copyright © Ahmed Sidky, Santeon Group, www.santeon.com

Let's break down this diagram.

» The correct way to adopt agile is shown by the green arrow on the left. Here we start by internalizing the agile mindset (welcoming change, small increments, etc.), and then we use those principles to guide our selection and implementation of agile practices. We start with a good understanding of *why* we are using the practices, which in turn helps us understand *how* to use them most effectively.

» The gray arrow on the right represents a team that decides to adopt agile practices (such as daily stand-up meetings and short iterations), without taking the time to understand what these practices are designed to accomplish. Here, we jump directly into the *how* of agile without first understanding the *why*. This is a common problem in agile adoption.

Based on this diagram, people sometimes use the term "left-to-right adoption" as a shorthand way of saying "teach agile values first."

Personal, Team, and Organizational Agility

Within an organization, people will develop an understanding of the agile mindset at different rates. To some, agile values will be intuitively easy to grasp—they seem to describe familiar beliefs and behavior patterns. To others, these values will have to be consciously learned and then actively practiced before they can be understood and accepted. Regardless of how this understanding develops, the more people there are in an organization who embrace and act upon agile principles, the more effective agile practices will be.

» If just one member of an organization adopts an agile mindset, it can help that person become more effective. However, they will feel continually frustrated that others in the organization don't seem to realize what is important, or are focused on the wrong goals and metrics.

» If one team in the organization adopts agile principles and practices, it can help the team members become more effective at delivering their project work. However, they will feel inhibited or misunderstood by other groups or systems in the organization, such as the project management office (PMO) or functional silos.

» If the entire organization adopts the agile way of thinking, then everyone will be working together to improve agility and the delivery of value. By adopting common goals and values, such as continuous improvement and welcoming change, everyone's effectiveness will be enhanced.

Although organizational agility is the ideal goal, today most organizations are not there yet. So how can we help our organizations get there? The way organizations change is through influence exerted by individuals. The diagram below depicts the steps involved in this process as the layers of an onion skin.

Figure 1.2: Creating Organizational Change

Let's look more closely at each of these layers:

Think
First, we need to *think*—this means individually learning and internalizing agile principles. Only once we fully understand the agile mindset can we move on to the next step.

Do
Doing is the practice of agile. For example, this might involve visualizing work items, using short iterations, or building in feedback and improvement steps. Once we understand and can practice these steps ourselves, we can move on to the next step, where we begin influencing others.

Encourage Others
This final step involves *encouraging others* to become agile. Although this may appear to be exponentially harder to achieve than the first two steps, it is also the most worthwhile when accomplished. That's because persuading others to adopt the agile mindset and practices will magnify agile learning and effectiveness across the entire organization. Also, the more people in your circle of work you can successfully educate about the merits of agile, the more allies you will have in advocating the cause. The end result of this process can be a complete transformation of the organization based on agile principles.

Exam tip

Like the "Think-Do-Encourage others" model described above, you'll see many process steps and bulleted lists in this book. Don't approach this material by trying to cram it, as if you were studying for a college exam. That won't work because the PMI-ACP exam doesn't test information—most of the questions will be situational. So your study goals should be, first, to fully absorb the agile mindset and then, second, to become generally familiar with agile tools and practices. So instead of memorizing details, focus on the big picture: "What is important about this list or process for agile?" "How does each item listed fit into the agile way of doing things?" This is what you need to understand for the exam.

The Agile Triangle

Another key difference between the agile mindset and traditional project management is the agile or "inverted" triangle of constraints. This triangle, shown below, was introduced in the first edition of the DSDM Manual, published in 1994.

Figure 1.3: Inverted Triangle Model

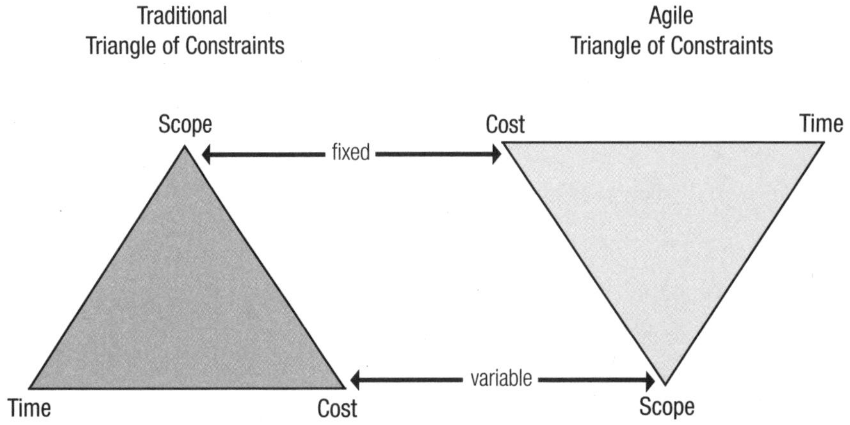

Traditional Triangle of Constraints

Agile Triangle of Constraints

Scope — fixed → Cost ... Time

Time ← variable → Cost ... Scope

This reversal of the traditional triangle means that agile teams allow scope to vary within the fixed parameters of cost and time. In other words, we aim to deliver the most value we can by X date within X budget. Although we'll begin with a high-level vision of the end product, we can't define up front how much we'll be able to get done—that will emerge as we get closer to the target date.

You might hear criticism that this approach doesn't make sense, because it won't work for tangible, industrial work projects. For example, "If I'm paying to have my bathroom remodeled, I want the whole thing to be completed, not just most of it!" However, remodeling a bathroom should be a defined, repeatable process with little R&D or uncertainty. That isn't the kind of project that calls for an agile approach or an inverted triangle of constraints.

Knowledge work projects, on the other hand, are characterized by experimentation and uncertainty—and the end product can be refined forever. This is why we need to determine acceptable operating boundaries, which usually take the form of cost and time constraints. For example, let's say we are developing training materials for a course on short story writing. For a project such as this, we could continue adding and tweaking our lesson material and exercises indefinitely. Instead we need to get to an acceptable performance point, and then have the discipline to stop.

K&S The Agile Manifesto

The name of domain I ("Agile Principles and Mindset") highlights not only the agile mindset, but also the values and principles it is based on. For the exam, you'll need a thorough understanding of the most important statement of agile values and principles— a document called the Agile Manifesto.

The Agile Manifesto was created during a meeting in February 2001 that brought together a number of software and methodology experts who were in the forefront of the emerging agile methods. The people in attendance were:

» Kent Beck	» James Grenning	» Robert C. Martin
» Mike Beedle	» Jim Highsmith	» Steve Mellor
» Arie van Bennekum	» Andrew Hunt	» Ken Schwaber
» Alistair Cockburn	» Ron Jeffries	» Jeff Sutherland
» Ward Cunningham	» Jon Kern	» Dave Thomas
» Martin Fowler	» Brian Marick	

Exam tip

The PMI-ACP exam won't ask about the origin of the Agile Manifesto or its creators. We are including this information to set the context for the discussion that follows. The exam will focus on your ability to apply the values and principles set forth in this document in situational questions. It won't ask "What is the fourth value?" or "How is the second principle phrased?"

The Four Values

The Agile Manifesto includes a statement of four values and twelve guiding principles. We'll examine the values first. This section of the document reads as follows:[3]

> ### Manifesto for Agile Software Development
>
> *We are uncovering better ways of developing software by doing it and helping others do it. Through this work we have come to value:*
>
> **Individuals and interactions** *over processes and tools*
> **Working software** *over comprehensive documentation*
> **Customer collaboration** *over contract negotiation*
> **Responding to change** *over following a plan*
>
> *That is, while there is value in the items on the right, we value the items on the left more.*

While the Agile Manifesto is simple in structure and sparse in words, there is a lot of good stuff in these four values. Understanding the ideas conveyed in these values is important not just for the exam, but also for the application of an agile approach on any kind of knowledge work project.

Exam tip

The Agile Manifesto was written by software development experts, so it uses terms from that field to express its ideas; however, those ideas are applicable to any kind of knowledge work project. So although you are likely to see some IT-specific terms on the exam, you'll need to be able to recognize these ideas in other contexts as well. As you read about the Manifesto, look beyond the specific terms and think about how these concepts apply to other types of knowledge work. For example, we might rephrase the second value as "Working *systems* over comprehensive documentation," substituting "systems" for "software."

The format of the four values—A *over* B ("Individuals and interactions *over* processes and tools")—addresses intention, focus, and effort. This isn't as black and white as just saying, "Do A *instead* of B." Instead, it acknowledges that both A and B will be components of our projects, but that we should apply more of our focus, emphasis, and intention to A than to B.

The Agile Manifesto isn't a set of rules telling us to do one thing instead of another. It is both more subtle and more powerful—it guides us to consider projects from a value-based perspective. Yes, we will need processes, tools, documentation, and plans on our projects. Yet while dealing with these assets, we should remember that our focus must be on the people engaged, the product we are building, cooperation, and flexibility. Agility is the capacity to execute projects while focusing our efforts on the items on the left side of these value statements, rather than those on the right.

With this in mind, let's look at these four statements in more detail.

Value 1: Individuals and Interactions Over Processes and Tools

The message here is that while processes and tools will likely be necessary on our projects, we should try to focus the team's attention on the individuals and interactions involved. This is because projects are undertaken by people, not tools, and problems get solved by people, not processes. Likewise, projects are accepted by people, scope is debated by people, and the definition of a successfully "done" project is negotiated by people. Focusing early on developing the individuals involved in the project and emphasizing productive and effective interactions help set up a project for success.

This is not to say that processes and tools cannot help in successfully completing a project. They are certainly important assets, and for those of us who have an engineering background, we may naturally tend toward the logic and predictability of processes and tools. Yet projects are ultimately about people, so to be successful, we need to spend the majority of our time in what may be the less comfortable, messy, and unpredictable world of people. If you tend toward processes rather than people, the first value of "individuals and interactions over processes and tools" is a great reminder of where to focus your time, energy, and passion.

Value 2: Working Software Over Comprehensive Documentation

This value speaks to the need to deliver. As mentioned in the exam tip on the previous page, it can be rephrased more broadly as "Working *systems* over comprehensive documentation," where "system" refers to the product or service that the project is delivering. This value reminds us to focus on the purpose or business value we're trying to deliver, rather than paperwork. (Some highly regulated industries, such as medical devices, require a lot of documentation—that's just part of the work that needs to be done. This value refers to the kind of documentation that isn't included in getting the work done properly.)

The agile approach to documentation is "just enough, just in time—and sometimes, just because." This phrase is shorthand to remind us of three important concepts:

» Agile documentation should be "barely sufficient"—just enough to cover our needs. This keeps most of our efforts focused on the emerging system.

» Agile documentation is done "just in time"—also known as the "last responsible moment"—so we don't have to spend extra time to keep it updated as our requirements and designs change.

» Finally, "just because" reminds us that sometimes when documentation is required or requested, it's easier and preferable to just produce it than to face the consequences of not doing so.

That last point doesn't mean we just pander to unnecessary documentation requests; for example, the lean methodology includes the powerful concept of asking "Why?" five times to find out what we really need. However, it pays to be smart, too—given the limited time and effort available on a project, we need to decide where to best focus our energy, and which battles to pursue. So although we value working on the product over documentation, that doesn't mean we abandon the opposing value.

Software projects are typically initiated with the goal of creating valuable, high-quality software, yet they often get caught up on interim deliverables such as extensive documentation that doesn't support the ultimate goal of working software. Software without documentation is certainly problematic and hampers support and maintenance. But comprehensive documentation without software has no value in most organizations.

Many software developers are detail-oriented and process-driven; although these characteristics are often highly beneficial, they can also mean the developers' focus can easily be distracted from the real reason they are undertaking software projects—to write valuable software. So this emphasis on valuing working software over comprehensive documentation acts as a necessary and useful reminder of why these projects are commissioned in the first place—to build something useful. Documentation by itself, or at the expense of working software, is not useful.

Value 3: Customer Collaboration Over Contract Negotiation

This value reminds us to be flexible and accommodating, rather than fixed and uncooperative. It is similar to the difference between "being right" and "doing the right thing." We could build the product exactly as originally specified, but if the customer's preferences or priorities change, it would be better to be flexible and work toward the new goal, as opposed to the goal that was originally stated.

It is notoriously difficult to define an up-front, unchanging view of what should be built. This challenge stems from the dynamic nature of knowledge work products, especially software systems; software is intangible and difficult to reference, companies rarely build the same systems twice, business needs change quickly, and technology changes rapidly. Rather than subject the client to a change management process that is really more of a change suppression process, we should recognize at the start that things are going to change, and we'll need to work with the customer throughout the project to reach a shared definition of "done." This requires a more trusting relationship and more flexible contract models than we often see on projects, but—like the previous value—it moves the emphasis from nonvalue-adding activities (such as arguing about scope) to productive work.

Value 4: Responding to Change Over Following a Plan

In knowledge work projects, we know that our initial plans are inadequate, since they are based on insufficient information about what it will take to complete the project. So instead of investing effort in trying to bring the project back in line with our original plan, we want to spend more of our effort and energy responding to the changes that will inevitably arise. Now, this doesn't mean we should abandon planning and just react to changes. We still need to plan, but we also need to acknowledge that our initial plans were made when we knew least about the project (at the beginning) and will need to be updated as the work progresses.

The importance of responding to change over following a plan is particularly true for software projects, where high rates of change are common. Again, instead of suppressing changes and spending a lot of time managing and tracking a largely static plan, we need to acknowledge that things will change. Agile projects have highly visible queues of work and plans in the form of backlogs and task boards. The intent of this value is to broaden the number of people who can be readily engaged in the planning process by adjusting the plans and discussing the impact of changes.

© 2015 RMC Publications, Inc • 952.846.4484 • info@rmcls.com • www.rmcls.com

EXERCISE: MATCH THE AGILE MANIFESTO VALUES

Note: As you do the exercises in this book, you may want to have a sheet of paper handy to cover up the answers to the exercises, since they are just after the questions.

Draw arrows to match the start of the Agile Manifesto values on the left to their correct endings on the right.

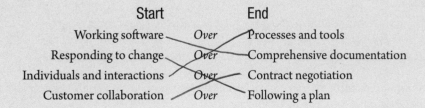

ANSWER

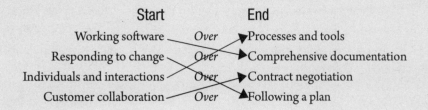

The Twelve Principles

In addition to the four agile values, the authors of the Manifesto created twelve guiding principles for agile methods. This part of the Manifesto reads as follows:[4]

Principles behind the Agile Manifesto

We follow these principles:

Our highest priority is to satisfy the customer through early and continuous delivery of valuable software.

Welcome changing requirements, even late in development. Agile processes harness change for the customer's competitive advantage.

Deliver working software frequently, from a couple of weeks to a couple of months, with a preference to the shorter timescale.

Business people and developers must work together daily throughout the project.

Build projects around motivated individuals. Give them the environment and support they need, and trust them to get the job done.

The most efficient and effective method of conveying information to and within a development team is face-to-face conversation.

Working software is the primary measure of progress.

Agile processes promote sustainable development. The sponsors, developers, and users should be able to maintain a constant pace indefinitely.

Continuous attention to technical excellence and good design enhances agility.

Simplicity—the art of maximizing the amount of work not done—is essential.

The best architectures, requirements, and designs emerge from self-organizing teams.

At regular intervals, the team reflects on how to become more effective, then tunes and adjusts its behavior accordingly.

Exam tip

Since these values and principles are fundamental to understanding agile, think of them when you are faced with two seemingly correct options on the exam. In such instances, look for the answer that best matches the Agile Manifesto values and principles.

Let's take a closer look at each of the Agile Manifesto's twelve principles. Again, although the principles may use software development terms, as you read about them, think about how these concepts apply to other types of knowledge work projects.

Principle 1: Our highest priority is to satisfy the customer through early and continuous delivery of valuable software.

This principle includes three main points. The first is to satisfy the *customer*. If we produce perfect plans and documentation, but only delight the project management office (PMO) or the quality assurance (QA) group, we have failed; our focus should be on the customer.

The second point is *early and continuous delivery*. We must structure the project and the development team to deliver value early and frequently. This can be a struggle if people are reluctant to share incomplete work, and it takes courage and support for everyone to become comfortable with this idea. However, we must achieve this point if we are to learn of problems while we still have time to fix them. It is better to get something wrong up front and have time to correct it than to discover the issue much later when so much more has been built on top of a faulty foundation.

The final point is that what we are delivering is *valuable software* (i.e., systems), not completed work products, WBS items, documentation, or plans. We need to stay focused on the end goal. For software projects, this is the software; for other types of projects, the end goal will be the product or service that the project was undertaken to deliver or enhance.

Principle 2: Welcome changing requirements, even late in development. Agile processes harness change for the customer's competitive advantage.

Changes can be great for a project—for example, if they allow us to deliver a late-breaking, high-priority feature. Yet in non-agile projects, changes are often seen in a negative light; they may be considered "scope creep" or blamed for the project deviating from the plan. Many non-agile projects have such rigorous change control procedures that only the highest priorities make it through. On such projects, much of the time and effort is spent logging and managing change requests.

This kind of rigorous change management is problematic for any project in a highly changeable environment, such as software development—so agile accepts that changes will occur. In fact, the XP methodology advocates that we "embrace change."

Instead of creating a high-overhead mechanism for suppressing or processing changes, agile methods use a lightweight, high-visibility approach—for example, continuously updating and prioritizing changes into the backlog of work to be done. Agile's well-understood, high-visibility methods for handling changes keep the project adaptive and flexible as long as possible. By welcoming the changes that will inevitably happen and setting up an efficient way to deal with them, we can spend more time developing the end product.

Principle 3: Deliver working software frequently, from a couple of weeks to a couple of months, with a preference to the shorter timescale.

Despite our best intentions and knowledge that early feedback is valuable, it's human nature to want our work to be as complete as possible before sharing it. However, we are doing ourselves a disservice by holding on to our work for too long. It's better to get feedback early and often, to avoid going too far down the wrong track.

This principle emphasizes the importance of releasing work to a test environment and getting feedback. Frequent testing and feedback are so important that software developers use continuous integration tools to provide feedback about any code they have written that breaks the build. Agile teams need feedback on what they have created thus far to see if they can proceed, or if a change of course is needed.

Delivering within a short timeframe also has the benefit of keeping the product owner engaged and keeping dialogue about the project going. With frequent deliveries, we will regularly have results to show the customer and opportunities to get feedback. Often at these demos, we learn of new requirements or changes in business priorities that are valuable planning inputs.

Principle 4: Business people and developers must work together daily throughout the project.

The frequent demos mentioned above are one example of how the business representatives and developers work together throughout the project. Daily face-to-face engagement with the customer is one of the most difficult principles to ensure from a practical standpoint, but it is really worth pushing for. Written documents, e-mails, and even telephone calls are less efficient ways of transferring information than face-to-face interactions.

By working with business representatives daily, we can learn about the business in a way that is far beyond what a collection of requirements-gathering meetings can ever achieve. As a result, we are better able to suggest solutions and alternatives to business requests. The business representatives also learn what types of solutions are expensive or slow to develop, and what features are cheap. They can then begin to fine-tune their requests in response.

When it isn't possible to have daily interactions between the business representatives and the development team, agile methods try to get the two groups working together regularly in some way, perhaps every two days or whatever type of frequent involvement will work. (Some teams use a "proxy customer," in which an experienced business analyst [BA] who is familiar with the business interests serves as a substitute, but this isn't an ideal option.)

Principle 5: Build projects around motivated individuals. Give them the environment and support they need, and trust them to get the job done.

As we'll see in chapter 4, software estimation data gathered by the COCOMO II® model has found that having the best people is significantly more important for a project than having the best processes and tools—by a factor of 10 to 1! So making sure we have smart and motivated people on the team is likely to make a big difference in whether our project is delivered successfully and efficiently.

While we may not always be able to pick our dream team, we can motivate and empower the team members we do have. Since the development team is such an important factor, agile methods promote empowered teams. People work better when they are given the autonomy to organize and plan their own work. Agile methods advocate freeing the team from the micromanagement of completing tasks on a Gantt chart. Instead, the emphasis is on craftsmanship, peer collaboration, and teamwork, which result in higher rates of productivity.

Knowledge work projects involve team members who have unique areas of expertise. Such people do their best work when they are allowed to make many of the day-to-day decisions and local planning for the project. For leaders, this doesn't mean abdicating involvement or abandoning the team to fend for itself; instead, we recognize that our team members are experts in what they do, and we provide the support they need to ensure they are successful.

Principle 6: The most efficient and effective method of conveying information to and within a development team is face-to-face conversation.

Written documents are great for creating a lasting record of events and decisions, but they are slow and costly to produce. In contrast, face-to-face communication allows us to quickly transfer a lot of information in a richer way that includes emotions and body language.

In a face-to-face conversation, questions can be immediately answered, instead of "parked" with the hope that there will be a follow-up explanation, or the answer will become clear later. For example, if instead of reading this book, you were talking to me directly, we could quickly skip all the stuff you already know and focus on the areas you want to learn more about. Written documents have to assume a lower starting point so as not to confuse their broader audience.

Of course, this recommendation for face-to-face conversations can't be applied to all project communications, but agile teams aim to follow it whenever possible. This is one example of how agile methods need to be customized or scaled for each project. As team sizes grow, it becomes harder to rely on face-to-face communications, and we have to introduce an appropriate level of written documentation.

Principle 7: Working software is the primary measure of progress.

By adopting "working software" (or "working systems") as our primary measure of progress, we shift our focus to working results rather than documentation and design. In agile, we assess progress based on the emerging product or service we are creating. Questions like "How much of the solution is done and accepted?" are preferred over "Is the design complete?" since we want to focus on usability and utility rather than conceptual progress.

Expressions like "What gets measured gets done" and "You get what you measure" apply here, since measurement reinforces activity. If a feature can't be measured or tested—in other words, if it doesn't "work"—it isn't considered complete. This emphasis on a working product helps ensure that we get features accepted by the customer, rather than marking items as "completed development" when they haven't yet been accepted as "done."

This definition of progress as "working systems" creates a results-oriented view of the project. Interim deliverables and partially completed work will get no external recognition. So we want to focus instead on the primary goal of the project—a product that delivers value to the business.

Principle 8: Agile processes promote sustainable development. The sponsors, developers, and users should be able to maintain a constant pace indefinitely.

Agile methods strive to maximize value over the long term. Some of the rapid application development (RAD) techniques that preceded agile promoted—or at least accepted—long, intense periods of prototyping prior to demos. The trouble with the RAD approach of working teams for long hours over an extended period of time is that people start to burn out and make mistakes. This is not a sustainable practice.

Instead of long, intense development periods, agile methods recognize the value of a sustainable pace that allows team members to maintain a work-life balance. A sustainable pace is not only better for the team; it benefits the organization as well. Long workdays lead to resignations, which means the organization loses talent and domain knowledge. Hiring and integrating new members into a team is a slow and expensive process.

In contrast, working at a pace that can be maintained indefinitely leads to a happier and more productive team. Happy teams also get along with the business representatives better than overworked teams. There is less tension, and work relationships improve. Some agile methods, like XP, previously recommended maintaining a regular 40-hour workweek as a guiding principle. Today most methods don't establish specific limits, but they do recommend paying close attention to the level of effort the team is putting forth to ensure a sustainable pace.

Principle 9: Continuous attention to technical excellence and good design enhances agility.

While we want the development team to work hard and deliver a lot of value, we also have to be mindful of keeping the design clean, efficient, and open to changes. Technical excellence and good design allow the development team to understand and update the design easily.

An agile team needs to balance its efforts to deliver high-value features with continuous attention to the design of the solutions. This balance allows the product to deliver long-term value without becoming difficult to maintain, change, or extend—cleaning and preventative maintenance are preferable to fixing problems. This helps the project run more smoothly and speeds up the team's progress.

In the software world, once the code base becomes tangled, the organization loses its ability to respond to changing needs. In other words, it loses its agility. So we need to give the development team enough time to undertake refactoring. Refactoring is the housekeeping, cleanup, and simplifications that need to be made to code to ensure it is stable and can be maintained over the long term.

Principle 10: Simplicity—the art of maximizing the amount of work not done—is essential.

The most reliable features are those we don't build—since there is nothing that could go wrong with them. And, in the software world, up to 60 percent of features that are built are used either infrequently or never.[5] Because so many features that are built are never actually used, and because complex systems have an increased potential to be unreliable, agile methods focus on simplicity. This means boiling down our requirements to their essential elements only.

Complex projects take longer to complete, are exposed to a longer horizon of risk, and have more potential failure points and opportunities for cost overruns. Therefore, agile methods seek the "simplest thing that could possibly work" and recommend that this solution be built first. This approach is not intended to preclude further extension and elaboration of the product—instead, it simply says, "Let's get the plain-vanilla version built first." This approach not only mitigates risk but also helps boost sponsor confidence.

Principle 11: The best architectures, requirements, and designs emerge from self-organizing teams.

This principle may sound odd, depending on how literally you read it. Essentially it is saying that to get the best out of people, we have to let them self-organize. People like self-organizing; it allows them to find an approach that works best for their methods, their relationships, and their environment. They will thoroughly understand and support the approach, because they helped create it. As a result, they will produce better work.

However, you may well ask why this principle says that the best architectures, requirements, and designs come from the project team—rather than the organization's best architects, business analysts, and designers, who may not be on the team. In my experience, the answer to this question is that our architectures, requirements, and designs work best when they are implemented by those who originate them. Although external recommendations may have more technical merit on paper, if they are implemented differently than originally envisioned, or without conviction by the team, they will ultimately be less successful.

Self-organizing teams that have the autonomy to make local decisions have a higher level of ownership and pride in the architectures, requirements, and designs they create than in those that are forced on them or "suggested" by external sources. Ideas created by the team have already gone through the team vetting process for alignment and approval—so they don't need to be "sold" to the team. In contrast, ideas that come from outside sources need to be sold to the team for the implementation to be successful, and this is sometimes a challenging task.

Another factor that supports this principle is that the members of a self-organizing project team are closest to the technical details of the project. As a result, they are best able to spot implementation issues, along with opportunities for improvements. So instead of trying to educate external people about the evolving structure of the project, agile methods leverage the capacity of the team to best diagnose and improve the project's architectures, requirements, and designs. After all, the team members are the most informed about the project and have the most vested in it.

Principle 12: At regular intervals, the team reflects on how to become more effective, then tunes and adjusts its behavior accordingly.

Gathering lessons learned at the end of a project is, frankly, too little, too late. Instead, we need to gather our lessons learned while they are still applicable and actionable. This means we need to gather them *during* the project and—most importantly—make sure we do something about what we've learned to adjust how we complete the remainder of the project.

Agile methods employ frequent lookbacks, called "retrospectives," to reflect on how things are working on the project and identify opportunities for improvements. These retrospectives are typically done at the end of each iteration, ensuring that the team has regular opportunities to review their process. One advantage of doing retrospectives so frequently is that we don't forget about problems and issues. Compare this to conducting a single lessons learned review at the end of a project, in which team members are asked to think back over a year or more to recall what went well and where they ran into issues.

Another disadvantage of only gathering lessons learned at the end of a project is that the lessons won't be really helpful to the organization until another project with a similar business or technical domain or team dynamics comes along. And at that point, it is easy to dismiss the lessons learned from an earlier project as not applicable to the current situation. On an agile project, we capture the lessons learned as the project progresses, so we can't pretend they aren't applicable. We know they are relevant, and we are motivated to tune and adjust our process accordingly.

EXERCISE: SEVERE SUMMARIES

To make sure you understand the intent behind each of the Agile Manifesto principles, shorten them to just five words or less that describe the essence of the idea. You can use these shortened descriptions as a memory aid to help you recall these concepts. The first two have been done for you as an example.

	Principle	Shortened Version
1	Our highest priority is to satisfy the customer through early and continuous delivery of valuable software.	*Satisfy customer with great systems*
2	Welcome changing requirements, even late in development. Agile processes harness change for the customer's competitive advantage.	*Welcome change*
3	Deliver working software frequently, from a couple of weeks to a couple of months, with a preference to the shorter timescale.	Deliver functioning software often
4	Business people and developers must work together daily throughout the project.	Daily crossfunctional collaboration

	Principle	Shortened Version
5	Build projects around motivated individuals. Give them the environment and support they need, and trust them to get the job done.	Foster a motivational environment
6	The most efficient and effective method of conveying information to and within a development team is face-to-face conversation.	Communicate face to face
7	Working software is the primary measure of progress.	Working software equals progress
8	Agile processes promote sustainable development. The sponsors, developers, and users should be able to maintain a constant pace indefinitely.	Encourage a sustainable pace
9	Continuous attention to technical excellence and good design enhances agility.	Keep code clean and agile
10	Simplicity—the art of maximizing the amount of work not done—is essential.	Keep it simple
11	The best architectures, requirements, and designs emerge from self-organizing teams.	Trust teams to generate requirements
12	At regular intervals, the team reflects on how to become more effective, then tunes and adjusts its behavior accordingly.	Reflect and adapt

© 2015 RMC Publications, Inc • 952.846.4484 • info@rmcls.com • www.rmcls.com

ANSWER

Your description may vary, depending on what part of the principle stands out most for you, but the following are possible abbreviations.

	Principle	Shortened Version
1	Our highest priority is to satisfy the customer through early and continuous delivery of valuable software.	*Satisfy customer with great systems*
2	Welcome changing requirements, even late in development. Agile processes harness change for the customer's competitive advantage.	*Welcome change*
3	Deliver working software frequently, from a couple of weeks to a couple of months, with a preference to the shorter timescale.	*Deliver frequently*
4	Business people and developers must work together daily throughout the project.	*Work with business*
5	Build projects around motivated individuals. Give them the environment and support they need, and trust them to get the job done.	*Motivate people*
6	The most efficient and effective method of conveying information to and within a development team is face-to-face conversation.	*Face-to-face communications*
7	Working software is the primary measure of progress.	*Measure systems done*
8	Agile processes promote sustainable development. The sponsors, developers, and users should be able to maintain a constant pace indefinitely.	*Maintain sustainable pace*
9	Continuous attention to technical excellence and good design enhances agility.	*Maintain design*
10	Simplicity—the art of maximizing the amount of work not done—is essential.	*Keep it simple*
11	The best architectures, requirements, and designs emerge from self-organizing teams.	*Team creates architecture*
12	At regular intervals, the team reflects on how to become more effective, then tunes and adjusts its behavior accordingly.	*Reflect and adjust*

K&S Agile Methodologies

There are over a dozen actively used agile methodologies. The most common approaches are Scrum, Extreme Programming (XP), lean product development, Kanban, Feature-Driven Development (FDD), Dynamic Systems Development Method (DSDM), and the Crystal family of methods.

In the discussion that follows, we'll describe each of these methodologies in the level of detail you'll need for the exam. The two most widely used agile methods are Scrum and XP, which are prominently featured on the PMI-ACP exam. We'll explore these two approaches in the most detail, covering their key concepts, teams, activities, and deliverables. The exam may also include multiple questions about the lean and Kanban approaches. Although these sections aren't as long, it's important to understand all the information provided about these methods. Finally, the exam might not have any questions about FDD, DSDM, or Crystal— or perhaps just one question about them. However, it's still helpful to have a topline understanding of these methods, since they might appear as incorrect answer options that you need to rule out.

We've said that agile is a mindset, not a specific set of practices, and many agile teams end up tailoring their agile methodology for their own circumstances. This is called process tailoring, and we'll discuss its pros and cons in more detail in chapter 7. At this point, just know that some of these methodologies work best when used "as is" (for teams new to agile), while others lend themselves quite easily to adjustment and adaptation.

For example, Kanban encourages modification and adaptation; Crystal offers a range of methodologies for different kinds of projects; and DSDM offers suitability filters for assessing how well an approach fits a project. On the other hand, both Scrum and XP consist of a balanced set of practices that works best when used "out of the box" as an integrated system, at least until the team has gained enough agile experience to understand how all the pieces fit together.

Exam tip

The different agile methods, and their inherent flexibility, presented a dilemma for PMI in creating the exam, since there is no single "right" way to do agile. When the PMI-ACP Steering Committee originally developed the exam content outline, we recognized the irony of creating a prescriptive exam outline for a scalable, adaptive process. However, we had to make some assumptions about the team size, composition, and project complexity that would be tested. In the end, we decided to focus on the practices that are most commonly used on small- to medium-sized projects. So if you are struggling with the context of an exam question, think of a project that has a development team of up to 12 people and is using discretionary or essential funds.

In this chapter, we'll be describing each agile approach as a standalone methodology, since that information will be tested on the exam. However, it's not uncommon for agile teams to use a customized hybrid approach that incorporates elements from more than one methodology. For example, a team might combine some of XP's team practices, such as refactoring, pair programming, and simple design, with Scrum's approach for planning sprints. We'll talk more about these "hybrid models" in chapter 7, providing some examples of how different approaches can be used together.

Exam tip

I've gotten feedback that some people who are steeped in the Scrum methodology are having difficulty passing the PMI-ACP exam. These students believe that "Scrum *is* agile"—and so by extension, agile is Scrum, and *only* Scrum—and since they already know Scrum, they don't study properly for the exam. For example, they assume that all agile teams have a product owner and work in timeboxed sprints followed by sprint review meetings. However, the PMI-ACP exam tests a much broader picture of agile—it also encompasses Kanban and lean methods that don't have a product owner, timeboxed sprints, or scrum meetings. The underlying principles of all the agile methodologies are still compatible. But if you're mostly familiar with Scrum, you'll need to think of agile more broadly and learn about other agile approaches to pass the exam.

Scrum

Scrum is a popular agile model that is lightweight and easy to understand—but like all agile methods, it is difficult to truly master. The methodology documented in the "Scrum framework" is a set of practices, roles, events, artifacts, and rules that are designed to guide the team in executing the project.

Scrum Pillars and Values

The theory behind Scrum is based on the three pillars of transparency, inspection, and adaptation. These principles guide all aspects of the Scrum methodology:

» **Transparency:** This involves giving visibility to those responsible for the outcome. An example of transparency would be creating a common definition of what "done" means, to ensure that all stakeholders are in agreement.

» **Inspection:** This involves doing timely checks of how well the project is progressing toward its goals, looking for problematic deviations or differences from the goals.

» **Adaptation:** This involves adjusting the team's process to minimize further issues if an inspection shows a problem or undesirable trend.

In addition to the three pillars, Scrum also recognizes five fundamental values—focus, courage, openness, commitment, and respect. These values reflect the concepts laid out in the Agile Manifesto, and they are also similar to the core values of XP (as we'll see when we get to that methodology).

The Scrum process is shown on the diagram below. As you continue reading about Scrum, refer back to this diagram to see how each component fits into the Scrum process.

Figure 1.4: Scrum Process

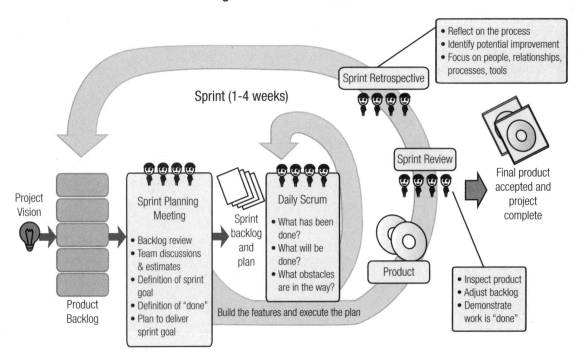

In this section, we'll discuss the key terms and components of Scrum, including the roles on a Scrum team, the activities that occur during a sprint, and the sprint deliverables or artifacts. We'll kick off this discussion by examining a key Scrum concept—sprints.

Sprints

A sprint is a timeboxed (time-limited) iteration of one month or less in which the team builds a potentially releasable product. (The Scrum term "sprint" is equivalent to the generic agile term "iteration.") Most Scrum sprints are one or two weeks long. Each sprint is like a mini-project. During a sprint, no changes are made that would affect the sprint goal. The scope might be clarified or renegotiated as new information becomes available, but if (for example) a change has to be made to the members of the development team, that wouldn't be done during a sprint.

A sprint can be cancelled before the timebox is over if the sprint goal becomes obsolete because of a change in business direction or technology conditions. However, only the product owner can cancel the sprint. When that happens, any incomplete product backlog items are re-estimated and returned to the product backlog.

The sequence of activities within each sprint consists of a sprint planning meeting, a development period that includes daily scrums, a sprint review meeting, and a sprint retrospective meeting. (We'll be explaining these terms shortly.)

Scrum Team Roles

Scrum teams are made up of the development team, the product owner, and the ScrumMaster. Let's look at each of these roles to see what they are responsible for on a Scrum project.

Development Team

The development team is the group of professionals who build the product increments in each sprint. The members of the development team are self-organizing—that is, they are empowered to manage their own work. Scrum teams are also cross-functional; each team member can fulfill more than one of the roles needed to complete the work (such as analysis, build, and test on a software development team).

Product Owner

The product owner is responsible for maximizing the value of the product by managing the product backlog, or list of work to be done. This includes ensuring that the work items in the backlog are up to date and accurately prioritized based on business value. The product owner is also responsible for making sure that the business and the team have a shared understanding of the project vision, the project goals, and the details of the work to be done, so that the team can plan and build the work items.

Although the product owner is ultimately responsible for keeping the backlog prioritized and updated, the ScrumMaster or development team also assists in this process by sharing information about estimates, dependencies, technical work items, and so on.

ScrumMaster

The ScrumMaster is responsible for ensuring that the Scrum methodology is understood and used effectively. This person is a servant leader to the development team, removing any impediments to their progress, facilitating their events (meetings), and coaching the team members. The ScrumMaster also assists the product owner with managing the backlog and communicating the project vision, project goals, and the details of the backlog items to the development team. Finally, the ScrumMaster serves the organization by facilitating its adoption of Scrum, not just on one project, but on a wider scale throughout the organization.

Scrum Activities (Events, Ceremonies)

The Scrum methodology defines five "activities," which are actually meetings that are focused on a specific purpose. These include product backlog refinement, sprint planning meetings, daily scrums, sprint reviews, and sprint retrospectives. The last four of these meetings are the team's planned opportunities for inspection and adaptation (two of the three pillars) in each sprint.

Exam tip

On the exam, you may see the terms "Scrum events" or "Scrum ceremonies." These are older terms for the Scrum activities (meetings).

Backlog Refinement

The backlog refinement meetings are where "grooming the backlog" is done. This basically means that everyone involved in the project gathers to discuss and update the items in the backlog. (If you aren't sure what this means, keep reading—we'll explain this activity more fully when we describe the product backlog.)

Sprint Planning Meeting

In the sprint planning meeting, everyone gathers to determine what will be delivered in the upcoming sprint and how that work will be achieved. The product owner presents the updated backlog items, and the group discusses them to ensure they have a shared understanding. The development team forecasts what can be delivered in the sprint, based on their estimates, projected capacity, and past performance. With this forecast in mind, they define the sprint goal. The development team then determines how that functionality will be built and how they will organize themselves to deliver the sprint goal. (Sprint planning—i.e., iteration planning—is covered in more detail in chapter 5.)

Daily Scrum

The daily scrum is a 15-minute timeboxed meeting that is held at the same time and place every day while the team is working toward the sprint goal. The ScrumMaster makes sure the meeting happens every day, and follows up on any identified obstacles. The daily scrum is primarily for the members of the development team, who use it to synchronize their work and report any issues they are facing. The scope of this meeting is strictly limited—each member of the team briefly answers three questions about what they are doing to meet the sprint goal:

1. What have I done since the last daily scrum?
2. What do I plan to do today?
3. Are there any impediments to my progress?

These questions help the team members assess their progress toward the sprint goal. In answering the questions, they address each other, and any further discussion or problem solving takes place offline. (We'll cover these meetings in more detail in chapter 5, where we talk about daily stand-ups—the generic agile term for "daily scrum.")

As Scrum projects get larger, multiple Scrum teams might need to coordinate their work. To do this, they commonly use an approach called "scrum of scrums." As the name suggests, this is a meeting in which a representative from each team reports their progress to the representatives of the other teams. Just like a normal daily scrum, the participants answer three questions, and they might also address a fourth question to help surface potential conflicts between teams: "Are you about to put something in another team's way?" This process is illustrated below.

Figure 1.5: Scrum of Scrums

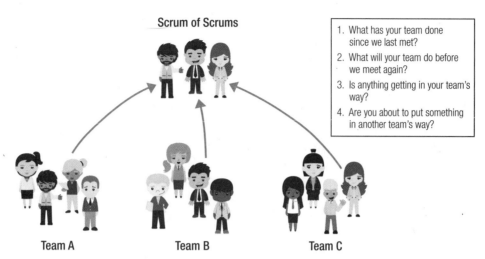

1. What has your team done since we last met?
2. What will your team do before we meet again?
3. Is anything getting in your team's way?
4. Are you about to put something in another team's way?

Like daily scrums, the scrum of scrums meetings take place on a regular schedule, but not necessarily daily. Let's say the individual teams have their daily scrum meeting at 9:00 a.m. After those meetings, at 9:30—perhaps daily or twice a week, as needed—someone from each team will take part in the scrum of scums meeting. If a participant learns about any issues that impact their team, they will report them to the rest of their team after the meeting.

Larger endeavors may even warrant a "scrum of scrum of scrums" where the teams repeat this pattern with a third-level meeting. Here, a representative from each scrum of scrums will attend a "scrum of scrum of scrums" to coordinate the work across a larger set of teams.

Figure 1.6: Scrum of Scrum of Scrums

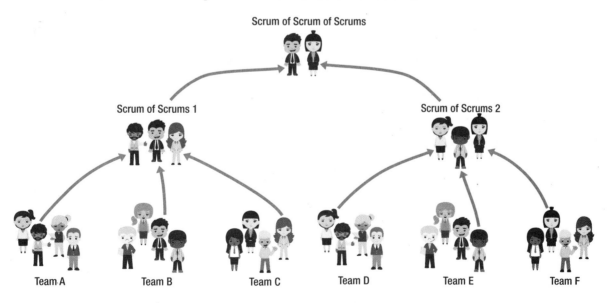

Sprint Review

The sprint review meeting is held at the end of the sprint and includes the development team, the product owner, and the ScrumMaster (and potentially other stakeholders). In this meeting, the team demos the increment, or evolving product, that they built in the sprint to the product owner. The product owner inspects the work to see whether it is acceptable—deciding if it is "done" or explaining what is missing. The team and the product owner discuss the increment and the remaining items in the product backlog. Together, they make any changes needed to the backlog and decide what to work on next.

Sprint Retrospective

After the sprint review, but before the next sprint planning meeting, the development team holds their final "inspect and adapt" activity for the sprint—the sprint retrospective. Although others may be invited to participate, this meeting is primarily for the development team and their ScrumMaster. This is their opportunity to gather lessons learned and look for opportunities for improvement. The timing of this meeting—after the sprint review, but before the next planning meeting—allows them to consider the product owner's feedback from the sprint review in their deliberations and also factor any improvements they identify during the retrospective into their plan for the next sprint.

The reflection that takes place during the retrospective might cover anything that occurred during the sprint, including the areas of people, process, and product. The team members explore what went well, look for opportunities for improvement, and decide what changes to implement in the next sprint.

Scrum Artifacts

Now that we've discussed the Scrum roles and activities, we'll turn to the scrum artifacts. These are three tangible items that are produced or used by the team during a sprint—the product increment, the product backlog, and the sprint backlog.

Product Increment

During each sprint, the members of the development team build an increment of the solution (the end product of the project). As we've seen, in the sprint review, they demo their latest increment to get feedback from the product owner and find out if that item is "done." To maximize the chances that the product increment will be acceptable, the team and the product owner need to agree upon a definition of done before the team begins working on it so that everyone has a shared understanding of what "done" will look like for that increment.

Product Backlog

The product backlog is a prioritized list of all the work that needs to be done to build the product—it serves as the single source for all the product requirements. The items in the backlog may include features to be built, functions, requirements, quality attributes (i.e., nonfunctional requirements), enhancements, and fixes. This list is dynamic; it evolves as the product evolves and needs to be continually updated.

The items in the backlog are prioritized by the product owner and sorted so that the highest-priority work is at the top of the list. The team will always try to work on these top-priority items next—or as soon as possible, given dependencies and other factors. The work items are progressively refined as they get closer to being worked on, so the highest-priority items are the most detailed, and the team's estimates for those items are more precise. The lowest-priority items might not get developed at all, since they may be continually deferred in favor of higher-priority work.

© 2015 RMC Publications, Inc • 952.846.4484 • info@rmcls.com • www.rmcls.com

The backlog is constantly being refined, and the process of adding more detail to the backlog items and refining their estimates is the "backlog refinement" Scrum activity described earlier, which is also known as "grooming the backlog." This activity is done by the development team and the product owner working together. The team estimates the work items, and product owner prioritizes them. For example, each time the team refines their estimates with a higher level of detail, the product owner might need to adjust the priority of those items in the backlog.

Sprint Backlog

The sprint backlog is the subset of items in the product backlog that have been selected as the goal of a specific sprint (the sprint backlog isn't a separate list from the product backlog). Along with the sprint backlog, the team develops a plan for how they will achieve the sprint goal—this is their commitment for the functionality that they will deliver in the sprint. The sprint backlog serves as a highly visible view of the work being undertaken and may only be updated by the development team.

EXERCISE: SCRUM OWNERSHIP/RESPONSIBILITY

In the following table, place a check mark in the appropriate column(s) for the development team, product owner, or ScrumMaster to indicate who owns or is responsible for each item.

Item	Development Team	Product Owner	ScrumMaster	
Estimates	✗			
Backlog priorities		✗		
Agile coaching			✗	
Coordination of work	✗			
The definition of "done"	✗	✗	✗	All
Process adherence			✗	
Technical decisions	✗			
Sprint planning	✗	✗	✗	All

ANSWER

Item	Development Team	Product Owner	ScrumMaster
Estimates	×		
Backlog priorities		×	
Agile coaching			×
Coordination of work	×		
The definition of "done"	×	×	×
Process adherence			×
Technical decisions	×		
Sprint planning	×	×	×

Extreme Programming (XP)

Extreme Programming—commonly referred to as "XP" based on the initials eXtreme Programming—is an agile method that is focused on software development. While Scrum focuses at the project management level on prioritizing work and getting feedback, XP focuses on software development best practices. Therefore, there will be a lot of software references in the following discussion. As with the Agile Manifesto values and principles, as you read this section see if you can think of ways that XP values and practices can be applied to other kinds of knowledge work projects.

XP Core Values

The core values of XP are simplicity, communication, feedback, courage, and respect; these values manifest themselves in the practices undertaken throughout the XP life cycle.

» **Simplicity**: This value focuses on reducing complexity, extra features, and waste. XP teams keep the phrase "Find the simplest thing that could possibly work" in mind, and build that solution first.

» **Communication**: This value focuses on making sure all the team members know what is expected of them and what other people are working on. For example, the daily stand-up meeting is a key communication tool (these meetings will be explained in chapter 5).

» **Feedback**: The team should get impressions of suitability early. Failing fast can be useful, especially if in doing so we get new information while we still have time to improve the product.

» **Courage**: It takes courage to allow our work to be entirely visible to others. In pair programming (described below under "XP Core Practices"), team members share code and often need to make bold simplifications and changes to that code. Backed up by automated builds and unit tests, developers need to have the confidence to make important changes.

» **Respect**: Respect is essential on XP projects, where people work together as a team and everyone is accountable for the success or failure of the project. This value also relates to pair programming; team members need to recognize that people work differently, and respect those differences.

The figure below illustrates the XP process. If some of the terms on this diagram are Greek to you, don't worry; we'll revisit the technical terms used in this diagram and in the discussion of XP that follows in more detail in later chapters.

© 2015 RMC Publications, Inc • 952.846.4484 • info@rmcls.com • www.rmcls.com

Figure 1.7: XP Life Cycle

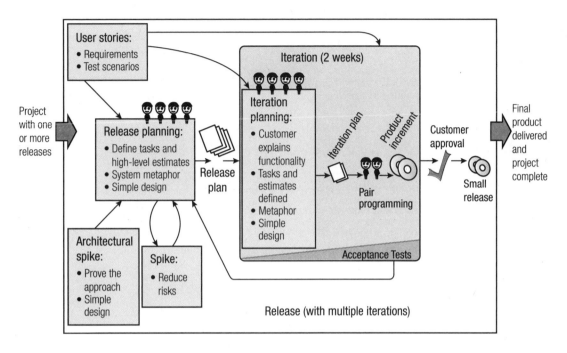

As shown in this diagram, XP teams use lightweight requirements called "user stories" to plan their releases and iterations. Iterations are typically two weeks long, and developers work in pairs to write code during these iterations. All software developed is subjected to rigorous and frequent testing. Then, upon approval by the on-site customer, the software is delivered as small releases.

"Spikes" are periods of work undertaken to reduce threats and issues, and "architectural spikes" are iterations used to prove a technological approach. The spikes are blended into the release planning processes.

XP Team Roles

XP defines the team roles differently than Scrum—the XP roles are coach, customer, programmer, and tester. Let's see how each of these roles participates in an XP project.

Coach

The coach acts as a mentor to the team, guiding the process and helping the team members stay on track. The coach is also a facilitator—helping the team become more effective—and a conduit, reinforcing communication both within the team and across teams. This role shares many responsibilities with a ScrumMaster. Although the formal definitions of these two terms differ, they are often used interchangeably.

In addition to the coach, there may also be a manager who facilitates external communications and coordinates the team's activities, but this isn't a formal XP role.

Customer

On an XP team the "customer" is the business representative who provides the requirements, priorities, and business direction for the project. This person defines the product that will be built, determines the priority of its features, and confirms that the product actually works as intended. This role is similar to the product owner in Scrum.

Programmers

The programmers are the developers who build the product by writing and implementing the code for the requested user stories.

Testers

The testers provide quality assurance and help the customer define and write acceptance tests for the user stories. This role may also be filled by the developers (programmers), if they have the required skills.

> ### Exam tip
>
> Think about the XP and Scrum terms for team roles—you might encounter an exam question that requires you to recognize a methodology based on the terms being used. You should also understand that from a broader agile perspective, "product owner" and "customer" are roughly equivalent roles, as are "ScrumMaster" and "coach."

XP Core Practices

The XP method draws upon 13 simple but powerful core practices, as shown below. We'll examine each of these practices in more detail, beginning with the outer ring and working our way in.

Figure 1.8: XP Core Practices

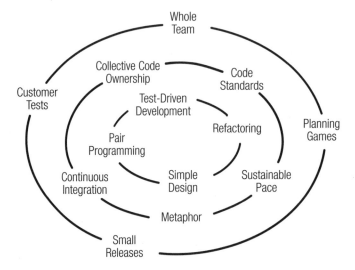

Whole Team

The whole team practice is the idea that all the contributors to an XP project sit together in the same location, as members of a single team. XP emphasizes the notion of generalizing specialists, as opposed to role specialists. In other words, anyone who is qualified to perform a role can undertake it—the roles are not reserved for people who specialize in one particular area. This practice helps optimize the use of resources, since people who can perform multiple jobs are able to switch from one role to another as the demand arises. The practice also allows for more efficient sharing of information and helps eliminate the possibility that people in certain roles will be idle or overstretched at certain points in the project.

Planning Games

XP has two primary planning activities, or planning games—release planning and iteration planning.

A release is a push of new functionality all the way to the production user. A project typically has one or more releases, with no more than one or two releases in a single year. During release planning, the customer outlines the functionality required, and the developers estimate how difficult the functionality will be to build. Armed with these estimates and priorities, the customer lays out the plan for the project delivery. Since the initial attempts at estimating will likely be imprecise, this process is revisited frequently and improved as the priorities and estimates evolve.

Iterations are the short development cycles within a release that Scrum calls "sprints." Iteration planning is done at the start of every iteration. The customer explains what functionality they would like to see in the next iteration, and then the developers break this functionality into tasks and estimate the work. Based on these estimates (which are more refined than the release planning estimates) and the amount of work accomplished in the previous iteration, the team commits to the work items they think they can complete in the two-week period.

Small Releases

Frequent, small releases to a test environment are encouraged in XP, both at the iteration level, to demonstrate progress and increase visibility to the customer, and at the release level, to rapidly deploy working software to the end users. Quality is maintained in these short delivery timeframes by rigorous testing and through practices like continuous integration, in which suites of tests are run as frequently as possible.

Customer Tests

As part of defining the required functionality, the customer describes one or more test criteria that will indicate that the software is working as intended. The team then builds automated tests to prove to themselves and the customer that the software has met those criteria.

Collective Code Ownership

In XP, any pair of developers can improve or amend any code. This means multiple people will work on all the code, which results in increased visibility and broader knowledge of the code base. This practice leads to a higher level of quality; with more people looking at the code, there is a greater chance defects will be discovered. There is also less of an impact to the project if one of the programmers leaves, since the knowledge is shared.

Code Standards

Although collective code ownership has its advantages, allowing anyone to amend any code can result in issues if the team members take different approaches to coding. To address this risk, XP teams follow a consistent coding standard so that all the code looks as if it has been written by a single, knowledgeable programmer. The specifics of the standard each team uses are not important; what matters is that the team takes a consistent approach to writing the code.

Sustainable Pace

XP recognizes that the highest level of productivity is achieved by a team operating at a sustainable pace. While periods of overtime might be necessary, repeated long hours of work are unsustainable and counterproductive. The practice of maintaining a sustainable pace of development optimizes the delivery of long-term value.

Metaphor

XP uses metaphors and similes to explain designs and create a shared technical vision. These descriptions establish comparisons that all the stakeholders can understand to help explain how the system should work. For example, "The billing module is like an accountant who makes sure transactions are entered into the appropriate accounts and balances are created."

Even if the team cannot come up with an effective metaphor to describe something, they can use a common set of names for different elements to ensure everyone understands where and why changes should be applied.

Continuous Integration

Integration involves bringing the code together and making sure it all compiles and works together. This practice is critical, because it brings problems to the surface before more code is built on top of faulty or incompatible designs.

XP employs continuous integration, which means every time a programmer checks in code to the code repository (typically several times a day), integration tests are run automatically. Such tests highlight broken builds or problems with integration, so that the problems can be addressed immediately.

Test-Driven Development

Testing is a critical part of the XP methodology. To ensure good test coverage so that problems can be highlighted early in development, XP teams often use the practice of test-driven development. With this approach, the team writes the acceptance tests prior to developing the new code.

If the tests are working correctly, the initial code that is entered will fail the tests, since the required functionality has not yet been developed. The code will pass the test once it is written correctly. The test-driven development process strives to shorten the test-feedback cycle as much as possible to get the benefits of early feedback.

Refactoring

Refactoring is the process of improving the design of existing code without altering its external behavior or adding new functionality. By keeping the design efficient, changes and new functionality can easily be applied to the code. Refactoring focuses on removing duplicated code, lowering coupling (dependent connections between code modules), and increasing cohesion.

Simple Design

By focusing on keeping the design simple but adequate, XP teams can develop code quickly and adapt it as necessary. The design is kept appropriate for what the project currently requires. It is then revisited iteratively and incrementally to ensure it remains appropriate.

XP follows a deliberate design philosophy that asks, "What is the simplest thing that could work?" as opposed to complex structures that attempt to accommodate possible future flexibility. Since code bloat and complexity are linked to many failed projects, simple design is also a risk mitigation strategy.

Pair Programming

In XP, production code is written by two developers working as a pair. While one person writes the code, the other developer reviews the code as it is being written—and the two change roles frequently. This practice may seem inefficient, but XP advocates assert that it saves time because the pairs catch issues early and there is a benefit in that the two people will have a larger knowledge base. Working in pairs also helps spread knowledge about the system through the team.

By taking a disciplined and rigorous approach to applying these practices, XP teams succeed in delivering high-quality software systems.

EXERCISE: XP CHUNK OR MADE-UP JUNK?

Separate the real XP practices and concepts (the XP Chunks) from the made-up terms (Made-up Junk) in the following table by placing a check mark in the appropriate column.

Item	XP Chunk	Made-up Junk
Sample design		×
Small revisions		×
Courage	×	
Passion		×
Execution game		×
Complexity		×
Refactoring	×	
Remarketing		×
Simplicity	×	
Architectural spoke		×
Iteration	×	
Sustainable pace	×	
Continuous interpretation		×

ANSWER

Item	XP Chunk	Made-up Junk
Sample design		×
Small revisions		×
Courage	×	
Passion		×
Execution game		×
Complexity		×
Refactoring	×	
Remarketing		×
Simplicity	×	
Architectural spoke		×
Iteration	×	
Sustainable pace	×	
Continuous interpretation		×

Lean Product Development

Strictly speaking, lean is not an agile methodology; however, the lean approach is closely aligned with agile, as we'll see. Lean originated in the Toyota Production System that was developed to improve upon Henry Ford's mass production system for building cars.[6] So lean began as a manufacturing approach, which was then applied to software development, and eventually adapted for other kinds of knowledge work.

When referring to lean in an agile context, we are really talking about a subset of lean called "lean product development." While the original lean production systems dealt with manufacturing products, lean product development deals with developing new and better products. The high-level principles of lean product development include:

» Using visual management tools
» Identifying customer-defined value
» Building in learning and continuous improvement

Background information

How Are Agile and Lean Related?

The high-level lean principles listed above are common to Kanban and all agile methods. Although opinions vary, I see lean as a superset of agile and Kanban methods. From this viewpoint, lean product development, Kanban, and agile methods are all specialized instances of lean thinking. They are offshoots that start with lean's general principles and add further guidance for their specific domains. While this viewpoint won't be tested on the exam, it may help you understand the relationship between lean, Kanban, and agile methods.

Figure 1.9: Relationship between Lean, Kanban, and Agile Methods

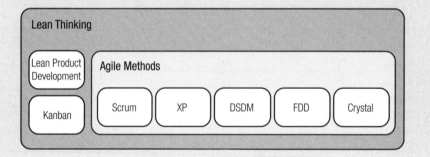

Lean Core Concepts

Lean focuses on seven core concepts, as shown below:

Figure 1.10: Lean Core Concepts

Eliminate Waste — Amplify Learning — Empower the Team — LEAN — Defer Decisions — Deliver Fast — Build Quality In — Optimize the Whole

Let's examine these seven concepts in more detail:

» **Eliminate waste**: To maximize value, we must minimize waste. In knowledge work, waste can take the form of partially done work, delays, handoffs, unnecessary features, etc. Therefore, to increase the value we are getting from projects, we must develop ways to identify, and then remove, waste.

» **Empower the team**: Rather than taking a micromanagement approach, we should respect the team members' superior knowledge of the technical steps required on the project and let them make local decisions to be productive and successful.

» **Deliver fast**: We can maximize the project's return on investment (ROI) by quickly producing valuable deliverables and iterating through designs. We will find the best solution through the rapid evolution of options.

» **Optimize the whole**: We aim to see the system as more than the sum of its parts. We go beyond the pieces of the project and look for how it aligns with the organization. As part of optimizing the whole, we also focus on forming better intergroup relations.

» **Build quality in**: Lean development doesn't try to "test in" quality at the end; instead, we build quality into the product and continually assure quality throughout the development process, using techniques like refactoring, continuous integration, and unit testing.

» **Defer decisions**: We balance early planning with making decisions and commitments as late as possible. For example, this may mean reprioritizing the backlog right up until it is time to do the work, or avoiding being tied to an early technology-bounded solution.

» **Amplify learning**: This concept involves facilitating communication early and often, getting feedback as soon as possible, and building on what we learn. Since knowledge work projects are business and technology learning experiences, we should start early and keep learning.

Exam tip

The exam questions may ask about these seven core concepts, either directly or indirectly. However, you should be aware that these ideas have been worded in different ways. This means that you can't just memorize the seven phrases listed above; you have to understand these concepts well enough to recognize them even if they are worded a bit differently.

The Seven Wastes of Lean

Although it's important to understand all seven of these concepts, the goal of eliminating waste is the primary driver for the lean approach. Lean uses the Japanese term *muda* to refer to the seven kinds of wastes that should be eliminated. Lean experts Mary and Tom Poppendieck, who have written extensively on the use of lean in software projects, have converted the seven traditional manufacturing wastes into seven comparable software development wastes, as shown below. As you read these examples, think about how these wastes could apply to other kinds of knowledge work.[7]

Waste	Description	Example
Partially done work	Work started, but not complete; partially done work can entropy	Code waiting for testing Specs waiting for development
Extra processes	Extra work that does not add value	Unused documentation Unnecessary approvals
Extra features	Features that are not required, or are thought of as "nice-to-haves"	Gold-plating Technology features
Task switching	Multitasking between several different projects when there are context-switching penalties	People assigned to multiple projects
Waiting	Delays waiting for reviews and approvals	Waiting for prototype reviews Waiting for document approvals
Motion	The effort required to communicate or move information or deliverables from one group to another; if teams are not co-located, this effort may need to be greater	Distributed teams Handoffs
Defects	Defective documents or software that needs correction	Requirements defects Software bugs

EXERCISE: CATEGORIZE WASTES

Using the Poppendiecks' table of wastes as a reference, categorize the following day-to-day office wastes shown in the table below. The first one is done for you as an example.

Activity	Type of Waste
Queuing for elevator	*Waiting*
Rebooting a computer after a program crash	Defects, waiting
Saving documents in old formats for compatibility	Extra processeses, motion
Creating notices in French and Spanish to comply with company standards, even though nobody at your location speaks these languages	Extra processes, extra features
Submitting stationery and letterhead orders for approval	Extra processes, motion, partially done work

ANSWER

Activity	Type of Waste
Queuing for elevator	Waiting
Rebooting a computer after a program crash	Defects, waiting
Saving documents in old formats for compatibility	Extra processes, motion
Creating notices in French and Spanish to comply with company standards, even though nobody at your location speaks these languages	Extra processes, extra features
Submitting stationery and letterhead orders for approval	Extra processes, motion, partially done work

Of course, these types of activities aren't always waste. Waste occurs only if there aren't any benefits to be gained from the activity. So while safety notices should be provided for anyone likely to need them, unnecessary translations of such notices can be considered waste. Likewise, approvals for stationery and letterhead orders might be useful to prevent abuse or petty theft, but for people who just need some basic supplies, this process adds no value and can be considered waste.

Lean has contributed important techniques and concepts to agile, including the seven forms of waste, pull systems, value stream mapping, and work in progress, or WIP. (These concepts will be described as we continue.) As we've seen, lean is also the source of the Kanban methodology, which will be discussed next. However, before moving on, try your hand at the next exercise, which is designed to reinforce your understanding of lean principles.

EXERCISE: MATCH AGILE PRACTICES TO LEAN PRINCIPLES

In the following table, match the agile practice to the most relevant lean principle.

Agile Practice	Lean Principle						
	Eliminate waste	Empower the team	Deliver fast	Optimize the whole	Build quality in	Defer decisions	Amplify learning
Teams make their own decisions		X					
Just-in-time iteration planning						X	
Team retrospectives							X
Two-week iterations			X				
Unit test as we go					X		
Shadow the users to learn what they do				X			
The evolving prototype is the specification	X						

ANSWER

While some of these agile practices map to multiple lean principles, the most direct matches are indicated below:

Agile Practice	Lean Principle						
	Eliminate waste	Empower the team	Deliver fast	Optimize the whole	Build quality in	Defer decisions	Amplify learning
Teams make their own decisions		×					
Just-in-time iteration planning						×	
Team retrospectives							×
Two-week iterations			×				
Unit test as we go					×		
Shadow the users to learn what they do				×			
The evolving prototype is the specification	×						

Kanban

The Kanban method is derived from the lean production system developed at Toyota. "Kanban" is a Japanese word meaning "signboard." The signboard, or Kanban board as it is called, plays an important role in the Kanban methodology. This board shows the work items in each stage of the production process, as defined by the team. Here is a simple example of such a board:

Figure 1.11: Kanban Board Example

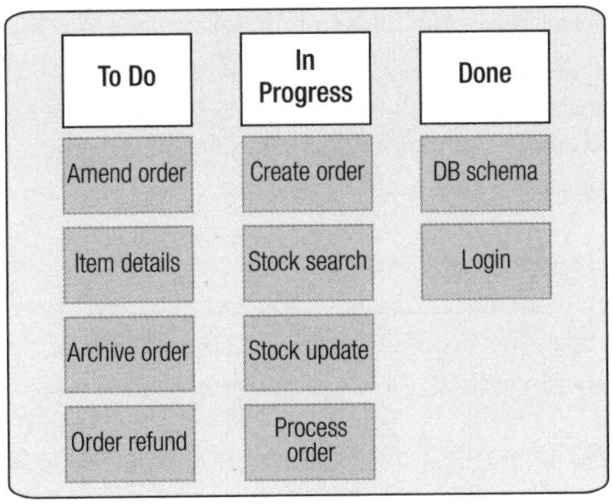

Five Principles of Kanban

Kanban development operates on five core principles:

» **Visualize the workflow**. Knowledge work projects, by definition, manipulate knowledge, which is intangible and invisible. Therefore, having some way to visualize the workflow is very important for organizing, optimizing, and tracking it.

» **Limit WIP (work in progress)**. Restricting the amount of work in progress improves productivity, increases the visibility of issues and bottlenecks, and facilitates continuous improvement. This makes it easier for the team to identify issues and minimize the waste and cost associated with changes. It also results in a steady "pull" of work through the development effort, since new work can only be moved forward as existing work is completed.

» **Manage flow**. By tracking the flow of work through a system, issues can be identified and changes can be measured for effectiveness.

» **Make process policies explicit**. It is important to clearly explain how things work so the team can have open discussions about improvements in an objective, rather than an emotional or subjective, way.

» **Improve collaboratively**. Through scientific measurement and experimentation, the team should collectively own and improve the processes it uses.

Kanban's Pull System

Kanban has some distinct features that differentiate it from Scrum, XP, and generic agile. The main difference you should understand for the exam is that Kanban teams employ a "pull system" to move work through the development process, rather than planning their work in timeboxed iterations. Each time a Kanban team completes an item of work, it triggers a "pull" to bring in the next item they will work on.

As we'll see, only a certain number of slots are available for each column on the Kanban board, and whenever there is an empty slot on the board, that's a signal for the team to pull work from the previous stages, if there are any items available. So work is continuously being pulled from the left side of the board to the right side.

(Note: This might look the same as moving work items through the process from left to right on an agile task board, but it's not. On a Kanban board, capacity is what signals the team to pull work items into the next stage of the process. On an agile task board, the team simply moves the work items from left to right to show their status in the process.)

This pull system means that Kanban has much less emphasis on iterations than agile methods. Although a Kanban team may use iterations if they wish, the pull mechanism just described means that there's no need for it. Of course, most organizations will want the team to commit to a delivery cadence that defines how often the work increments will be available for consumption. However, the team's development cadence doesn't need to be coupled to that delivery cadence, or any aggregation of it.

For example, an agile team might decide to use two-week iterations with deployments into a test environment as well as live deployments to the customers every quarter. So within each quarterly live deployment timeframe (13 weeks) they would typically get six iterations of work done. So, the team knows they will have six iterations between each major release and will plan their features and stories accordingly.

A Kanban team, on the other hand, might still have quarterly customer releases, but they wouldn't use any internal iterations. Instead, any accepted work packages they have ready would be candidates for the next release. They can still plan and track their work for the releases, but instead of using velocity metrics such as points accepted per iteration, they would use cycle time, lead time, and throughput metrics (as described in chapter 6).

WIP Limits in Kanban

WIP limits will be explained more fully in the next chapter—but basically, this term refers to capping the number of items that can be in a given state of progress, as defined by the columns on the team's Kanban board. Once the limit at the top of a column is reached, no new items may be moved into that column until another item is moved out. Here's an example of a Kanban board with WIP limits:

Figure 1.12: Kanban Board with WIP Limits

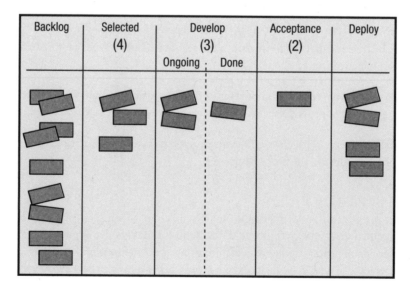

In this figure, we see the team's backlog in the first column. The second column shows that there are three user stories currently selected for development (as shown by the green index cards), which is fine, since the WIP limit at the top of that column is four. The Develop and Acceptance columns also have WIP limits—three and two, respectively.

Why is limiting WIP so important? The reason is that lowering WIP actually increases a team's productivity—it speeds up the rate at which the work is completed. If that claim seems debatable, the relationship between work in progress and productivity has actually been proven mathematically by Little's Law, as illustrated below. This law demonstrates that the *duration* of a queue (how long it will take to complete the work) is proportional to its *size* (how much work is in progress). In other words, teams that limit WIP will get their work done faster.

Figure 1.13: Little's Law

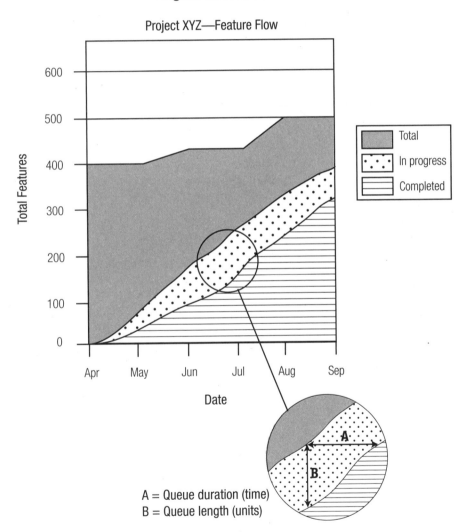

A = Queue duration (time)
B = Queue length (units)

The Kanban development methodology has helped popularize task boards with WIP limits and pull systems created by limiting WIP. It also gives permission and encourages team members to try new approaches and change processes. With a continuous pull model, iterations may not be required, and therefore activities like creating estimates can be considered waste and reduced or eliminated entirely.

Feature-Driven Development (FDD)

Feature-Driven Development (FDD) is a simple-to-understand yet powerful approach to building products or solutions. A project team following the FDD method will first develop an overall model for the product, build a feature list, and plan the work. The team then moves through design and build iterations to develop the features.

Figure 1.14: Feature-Driven Development Process

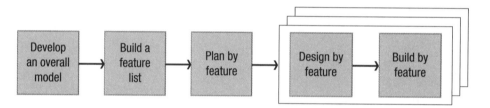

FDD recommends a set of good practices, derived from software engineering. These practices include:

» **Domain object modeling**: In this practice, teams explore and explain the domain (or business environment) of the problem to be solved.

» **Developing by feature**: This involves breaking functions down into two-week or shorter chunks of work and calling them features.

» **Individual class (code) ownership**: With this practice, areas of code have a single owner for consistency, performance, and conceptual integrity. (Note that this is quite different from XP's collective code ownership idea that aims to spread the knowledge to other team members.)

» **Feature teams**: These are small, dynamically formed teams that vet designs and allow multiple design options to be evaluated before a design is chosen. Feature teams help mitigate the risks associated with individual ownership.

» **Inspections**: These are reviews that help ensure good-quality design and code.

» **Configuration management**: This involves labeling code, tracking changes, and managing the source code.

» **Regular builds**: Through regular builds, the team makes sure the new code integrates with the existing code. This practice also allows them to easily create a demo.

» **Visibility of progress and results**: This practice tracks progress based on completed work.

Feature-Driven Development is the agile methodology that popularized cumulative flow diagrams (discussed in chapter 2) and parking lot diagrams, which are one-page summaries of project progress. Both are useful tracking and diagnostic tools that are now used by other agile approaches.

Dynamic Systems Development Method (DSDM)

DSDM was one of the earlier agile methods, and it started out quite prescriptive and detailed. Its coverage of the project life cycle is broad, encompassing aspects of an agile project ranging from feasibility and the business case to implementation. The figure below illustrates the DSDM life cycle.

Figure 1.15: DSDM Life Cycle

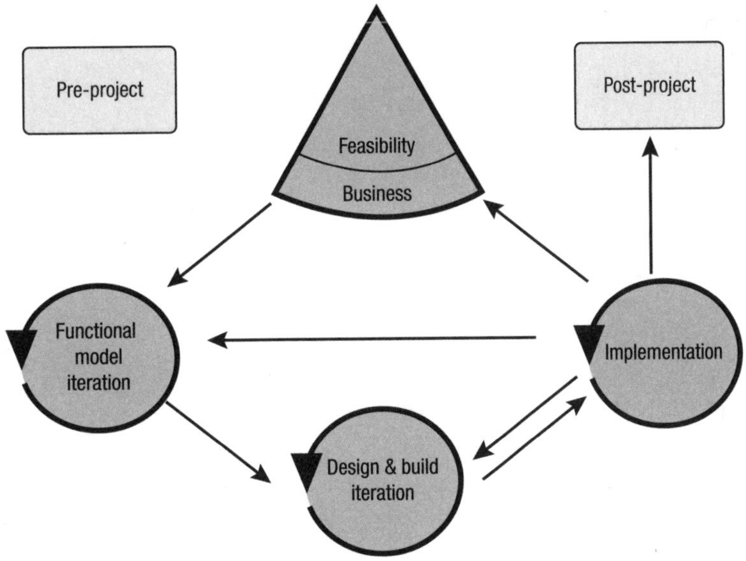

DSDM is centered on eight principles. Although these principles were created before the Agile Manifesto was written, they are closely aligned to the Manifesto. The eight principles are:

1. Focus on the business need
2. Deliver on time
3. Collaborate
4. Never compromise quality
5. Build incrementally from firm foundations
6. Develop iteratively
7. Communicate continuously and clearly
8. Demonstrate control

DSDM has influenced the development of agile by helping to popularize early architectural considerations, agile suitability filters, and agile contracts.

Crystal

Crystal isn't just one method; it consists of a family of situationally specific, customized methodologies that are coded by color names. Each methodology is customized by criticality and team size, which allows Crystal to cover a wide range of projects, from a small team building a low-criticality system (Crystal Clear) to a large team building a high-criticality system (Crystal Magenta).

Figure 1.16: Agile Process Overview

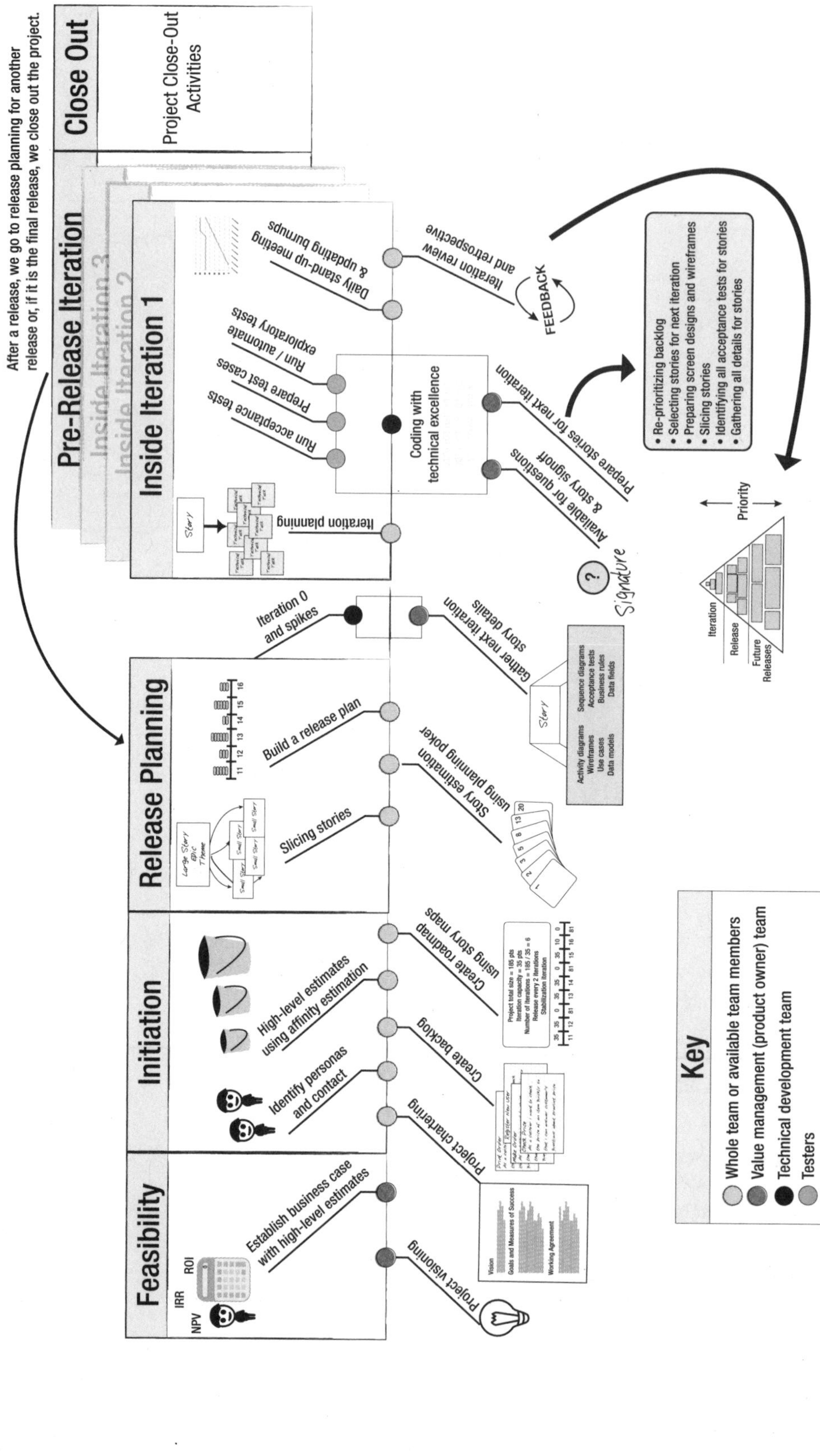

Image credit: Ahmed Sidky, Santeon Group, www.santeon.com

The Crystal framework provides a great example of how agile methods can be tailored to match the specific characteristics of a project. For example, the figure below shows the first five "colors" (methodologies) of Crystal.

Figure 1.17: Crystal Methods

Criticality	Crystal Clear	Crystal Yellow	Crystal Orange	Crystal Red	Crystal Magenta
Life	L6	L20	L40	L100	L200
Essential funds	E6	E20	E40	E100	E200
Discretionary funds	D6	D20	D40	D100	D200
Comfort	C6	C20	C40	C100	C200
	1–6	*7–20*	*21–40*	*41–100*	*101–200*

Team Size

Note: This matrix shows all the potential versions of Crystal—however, the unshaded options aren't recommended because of the project criticality.

In this diagram, team size is shown on the horizontal axis and project criticality is shown on the vertical axis. A project's "criticality" is based on the potential impact of a product defect—whether it could cause loss of comfort, discretionary funds, essential funds, or life. For example, we can see that Crystal Clear is designed for small projects with a core team size of one to six people, in which failure might result in a loss of comfort (for example, a video game crash) or a low-level financial loss (for example, a word processor crash resulting in lost business time). As projects engage more people and develop increasingly critical applications, we need to scale our processes beyond face-to-face communications and introduce additional validation and traceability measures.

For the exam, you don't need to understand the details of how the different Crystal methods work, but you should understand how Crystal uses the factors of criticality and team size to classify projects, as shown above. Crystal's use of this scaling method to determine the best methodology for a project has significantly influenced agile thinking. Another influential concept that Crystal has contributed to agile is the concept of osmotic communication (covered in chapter 4). Crystal methodologies also embrace and promote many other agile principles.

Agile Process Overview

The remaining chapters explain the agile tools, practices, and concepts that are covered in the remaining exam domains. The full-page figure on the facing page provides a high-level, timeline-oriented view of when many of the activities we will be discussing typically occur in a simple agile project, including which activities are iterated (repeated). As you read this book, revisit this diagram as needed to see how the different concepts fit into the overall agile process.

There is one thing to bear in mind, however. This diagram shows the typical process for a generic agile project, but that process isn't the same for lean and Kanban projects. Lean and Kanban teams might not use iterations, instead using a "pull model" to move work through the system. They would still use demonstrations, reviews, and retrospectives, but they wouldn't directly link the timing of these meetings to the delivery of an increment of functionality, as generic agile does at the end of each iteration.

K&S **Agile Leadership**

Agile is more humanistic than mechanistic, as evidenced by the agile value of "Individuals and interactions over processes and tools." The concept of valuing people over processes goes beyond how we manage the work to be done on the project; it also impacts how we organize and motivate our team members, as well as how we assume our role as leaders. Leadership is about tapping into people's intrinsic motivations. To be effective leaders, we need to discover why team members want to do things, understand what motivates them, and then align their project tasks and goals accordingly. It is by aligning project objectives with personal objectives that we can get higher levels of productivity.

Management versus Leadership

The leadership techniques employed on agile projects involve taking an interpersonal approach, rather than directing, command-and-control methods. There is a famous quote from Warren Bennis that speaks to the difference between these two approaches: "Management is getting people to do what needs to be done. Leadership is getting people to want to do what needs to be done."[8]

Instead of telling people what to do, we need to create an environment where people want to do what needs to be done. The difference between the two environments is like the difference between pulling a rope (when people want to do what needs to be done) and pushing the rope (when people are simply told what to do).

Management has a more mechanical focus than leadership; it is concerned with tasks, control, and speed. In contrast, leadership assumes a humanistic focus on people and purpose; it is more concerned with empowerment, effectiveness, and doing the right things. The following table illustrates the differences between a management focus and leadership focus:

Management Focus	Leadership Focus
Tasks/things	People
Control	Empowerment
Efficiency	Effectiveness
Doing things right	Doing the right things
Speed	Direction
Practices	Principles
Command	Communication

So does this mean that leadership is better than management? Can we have just leadership without management? No; we definitely need the mechanics of management in place. But to be truly effective, we need to layer leadership on top of those mechanics. We can best amplify team productivity through a combination of management and leadership.

EXERCISE

Review the activities in the table below, and determine whether they are mainly leadership- or management-based.

Activity	Mainly Leadership or Management?
Human resource management	*mangement*
Career planning	*leadership*
Team time tracking	*Management*
Team member recognition	*leadership*
Task assignment	*Management*
Team brainstorming	*leadership*
Planning workshops	*leadership*
Creating Gantt charts	*management*

ANSWER

Activity	Mainly Leadership or Management?
Human Resource management	Management
Career planning	Leadership
Team time tracking	Management
Team member recognition	Leadership
Task assignment	Management
Team brainstorming	Leadership
Planning workshops	Leadership
Creating Gantt charts	Management

T&T **Servant Leadership**

Agile promotes a servant leadership model that recognizes it is the team members, not the leader, coach, or ScrumMaster, who get the technical work done and achieve the business value. The servant leadership approach redefines the leader's role in relation to the team. It focuses the leader on providing what the team members need, removing impediments to their progress, and performing supporting tasks to maximize their productivity.

There are four primary duties a leader performs in this role of serving the team:

1. **Shield the team from interruptions**. Servant leaders need to isolate and protect the team members from diversions, interruptions, and requests for work that aren't part of the project.

   When business representatives are closely involved in a project, it can be tempting for them to make side requests for changes or enhancements directly to the developers that would sidetrack the planned development effort. While agile projects positively encourage these business insights and requests, they need to go through the proper channels. Business representatives should make such requests during the iteration planning meeting or submit them to the product owner for inclusion in the backlog. Part of being a servant leader involves reminding people about the designated channels so the team can maintain their focus on the iteration and establish a reliable velocity that allows the team's progress to be used to help plan future work.

   Although it is also important to shield the team from internal diversions, the leader must be especially vigilant in protecting the team from external diversions. Breaking people away from focusing on the project work and moving them back and forth between initiatives saps productivity. If the leader is able to shield the team from noncritical external demands, their productivity will be enhanced. Physically co-locating team members is an effective way to prevent external interference. If people are still located in their functional departments or workspaces, it is too easy for them to be drawn into nonproject work.

2. **Remove impediments to progress**. Servant leaders need to clear obstacles from the team's path that would cause delay or waste. These obstacles may include documentation or compliance activities that divert the team from completing the objectives of the current iteration. In lean terms, compliance work refers to efforts that do not directly contribute toward delivering business value. This might include duplicated time recording tasks, nonproject meetings, and other administrative activities.

   At the daily stand-up meeting where the team reports on its progress, planned work, and issues, the leader needs to note the issues and work to resolve them that same day, if possible. Removing or easing such impediments will allow the team to work faster and ultimately deliver more value to the business.

   Some agile project management tools now support impediment backlogs. These backlogs are a kind of prioritized obstacle removal list. Servant leaders can use such tools to track their impediment removal work.

3. **Communicate (and re-communicate) the project vision.** This may seem like an odd duty to place in the category of servant leadership, but communicating and re-communicating the project vision is critical to successfully leading a team. Only if stakeholders have a clear image of the goals for the completed product and project can they align their decisions with, and work toward, the common project objective. In their best-selling book *The Leadership Challenge*, James Kouzes and Barry Posner say that leaders need to reveal a "beckoning summit" toward which others can "chart their course."[9] Simply stated, a common vision helps to keep people all pulling in the same direction.

   Divergent views commonly develop between well-intentioned team members. In software projects, for example, a developer's desire for simplicity or a new technology can cause his or her work to diverge from the user's requirements. An analyst's or quality assurance specialist's desire for completeness and conformance may diverge from the sponsor's requirements for progress and completion. Communicating and re-communicating the project vision helps stakeholders recognize these divergences and bring them back in line with the project's objectives.

   Kouzes and Posner found that the most effective leaders dedicate a much higher percentage of their work time to communicating and re-communicating the project and corporate vision than people in the lower leadership levels. They believe it is almost impossible for leaders to overcommunicate project vision, and state that it is a critical step for effective leadership.

   So agile projects should not just have a vision exercise at the project kickoff or when developing the iteration goals. Such a limited focus is not enough. Instead, servant leaders need to continually look for opportunities to communicate the project vision and find new ways to illustrate and reinforce that vision.

4. **Carry food and water.** This duty isn't literally about food and water; it's about providing the essential resources a team needs to keep them nourished and productive. Such resources could include tools, compensation, and encouragement. People who are fueled by professionalism and duty alone can't continue to contribute to the best of their ability, iteration after iteration. Leaders need to learn what motivates their team members as individuals and find ways to reward them for good work. As a simple measure, a great place to start is a sincere "thank you" to someone for their hard work.

   Leaders also need to celebrate victories—the large ones, of course, but also the small ones—as the project progresses. It is often tempting to save the project celebrations for the end, but if the team members aren't receiving some regular recognition, the project may never reach a successful conclusion. Celebrations and recognition help build momentum, and leaders need to nourish their teams with such rewards frequently to keep the project moving forward productively.

   Training and other professional development activities are also examples of resources the team may need to be productive. The ScrumMaster or coach should take an interest in and arrange appropriate training for the individuals on the team. By building the team members' skills, the project will gain the benefits of their new knowledge. Such actions also show that we're not just trying to extract work and information from the team members, we also want them to grow as individuals.

Figure 1.18: Servant Leadership Approach

Image copyright © Leading Answers, Inc. Reproduced with permission from Leading Answers, Inc. www.leadinganswers.com

Twelve Principles for Leading Agile Projects

In addition to the four core duties we've just discussed, there are other activities that servant leaders should keep in mind. Jeffrey Pinto, in *Project Leadership: from Theory to Practice*, offers the following great list of principles for leaders to follow:[10]

» **Learn the team members' needs.** Find out what motivates and demotivates people.

» **Learn the project's requirements.** Talk to the customers and sponsors; find out their priorities.

» **Act for the simultaneous welfare of the team and the project.** Balance and promote the needs and desires of both the team and the other project stakeholders.

» **Create an environment of *functional accountability.*** Make sure people know what success and failure look like, and empower the team to self-organize to reach the goal. Be proud of accomplishments, but don't hide or shy away from failures—instead, examine them, learn from them, and adapt.

» **Have a vision of the completed project.** Create a "beckoning summit" to which others can chart their own course. When we are head-down in the weeds, it's good to know where we are trying to get to, so we can navigate our own course.

» **Use the project vision to drive your own behavior.** Model action toward the project goals.

» **Serve as the central figure in successful project team development.** Model desired behavior for the team.

» **Recognize team conflict as a positive step.** Unfiltered debate builds strong buy-in for well-discussed topics.

» **Manage with an eye toward ethics.** Be honest and ethical, because people don't want to be associated with goals or missions they feel are unethical.

» **Remember that ethics is not an afterthought, but an integral part of our thinking.** You cannot add trust in later—like quality, it has to be a core ingredient.

» **Take time to reflect on the project.** Review, diagnose, and adapt; improve through progressive change and learning.

» **Develop the trick of *thinking backwards*.** This means imagining that we have reached the end goal and then working backwards to determine what had to happen to get there and what problems and risks we were able to avoid. So first discuss and decide what "done" will look like; then chart the path to get there, and plan how you will avoid any obstacles in your way.

We've already discussed some of these principles in other parts of this chapter, but this list provides a nice summary to keep in mind as we strive to be effective servant leaders.

Exam tip

You don't need to memorize these leadership principles, or the leadership practices we'll discuss next—the exam won't test these concepts directly. Instead, these are the unstated assumptions behind the exam questions that deal with leadership scenarios. So as you read this section, think about the implications of this approach to leadership, and aim to integrate these ideas into your agile mindset.

Agile Leadership Practices

As leaders, we can help create a productive project environment by using practices like modeling the behavior we want the team to follow, communicating the project vision, enabling stakeholders to act, and being willing to challenge the status quo.[11]

Model Desired Behavior

In *The Leadership Challenge*, Kouzes and Posner describe a 10-year study that asked more than 75,000 people, "What values, personal traits, or characteristics do you look for or admire in your leader?"[12] The following were the highest-ranked values:

» **Honesty**: People will not follow leaders they know are deceptive, since doing so undermines their own feelings of self-worth. Therefore, we should pay special attention to transparency and make sure we follow through on what we say we will do. We shouldn't hide our mistakes—we should admit them openly. This is not only a healthy approach for us as leaders, but it also sets an example for how we want our team to operate. We shouldn't do things like ask our team members for estimates and then say we will double the estimates before we give them to management. Such statements hurt our credibility with the team and give them reason to doubt our integrity. (A better approach would be to explain the concept of adding a contingency to estimates and discuss how to base that contingency on realistic expectations of the risks involved in the project.)

» **Forward-looking**: People expect those who lead them to understand where they are going. Leaders should be able to paint the picture for the team so everyone understands what they are ultimately aiming for.

» **Competent**: Leaders don't need to have the strongest technical skills on the team, since team members are typically happy to provide specialist knowledge when required. However, leaders should be competent and not be an embarrassment or liability to the group.

» **Inspiring**: People want to be inspired in their work, rather than be met each day with a sense of doom and gloom. Therefore, leaders need to find ways to explain the project's vision and journey with genuine enthusiasm and spirit.

When we embody these traits as leaders, not only do we encourage people to follow us, we also model the behaviors that we want our team members to emulate. We are, in effect, leading by example.

Communicate the Project Vision

In our earlier discussion of the duties of a servant leader, we talked about the importance of communicating and re-communicating the project vision to keep stakeholders aligned with the project objectives. A leader can use a variety of practices to achieve this, based on what is most effective for the particular team. For example, XP teams use metaphors, some teams develop mantras, and other teams create elevator pitches or project Tweets, in which they explain the project purpose in 140 characters or less. Whatever the method used, it is important to frequently communicate the project's goals and objectives to ensure that all stakeholders are aware of and aligned with the vision.

Enable Others to Act

In order to enable our team members to feel confident in making decisions and taking actions that move the project forward in a productive way, we need to foster a collaborative environment. This involves building trust among team members and strengthening others by sharing power. We also need to create a safe work environment where people are not afraid to ask what they may think are dumb questions. People learn much more quickly when they can raise questions without fear of reprisal or ridicule.

Strengthening others by sharing power means the project manager, ScrumMaster, or leader does not keep the project plan or estimates to him- or herself. Instead, the leader makes sure that information and knowledge are spread throughout the team. For example, this might involve switching from planning tools like Gantt charts that only one or two people can update and maintain to using a task board that the whole team can engage with. In doing so, the planning and status information will be more accessible to the team, and the project will benefit because more people are able to vet, update, and optimize the plan.

Figure 1.19: Switching from Exclusive Tools Like Gantt Charts to Inclusive Tools Like Task Boards

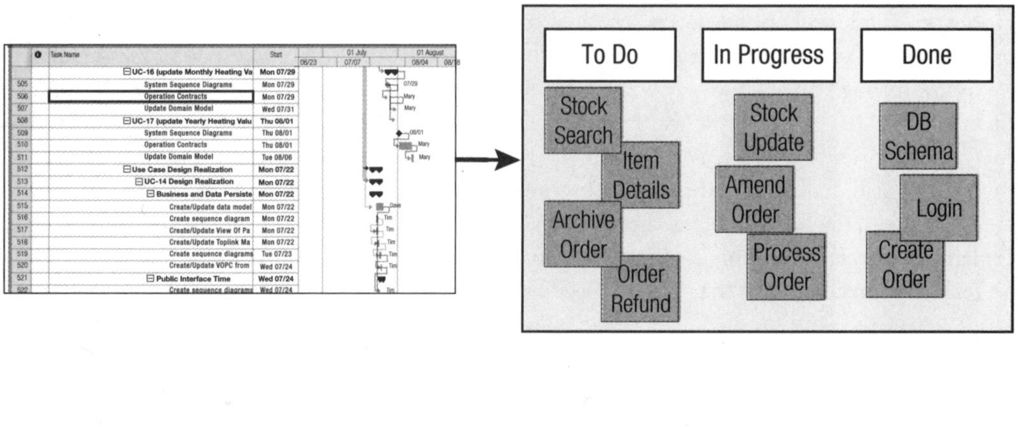

Be Willing to Challenge the Status Quo

Challenging the status quo means we search for innovative ways to change, grow, and improve—and then experiment and take risks by constantly generating small wins and learning from our mistakes. Iterations are perfect microcosms for experimentation. We can try new ideas for one or two iterations before committing to them. If the ideas work, we can institutionalize them. If they do not work, it's no big loss; at least we tried, and we learned something from the experience.

Allowing stakeholders to suggest new ideas for improvement and then giving them a chance to try out those ideas is one way to cement the concept that everyone's ideas have value. There is nothing more disheartening than having a good idea fall on deaf ears. If this happens, people will soon stop trying to make suggestions and will no longer care about the project. So to keep our stakeholders engaged, we should take advantage of the opportunities agile projects present us for small-scale, localized experiments in a supportive, low-risk environment.

As leaders, we need to encourage our team to challenge the status quo of how we operate, not only because the team members are in a great position to suggest process improvements, but also because doing so helps to motivate them. To be successful in this effort, we need to have analytical thinking skills to help our team brainstorm ideas and solutions and active listening skills to make sure we accurately understand their suggestions.

Leadership Tasks

Before moving on to the next chapter, we'll conclude our discussion of the agile principles and mindset by examining some of the leadership tasks in the exam content outline for this domain. Although these concepts will be discussed in more detail in later chapters, we'll introduce them here to help you understand not only these specific Domain I tasks, but also the underlying goals and values of the agile approach to leadership.

Practice Transparency through Visualization

"Transparency" means being open and honest, not only about our progress and achievements, but also about our issues and setbacks. Agile teams demonstrate transparency by openly displaying their work, progress, and review findings for other stakeholders to see. When walking into an agile team room, it's normal to see graphs showing velocity, defect rates, and the results of their last retrospective—including what is working well and what needs improvement. When there is nothing to hide and everything is shared, conversations can be more frank, and people can be less guarded and focus on improvements. People won't be scared or reluctant to discuss problems or mistakes, because the transparent environment all around them sets the example that it is accepted and expected to talk about issues as well as progress.

Create a Safe Environment for Experimentation

When people believe that it's okay to try new approaches and they won't be criticized if they don't succeed the first time, they are more likely to have a go and try new things. It is during these experiments that the most useful innovations and breakthroughs are often made. It's a common pattern that as people get more established in their careers, they grow more concerned about preventing failure and looking inferior to their peers, which can limit their willingness to experiment.

If a friend asks us to help out over the weekend as an entertainer at a children's party, we might try a number of different games to keep the kids happy. If the children don't like the first one, it's not a big problem—we'll just try something else. However, if we happen to have a PhD in child psychology, many published papers on child happiness, and a counseling practice at a high hourly rate, then fears for our credibility and reputation might limit our willingness to try new approaches.

On agile teams we need to create a safe environment for experimentation. This includes encouraging trials and tests, not dwelling on experiments that have failed, and encouraging the generation of new ideas. At Toyota they have a suggestion box for employees to submit ideas. Now, this is nothing new; most organizations have such a box. In most companies management will select the best one or two ideas to implement and then give the submitter a big bonus if their suggestion works out. However, Toyota gives small rewards to everyone who submits an idea. This results in *lots* of suggestions. In Matthew May's book, *The Elegant Solution*, he says that Toyota implements over one million suggestions each year! That's 3,000 suggestions or experiments per day. When we're innovating at that speed, it's really hard for competitors to catch up.[13]

Experiment with New Techniques and Processes

Not all suggestions for improvements are successful. Some likely-sounding ideas fail, and some unlikely ideas may work very well. The only real way to find out if a new process or technique is going to work for your project is to try it out. One important agile element that supports experimentation is short iterations. They provide perfect test cases for short experiments to see what works and should be institutionalized, and what does not and should be dropped.

Geneticists like to study inheritance patterns in mayflies because these insects reproduce in days, so they get their results quickly (also, it is illegal to breed people for scientific purposes). Short iterations offer the same short feedback cycles and provide ready-made test periods for new approaches.

Share Knowledge through Collaboration

It's important for teams to share knowledge about the product they are working on and how things work within their team. However, writing all this down takes a long time, and people often don't like to read it anyway. Engineers also have a deserved stereotype of not liking to produce documentation when they could be "doing real work" and even less of a penchant for "reading the instructions" when they could be tinkering.

This is why collaborative work—either through pairing workers or ad hoc knowledge transfer through co-location—has emerged as an agile best practice. By working with people we learn what they do, how they solve problems, and how to help them most effectively when they are stuck. By simply changing who people work with, project knowledge can be dispersed throughout the team, reducing the impact of information loss if a team member leaves.

Encourage Emergent Leadership via a Safe Environment.

Emergent leadership is when a team member takes the initiative and tries a new approach after gaining team approval. Agile team members are not only empowered to make decisions but also to lead the charge on improvements. If someone identifies a more efficient process, they are encouraged to try an experiment to prove if it can work.

Chapter Review

As explained in the study plans at the start of this book, it's best to wait to take the chapter review quizzes until you have read through the entire book at least once.

1. What is most important for your agile team to continuously focus on?

    A. Getting the right answers
    B. Understanding their tasks
    C. Defining their tasks
    D. Measuring their performance

2. What wouldn't be a key focus of your agile approach?

    A. Increasing return on investment
    B. Expecting change and uncertainty
    C. Measuring progress
    D. Working incrementally

3. You've been asked to recommend how a team should transition to using agile. How would you reply?

    A. Try out some agile practices first to see if they are helpful in your situation.
    B. Hire the best ScrumMaster you can afford and make that person accountable for the transition.
    C. Identify a successful agile team and copy what they are doing.
    D. Learn agile values and principles and use them to guide which practices to adopt in your situation.

4. On a Scrum team, whose responsibility is it to keep the priorities in the product backlog up to date as changes occur?

    A. Product owner
    B. ScrumMaster
    C. Sponsor
    D. Development team

5. As an agile leader, what would be your highest priority?

    A. Resolving conflicts and disagreements
    B. Keeping the team members well fed
    C. Making sure the team members understand what the project is trying to accomplish
    D. Controlling the team's performance

6. What isn't something that your agile team should focus on?

    A. Reflecting on their mistakes and how to improve
    B. Using feedback loops to discover the weaker performers
    C. Failing fast with learning
    D. Learning through trial and error

7. Your team is tasked with developing a breakthrough medical device, and they don't know what the final product will look like. How would you advise them to proceed?

   A. Try the most promising approaches in short iterations and learn as you go.
   B. Perform a risk-based spike.
   C. Ask the customer for more detailed information about the product.
   D. Follow the project charter as closely as possible.

8. The sponsor wants to know if the product you're building will be ready to demo at a trade show. What do you tell her?

   A. It will be ready when it's ready.
   B. We'll get your top-priority functionality done by then.
   C. We'll let you know as soon as our velocity has stabilized.
   D. That depends on your budget.

9. As ScrumMaster, you assess that the competitive market has shifted and the product the team is building is no longer viable. What should you do?

   A. Request that the product owner immediately re-prioritize the backlog.
   B. Cancel the project.
   C. Alert the team that they can expect to be assigned to a new project soon.
   D. Ask the product owner if the viability of the project has changed.

10. Which of the following would NOT an advantage of limiting your team's work in progress?

    A. It reduces the potential need to rework a large collection of flawed, partially completed items.
    B. It helps optimize throughput to make processes work more efficiently.
    C. It brings bottlenecks in the production process to the surface so they can be identified and resolved.
    D. It maximizes resource utilization to make processes work more efficiently.

11. The relationship between leadership and management in agile methods is:

    A. Leadership replaces all aspects of management.
    B. Leadership is subsidiary to management.
    C. Management and leadership are used together.
    D. Management and leadership are incompatible.

12. On a typical agile team, who has the best insight into task execution?

    A. Project manager
    B. Team members
    C. ScrumMaster
    D. Agile coach

13. The four primary roles of a servant leader include:

    A. Shielding team members from interruptions
    B. Resolving conflicts
    C. Determining which stories to include in an iteration
    D. Assigning tasks to the team members

14. In Scrum, the definition of done is created with the input of everyone except the:

    A. Development team
    B. Product owner
    C. ScrumMaster
    D. Process owner

15. In the lean approach, which of the following wouldn't be considered an example of one of the seven forms of waste?

    A. The handoff between coding and testing
    B. Testing the code
    C. Code that is waiting for testing
    D. Assigning a developer to work on two projects at the same time

16. Which of the following is one of the planned opportunities for inspection and adaptation in the Scrum method?

    A. Velocity review meeting
    B. Sprint risk meeting
    C. Daily scrum
    D. Retrospective planning meeting

17. The agile triangle of constraints is said to be inverted from the traditional triangle because it allows:

    A. Scope and time to vary instead of cost
    B. Cost and time to vary instead of scope
    C. Scope and cost to be fixed instead of time
    D. Scope to vary while time and cost are fixed

18. Which of the following Agile Manifesto principles reflects the agile focus on team empowerment?

    A. Working software is the primary measure of progress.
    B. Welcome changing requirements, even late in development.
    C. Simplicity—the art of maximizing the amount of work not done—is essential.
    D. Build projects around motivated individuals.

19. The Kanban pull system means that:

    A. Kanban team members "pull" work from each other, pairing up as needed.
    B. Each time a work item is completed, the next work item is "pulled" into that stage of the process.
    C. Iterations are "pulled" into the process as needed to keep the work organized.
    D. Kanban teams have shorter work queues, which means that the work takes longer to complete (as shown by Little's Law).

20. Which of the following isn't a core aspect of the agile mindset?

    A. Welcome change
    B. Learn through discovery
    C. Respect the process
    D. Deliver value continuously

Answers

1. **Answer**: B

   Explanation: This question tests your high-level understanding of knowledge work versus industrial work. Getting the right answers, defining tasks, and measuring performance are emphasized more in industrial work than knowledge work. Knowledge workers focus instead on understanding their tasks. Although all of these activities are performed on an agile project, notice that the question asks what the team should "continuously focus on." This is a clue that the question is looking at the team's high-level process, not specific parts of their work.

2. **Answer**: C

   Explanation: The best way to approach this question is by a process of elimination. You should be able to recognize that three of the options are essential elements of the agile mindset: increasing return on investment, expecting change and uncertainty, and working incrementally. Once you have eliminated those three options, then the remaining option (measuring performance) must be the correct answer. Although agile teams certainly measure their progress, this isn't one of the basic tenets of agile.

3. **Answer**: D

   Explanation: This question tests your understanding of "being" agile versus "doing" agile. We need to first understand and integrate the mindset behind agile practices before we can use them effectively. Therefore the best way to transition to using agile is to learn agile values and principles and use them to guide which agile practices to adopt in your situation.

4. **Answer**: A

   Explanation: One of the most basic things you should understand about agile teams (whether in Scrum or any other agile methodology) is that it is the product owner/customer who prioritizes the product backlog and keeps the priorities up to date. That's because the product owner represents the business and therefore understands the value the product is expected to deliver better than the other members of the team, who are focused on other areas.

5. **Answer**: C

   Explanation: Like many questions you will encounter on the PMI-ACP exam, you might find this question to be a bit tricky. When faced with a tricky question, try to rule out the incorrect answers first. "Keeping the team members well fed" wouldn't be the leader's responsibility except in an emergency situation, since one would assume that normally the team members could seek out food and feed themselves. "Resolving conflicts and disagreements" and "controlling the team's performance" are done by an agile leader as needed, but are they really the leader's HIGHEST priority? Probably not. (Also, the term "controlling" implies a command-and-control approach rather than servant leadership.) That leaves "making sure the team members understand what the project is trying to accomplish." If you think about it, this is simply another way of stating one of the four primary duties of a servant leader, to "communicate and re-communicate the project vision." So this is the correct answer.

6. **Answer**: B

    Explanation: All the choices are important aspects of the agile mindset except using feedback loops to discover the weaker performers on the team. Agile teams use feedback loops to improve their ability to deliver value, not to critique the performance of individual team members—that wouldn't be aligned with the agile values of respect and empowering the team.

7. **Answer**: A

    Explanation: In the uncertain environment of knowledge work projects, the agile method is to try the most promising approaches in short iterations and learn as we go. (This is a key agile theme that is emphasized throughout the book.) A risk-based spike and gathering more information might be good ideas, but they aren't the BEST answer. Agile charters usually don't provide much detail or technical guidance for the team.

8. **Answer**: B

    Explanation: Like many questions you will encounter on the PMI-ACP exam, this question tests your grasp of agile concepts in an indirect way. In this case, you need to know that agile teams will typically fix time and cost and allow scope to vary, as reflected in the agile triangle. When balancing constraints, the agile approach is to get the highest-priority product functionality done by the customer's deadline.

9. **Answer**: D

    Explanation: This question tests your understanding of agile team roles. In reading the scenario, you should have noticed that it implies the ScrumMaster is stepping outside of their role, since that role doesn't include assessing the value of the project. The ScrumMaster can't cancel the project because they don't make those kinds of decisions. They also wouldn't direct the product owner what to do when. The only person on a Scrum team who has the information needed to assess the viability of the project is the product owner. So the correct answer is "ask the product owner if the viability of the project has changed."

10. **Answer**: D

    Explanation: Since this question is looking for the answer that is NOT an advantage of limiting work in progress (WIP), if an option is true, that means it is not the answer we are looking for. Limiting WIP does reduce the potential need for rework. It also improves process efficiency and helps us find production bottlenecks. The only option listed here that is NOT an advantage of limiting WIP is the one that refers to maximizing resource utilization. Limiting WIP focuses on optimizing throughput, not resources, and we may actually decrease resource optimization to get more throughput.

11. **Answer**: C

    Explanation: Agile methods employ a combination of management and leadership. Leadership neither totally replaces nor is subsidiary to management—and since the two approaches can be used together, they aren't incompatible.

12. **Answer**: B

    Explanation: Agile's servant leadership approach recognizes that the "doers" of the work, the team members, are closest to the work and therefore have the best insight into its execution. This is why agile project managers, ScrumMasters, and coaches defer to the team's decisions about how best to execute the work.

© 2015 RMC Publications, Inc • 952.846.4484 • info@rmcls.com • www.rmcls.com

13. **Answer**: A

    Explanation: The four primary roles of a servant leader are shielding the team from interruptions, removing impediments to progress, communicating the project vision, and "carrying food and water." Although we haven't discussed the other three answer options in chapter 1, we'll see in later chapters that they aren't correct. Servant leaders don't take the lead in resolving conflicts—they first let the team try to resolve issues on their own. The team determines which stories to include in an iteration, rather than accepting the stories chosen by the team leader. Finally, task assignment isn't a servant leadership role; agile teams are encouraged to select their own work.

14. **Answer**: D

    Explanation: The whole team, including the development team, product owner, and ScrumMaster, is responsible for creating the shared definition of done. Since "process owner" is a made-up term, this is the correct choice for someone who would NOT be involved in defining done.

15. **Answer**: B

    Explanation: Testing the code is the only one of these options that wouldn't be considered waste in lean. The seven wastes of lean are partially done work, extra processes, extra features, task switching, waiting, motion, and defects. Handoffs are an example of motion, code that is waiting for testing is partially done work, and a developer who is working on multiple projects is an example of task switching.

16. **Answer**: C

    Explanation: The meetings that are Scrum's planned opportunities for inspection and adaptation are the sprint planning meeting, daily scrum, sprint retrospective, and sprint review. Velocity review meetings, sprint risk meetings, and retrospective planning meetings are not recognized Scrum events.

17. **Answer**: D

    Explanation: Unlike the traditional constraint triangle, in which scope is fixed and time and cost may need to bend to achieve that planned scope, agile teams typically allow scope to vary within fixed parameters of cost and time. In other words, they aim to deliver the most value they can by X date within X budget.

18. **Answer**: D

    Explanation: Agile Manifesto principle five, "Build projects around motivated individuals" addresses the importance of giving teams the environment and support they need, and trusting them to get the job done. Supporting and trusting the team members means recognizing that they are experts at what they do, and that they can work most effectively if they are empowered to plan and organize their own work.

19. **Answer**: B

    Explanation: In Kanban's "pull system," each time the team completes an item of work, it triggers a "pull" to bring in the next item to that stage. The other answer options are all incorrect. Although Kanban teams do tend to have shorter work queues (due to the pull system and limiting WIP), Little's Law actually demonstrates that shorter queues reduce how long it takes to complete the work, rather than lengthening it.

20. **Answer**: C

   Explanation: The concepts of welcoming change, learning through discovery, and continuous delivery of value are all core aspects of the agile mindset. That leaves the option "respect the process" as the outlier. Also, the first value of the Agile Manifesto—"individuals and interactions over processes and tools"—is another indication that "respect the process" isn't part of the agile mindset.

CHAPTER 2

Value-Driven Delivery

Domain II Summary

This chapter discusses domain II in the exam content outline, which is 20 percent of the exam, or about 24 exam questions. This domain deals with maximizing business value, including prioritization, incremental delivery, testing, and validation.

Key Topics

- » Agile contracting
- » Agile project accounting principles
- » Agile risk management
- » Agile tooling
- » Compliance/regulatory compliance
- » Cumulative flow diagrams (CFDs)
- » Customer-valued prioritization
- » Earned value management (EVM) for agile projects
- » Frequent verification and validation
- » Incremental delivery
- » Managing with agile KPIs
- » Minimal viable product (MVP)
  - – Minimal marketable feature (MMF)
- » Prioritization schemes
  - – Kano analysis
  - – MoSCoW
- » Relative prioritization/ ranking
- » ROI/NPV/IRR
- » Software development practices
  - – Continuous integration
  - – Exploratory and usability testing
  - – Red, Green, Refactor
  - – TDD/TFD/ATDD
- » Task/Kanban boards
- » Value-driven delivery
- » Work in progress (WIP)
  - – WIP limits

Tasks

1. Plan work incrementally.
2. Gain consensus on just-in-time acceptance criteria.
3. Tune process to organization, team, and project.
4. Release minimal viable products.
5. Work in small batches.
6. Review often.
7. Prioritize work.
8. Refactor often.
9. Optimize environmental, operational, and infrastructure factors.
10. Review and checkpoint often.
11. Balance adding value and reducing risk.
12. Reprioritize periodically to maximize value.
13. Prioritize nonfunctional requirements.
14. Review and improve the overall process and product.

© 2015 RMC Publications, Inc • 952.846.4484 • info@rmcls.com • www.rmcls.com

83

Delivering value, specifically business value, is a core component of agile methods. This concept is woven into the agile DNA in both the Agile Manifesto values ("Working software over comprehensive documentation"[1]) and the Agile Manifesto principles ("Deliver working software frequently" and "Working software is the primary measure of progress"[2]). In this chapter, we'll organize the agile practices involved in this domain into five themes: assessing value, prioritizing value, delivering incrementally, agile contracting, and finally, verifying and validating value. (The themes that we're using to organize the topics in each chapter aren't official terms and won't appear on the exam; we are simply using these categories to organize and set the context for our discussion of the agile concepts, tools, and practices covered in each chapter.)

What Is Value-Driven Delivery?

Let's start by defining value-driven delivery. The reason projects are undertaken is to generate business value, be it to produce a benefit or improve a service. Even safety and regulatory compliance projects can be expressed in terms of business value by considering the business risk and the impact of not undertaking them. So if delivering value is the reason for doing projects, then value-driven delivery must be our focus throughout the project.

In chapter 1, we answered the question "Why use agile?" in a historical context, based on the differences between industrial work and knowledge work. But we can also address this question from the standpoint of the value that agile methods are able to deliver in comparison to non-agile methods. This is called the "agile value proposition," and it is typically depicted visually, as shown below.

Figure 2.1: Agile Value Proposition

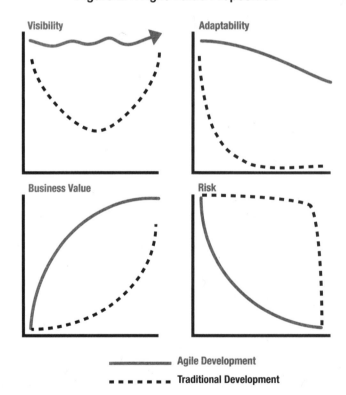

In this chapter, we'll discuss early delivery of business value (bottom left on the value proposition diagram) and begin a discussion of early reduction of risk (bottom right) that will be continued in chapter 6. We'll address visibility in chapter 3 and adaptability in chapter 5.

Maximizing value is an overarching theme, or mantra, for agile teams. When faced with a decision, they ask, "Which choice will add the most value for the customer?" The focus on delivering value drives many of the activities and decisions on an agile project and is the ultimate goal of many of the practices in the agile toolkit. This is such an essential component of agile methods that, out of all the domains, PMI has given the value-driven delivery domain the most weight for the exam. So this is one of the most important agile themes, tying together many fundamental agile concepts, such as prioritization, incremental delivery, and test-driven development.

Deliver Value Early (Eat Your Dessert First!)

One of the key ways agile teams try to maximize value is by delivering value early. This means they aim to deliver the highest-value portions of the project as soon as possible. There are some important reasons for this approach. First, life is short, stuff happens, and the longer a project runs, the longer the horizon becomes for risks that can reduce value such as failure, decreased benefits, erosion of opportunities, and so on. To maximize success, we have to try to deliver as many high-value components as soon as we can, before things change or go sideways.

The second major reason for focusing on early delivery of value is that stakeholder satisfaction plays a huge role in project success. Engaged, committed sponsors and product owners are vital for removing project obstacles and declaring success. Every team is on a trial period when a project starts, because the sponsors might not be sure whether the team can pull it off. By delivering high-value elements early, the team demonstrates an understanding of the stakeholders' needs, shows that they recognize the most important aspects of the project, and proves they can deliver. Tangible results help the team raise the confidence of stakeholders, build rapport with them, and get them on board early, creating a virtuous circle of support.

In short, value-driven delivery means making decisions that prioritize the value-adding activities and risk-reducing efforts for the project, and then executing based on these priorities.

Minimize Waste

Wasteful activities reduce value. This is why agile methodologies have adopted the lean concept of minimizing waste and other nonvalue-adding activities. For example, while the project is being executed, an agile team might reduce or postpone activities that are required by the organization but not directly focused on delivering value, such as formal time tracking and reporting. To maximize value, it is useful to revisit the Poppendiecks' list of seven wastes:[3]

» Partially done work
» Extra processes
» Extra features
» Task switching
» Waiting
» Motion
» Defects

Wherever we find project activities that are wasteful, we want to try to eliminate them.

Assessing Value

Business value is usually assessed in financial terms, starting with assessing the value of potential projects before they are approved to proceed. It's true that some projects are undertaken for safety or regulatory compliance purposes and don't have an easily determined monetary value. To assess the value of these projects, the organization might look at the financial ramifications of not undertaking the project, such as the risk that the business will be shut down, assessed a fine, or subjected to a lawsuit. Another option is to simply label the project as mandatory and not spend any additional time trying to quantify its value.

Financial Assessment Metrics

For business projects, value is commonly estimated using financial metrics such as return on investment (ROI), internal rate of return (IRR), and net present value (NPV). These metrics are used in basically the same way for agile projects as they are for non-agile projects. Let's start by looking at the benefits of these metrics.

Using financial formulas to assess the value of a project removes individual bias and emotion from the process of selecting and justifying projects. Instead, we can focus on comparing a common variable (financial return) across projects. Although there may be additional factors to consider that add value in other ways, the advantage of these financial tools is that their conclusions are "in the numbers," and are (in theory) more objective than other project selection models. (In practice, however, the inputs used for the calculations can be manipulated to sway the outcome.)

Exam tip

Financial metrics are unlikely to be a big topic for the exam, so you might not see any questions about ROI, NPV, or IRR. Any questions you do see are likely to involve interpreting the meaning of these metrics in a given scenario. You won't be asked to define these terms or perform any calculations. In this discussion, we'll be showing some example charts to help illustrate these concepts for those who might not be familiar with them. But don't let our examples and graphics intimidate you; the information you need to know for the exam is actually very simple, as pointed out in the exam tips below.

T&T ### Return on Investment (ROI)

Return on Investment, or ROI, measures the profitability of an investment by calculating the ratio of the benefits received from the investment to the money invested in it. ROI is expressed as a percentage, and the higher that percentage, the better the return that the project is expected to deliver.

> ROI = The ratio of the benefits received from an investment
> to the money invested in it, expressed as a percentage.

Exam tip

For the exam, you should understand the definition of ROI and how this metric differs from NPV and IRR. You should know that a higher ROI means you are getting a better return on your investment than a lower ROI. This metric might be mentioned in situational questions that deal with the return expected from investing in a project. You may be asked to interpret these numbers, but you won't be asked to calculate them.

Although ROI is a very helpful metric, it might not tell us the true value that a project will deliver. To see why, let's look at an example. The figure below shows a project that will run from January until June and then deliver a solution that will generate some financial return. From January until June, we will spend money on the project, paying hourly rates for the resources (shown by the green line indicating negative cumulative cash flow). In June, we will deploy our solution and start to get a positive income (shown by the black line).

Figure 2.2: Project Spending and Income

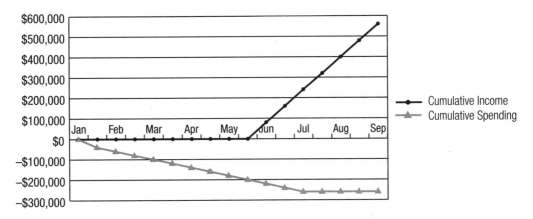

When we add the new income to the outflow figures, we can see the net cash flow (indicated below by the gray line). Based on this chart, when do you think our total return will equal our total investment?

Figure 2.3: Net Cash Flow

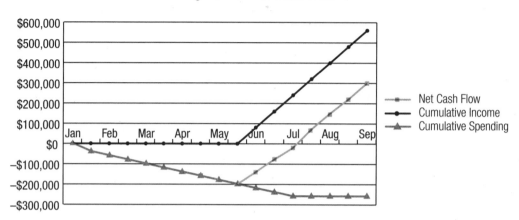

It is tempting to assume that our return will equal our investment in July, when our net cash flow reaches zero. In a way, this is true; at that point we will have received back as much money as we have spent. However, we need to consider the effects of inflation. It is a basic rule of finance that the money we expect to receive in the future is less valuable than the money we have available to invest today. This is especially true if we have to borrow money to generate that future return, since we'll need to pay back the borrowed funds with interest. So when determining our payback period, we need a way to take into account the effects of inflation and the cost of borrowing money.

Present Value

This is where the concept of present value comes in. Present value is a way of calculating the value of a future amount in today's terms, given an assumed interest rate and inflation rate.

Let's apply the concept of present value to our sample project. In the figure below, the darker gray line with the "x" markers indicates the present value of the money we will receive in the future, based on a projected interest rate of 2 percent (representing inflation). As you can see from this figure, the present value of the project investment doesn't go as far into the negative or the positive as the net cash flow, since its value is adjusted for the impacts of inflation.

Figure 2.4: Present Value

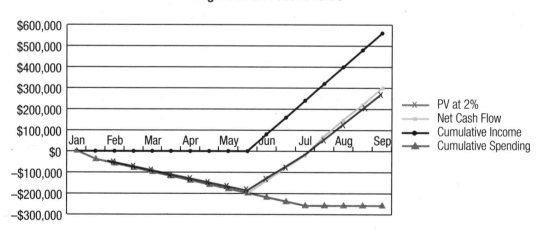

T&T Net Present Value (NPV)

We can extend the concept of present value to find the present value of the return expected for a project. This metric is called net present value, or NPV, and it can be defined as follows:

NPV = The present value of a revenue stream (income minus costs)
over a series of time periods

To assess the value of a project in terms of today's money, we calculate its NPV. This is the present value of its revenue stream, or income minus costs, over a series of time periods, such as months or years. Calculating the NPV of various potential projects allows a firm to compare them on an equal footing and select the best one. Generally, any project that has a positive NPV is a good investment, since we will make back more money than we have invested in today's terms. However, if we're using NPV to compare the value expected from multiple potential projects, then the higher the NPV, the better.

Exam tip

For the exam, you should understand the definition of NPV and how this metric differs from ROI and IRR. You should know that a project with a higher NPV is expected to deliver more value than a comparable project with a lower NPV. This metric might be mentioned in situational questions that deal with assessing the value of a project or comparing value between projects. You may be asked to interpret these numbers, but you won't be asked to calculate them.

Organizations have found NPV so valuable for comparing money going out and money coming in that this metric is widely used to evaluate project returns. It allows us to assess a project's cost and income stream and find—after adjustment for inflation—its true payback period. In large multiyear projects, the true payback period can be substantially longer, especially if the money to invest in these projects is borrowed at high interest rates.

NPV can be especially helpful for comparing projects that involve different timeframes or that are expected to start delivering value at different times. For example, what if we are trying to choose between a project that we expect will deliver a 2 percent ROI in 12 months and one that we expect will deliver a 4 percent ROI in 36 months? Do inflation and interest over the longer time period cancel out the higher ROI of the second project? To find out, we calculate the NPV of the two projects, and see which one yields more value in today's money.

In a case like this, the NPVs of the two projects will give us more information than their ROIs. However, the drawback of calculating NPV is that we have to estimate what inflation and interest rates will be in the future—and those guesses may not turn out to be correct.

T&T Internal Rate of Return (IRR)

To see how organizations address the problem of having to guess future inflation and interest rates, let's turn to the concept of internal rate of return, or IRR. To understand this concept, we have to use the financial term "discount rate," which means "the interest rate you need to earn on a given amount of money today to end up with a given amount of money in the future."[4]

The official definition of IRR is the discount rate at which "the project inflows (revenues) and project outflows (costs) are equal."[5] Another way of stating this is to ask, "What is the discount rate that would move the payback period to the end of the project?"

IRR = The discount rate at which the project inflows (revenues)
and project outflows (costs) are equal.

This metric helps simplify the evaluation of projects, since we don't have to guess what the future interest and inflation rates will be, as we do for NPV. Rather than using projected interest and inflation rates to calculate the value of a project in today's terms, we use our estimates of project duration and payback to calculate an effective interest rate (aka "discount rate") for the project.

The easiest way to think of IRR is to compare it to the interest rate paid by a savings account. When deciding where to put our savings, if we are savvy, we will shop around and try to select the account that will give us the highest return. In the same way, when a company is choosing which project to invest in, it will calculate the expected rate of return for each potential project and select the one that is projected to yield the highest IRR.

From an economic perspective, the higher the rate, the better the project.

Exam tip

If an exam question asks about the "rate of return" or "discount rate" for a project, it is referring to the project's IRR. For the exam, you should understand the definition of IRR and how this metric differs from NPV and ROI. You should know that a project with a higher IRR is a better investment than a comparable project with a lower IRR. This metric might be mentioned in situational questions that deal with decisions about which project to invest in. You may be asked to interpret these numbers, but you won't be asked to calculate them.

Note: It isn't meaningful to compare ROI, NPV, and IRR rates to each other. If the answers to an exam question include more than one of these metrics, don't just pick the option with the highest rate; instead, read the question carefully to determine which metric is relevant to the situation being described.

EXERCISE: UNDERSTANDING ROI, NPV, AND IRR

Test yourself! See how well you understand ROI, NPV, and IRR by answering the following questions:

1. Which of the following definitions best describes return on investment (ROI)?

    A. The point in time when the revenue received equals the costs expended for the project
    B. How much revenue the project will bring in once it is completed and operational, compared to its ongoing operating costs
    C. The ratio of the money we receive at the end of a project to the money we have invested in it
    D. The percentage of money the project will cost once all project expenditures are collected

2. A sponsor is trying to determine which project has the greatest business value. One project returns $5 million in three years, and another project returns $6 million in four years. The cost of borrowing capital to fund the project is 4 percent. Which of the following is the best approach to determine the project with the greatest value?

   A. Select the project that returns $6 million in four years, since it returns the highest amount.
   B. Select the project that returns $5 million in three years, since it has the shorter payback period.
   C. Calculate the NPV of the projects, and choose the project with the lowest cost.
   D. Calculate the NPV of the projects, and choose the project with the highest value.

3. A sponsor is considering the business value of two projects. Which of the following definitions best describes the approach for assessing and applying the concept of internal rate of return (IRR)?

   A. Calculate the internal rate of return, and choose the project with the highest rate.
   B. Calculate the internal rate of return, and choose the project with the lowest cost.
   C. Calculate the internal rate of return, and choose the project with the highest revenue.
   D. Calculate the internal rate of return, and choose the project with the lowest revenue.

ANSWER

1. **Answer**: C. ROI is the ratio of the money we receive at the end of a project to the money we have invested in it, expressed as a percentage. Looking at the other options, choice A describes the project payback or breakeven point, not ROI (which doesn't specify a point in time). Choice B focuses on revenue only, omitting any consideration of the money we have invested, and instead referring to operational costs after the project is over. Choice D describes project costs rather than ROI. Therefore, choice C is our best option.

2. **Answer**: D. To evaluate the value of two projects that will be completed at different times, we can use net present value (NPV) to level the amounts into today's values. So calculating the NPV and choosing the project with the highest NPV value is the way to go. Choice A points us to the project with the highest return, but the question is asking for the *best* approach. Likewise, options B and C do not address the best approach to take.

3. **Answer**: A. Internal rate of return (IRR) shows the earning potential for a project. Like comparing investment interest rates, the higher the rate, the better the investment proposition. So to use IRR to evaluate projects, we calculate the IRR for each and then select the project with the highest IRR value. Costs and revenue are rolled into the calculation of IRR and are not part of the final IRR evaluation.

T&T Earned Value Management (EVM)

Now that we've seen how organizations assess the value of potential projects, let's turn to the tools they can use to monitor the delivery of value while a project is in progress. The first tool we'll address is earned value management (EVM). To understand the benefits of EVM, we'll start by looking at some of the alternatives for tracking value.

One tool commonly used to track project spending is an "S-curve." This is simply a graph that tracks costs or some other variable against time. These graphs are called S-curves because a growth curve is usually shaped like an S, as in this example:

Figure 2.5: S-Curve Graph

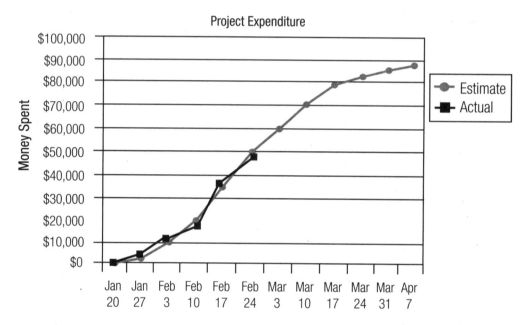

The advantages of S-curves are that they are simple to interpret and can quickly tell us whether our project is over or under budget. However, S-curves don't provide any information about the schedule. So our S-curve might show that we are doing fine spending-wise, but we might be behind schedule and not know it.

To monitor the status of the project schedule, people often use Gantt charts, like this example:

Figure 2.6: Gantt Chart

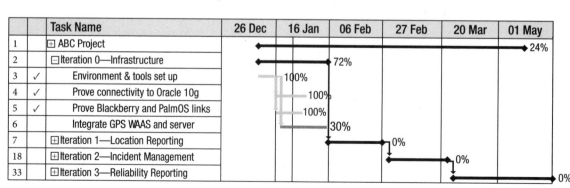

However, Gantt charts are also limited, since they lack the spending component of the project status—just like S-curves lack the schedule component. Since most projects will get either ahead of or behind budget and schedule at some point, it can be difficult to assess the project's overall health. In addition, because of the iterative nature of agile projects, Gantt charts can become deeply nested and confusing.

To assess the overall health of a project, what we really need is a single diagram that can show the project status in terms of both schedule and total value delivered to date. Earned value management was created to address this gap. This approach combines spending and schedule data to produce a comprehensive set of project metrics, including planned value (PV), earned value (EV), schedule variance (SV), cost variance (CV), schedule performance index (SPI), and cost performance index (CPI).

Using EVM for Agile Projects: Pros and Cons

People ask me if you can use earned value metrics on agile projects. The answer is yes. But while the math still works in the same way as it does on non-agile projects, we do need to be careful about what we are measuring against. Earned value compares actual project performance to planned performance at a particular point in time. So the quality of the baseline plan is a critical success factor in using this approach. If our initial plan is no longer accurate, this might be like tracking a road trip from Calgary to Salt Lake City on a map of France! On agile projects, we know that our initial plan will need to change, so the basis for effective EVM is quickly eroded as our plans evolve.

Another caution regarding earned value is that it doesn't truly indicate whether the project is successfully delivering value. We could be on time and on budget, but building a horrible, low-quality product that the customer doesn't like or need. Cost and schedule aren't the whole picture—our project might still be going badly even if it looks good from an earned value standpoint.

After reading about these issues that agile teams need to be wary of, you might be wondering, "Why use EVM at all?" Well, one of the key benefits of earned value metrics is that they are a leading indicator. Perfect rearview vision isn't much use to us. EVM looks forward, trying to predict completion dates and final costs. After all, imperfect leading metrics are generally more valuable than perfect trailing metrics, since leading metrics give us the opportunity to revisit our plan and change our approach.

Another benefit of earned value is that it is visual. People often forget the EVM diagrams and just focus on the numbers, but at the heart of this technique are some useful graphs. Visual representations of information engage the right side of our brain, helping us intuitively understand and interpret the data so we can plan an appropriate response. Visual depictions are also better for working collaboratively, since people can more easily mark up, point to, and extrapolate from pictures than words or numbers.

Constructing an Agile Earned Value Tool

How can agile teams take advantage of the forecasting and visual communication benefits of EVM, while reducing the downsides of that approach? To see how, we'll build up a graph that has all the metrics we need, one step at a time. We'll start with a double S-curve, like this:

Figure 2.7: Scope, Cost, and Schedule Performance (Double S-Curve)

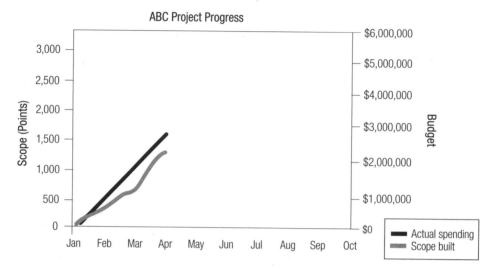

Here we have a familiar spending line, shown in black, which is being tracked against the dollar scale on the right side the graph. We also have a green line for completed scope (features built to date), which is being tracked against the story points scale on the left side.

The gradient of the green line shows the team's velocity. Where it rises steeply, the team was able to develop a lot of story points in a short period of time. Where it is flat, their progress was slow. (The agile concepts of velocity and story points will be explained in chapters 4 and 5.)

For the next step in building our graph, we'll add a background that shows the functional areas of the project, like this:

Figure 2.8: Scope, Cost, and Schedule Performance with Functional Areas

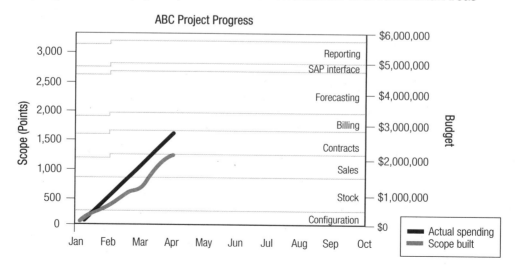

With this new background in place, we have a lot more information. We can see not only that just over 1,000 points' worth of functionality has been completed, but also the Configuration and Stock components of the system have been built, and the team is currently working on the Sales piece of the scope. Also, we can easily see that scope was added to the Sales portion of the system in February. Since this change increased the total project scope (in terms of story points), all the functional areas after Sales show a step up to reflect that change.

Now we have a graph that shows scope, schedule, and cost performance to date, but what we are lacking is any projections that tell us whether we are ahead of or behind our predicted budget and schedule. So in the next step we will add projections to the graph, so that it looks like this:

Figure 2.9: Scope, Cost, and Schedule Performance with Functional Areas and Projections

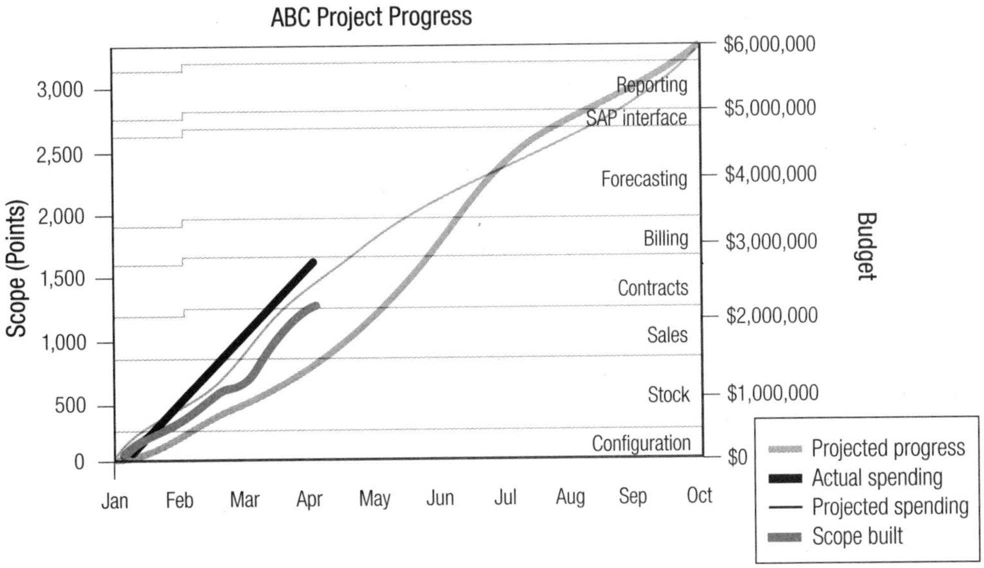

Now we can see how the project's actual progress compares to projected performance—at present in our example project we have overspent a bit, but we are ahead in building the scope. With this step, we now have a graph that can be used for agile earned value management, since it provides the same values and indices as EVM, but depicts them in a visual way, as shown in the next figure.

Figure 2.10: A Visual Tool for EVM Metrics

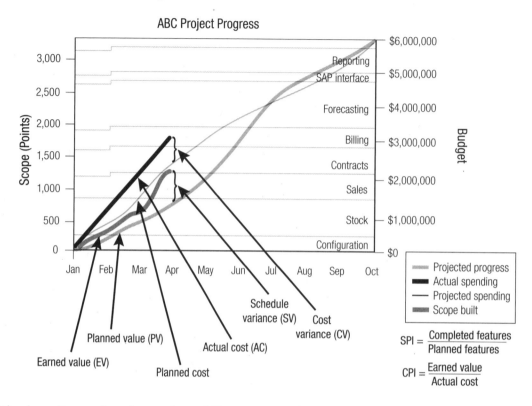

The above diagram shows how traditional EVM metrics such as schedule performance index (SPI) and cost performance index (CPI) can be translated into agile terms. For example:

» We planned to complete 30 story points in the last iteration, but we only completed 25 points. To find our SPI, we divide 25 by 30 for an SPI of 0.83. This tells us that we are working at only 83 percent of the rate planned.

» CPI is the earned value (EV, or value of completed features) to date divided by the actual costs (AC) to date. So in the above diagram, CPI = $2,200,000 / $2,800,000 = 0.79. This means we are only getting 79 cents on the dollar compared to what we had predicted.

Exam tip

The PMI-ACP exam usually doesn't include many figures or graphs, but you might be asked to interpret an EVM or S-curve graph such as those shown above. You won't have to do any calculations, but you'll need to know what the graph tells you about the project. The exam might also include questions about how EVM is used on an agile project, such as how an evolving project baseline impacts the use of earned value metrics for agile projects.

K&S Agile Project Accounting

Agile "accounting" refers to how the economic models of agile projects work. We will be covering the differences between agile and traditional life cycles from various perspectives, in relation to planning, estimating, delivering value, and so on—and these differences also have implications for project accounting. Prioritizing work by business value and delivering the highest-value components first can radically change the payback period and return on investment for a project. As we'll see, agile teams aim to deliver a minimal viable product (MVP) as quickly as possible. So there may be opportunities to get early benefits by making use of some of the product functionality while the remainder of the project is still being executed.

Another change associated with breaking down the work into a minimum viable set of features is a greater opportunity for incremental funding. For example, this can reduce financial risk when dealing with third-party vendors. Instead of having large statements of work and legal teams battling over the contracts and the meaning of the specifications, we can have smaller statements of work—freeing up the legal teams to argue over multiple contracts and definitions of done. (Just joking.) By chunking up the work, we reduce the economic value at stake, de-escalate the impact of disputes on future work, and can use agile contract models, such as the DSDM contract we'll discuss later, which speaks to products being fit for business purpose as opposed to meeting every requirement of a detailed specification.

K&S Key Performance Indicators (KPIs)

Sponsors of agile projects want to know the same things as sponsors of traditional projects—they want to know when the project will be done, and how much it will cost. In discussing the inverted agile triangle in chapter 1, we said that agile projects can fix time and cost if the scope can vary. However, usually the sponsor will still be asking, "Given the currently agreed-upon scope, when will you be done, and how much will it cost?" Fortunately, there are several key performance indicators (KPIs) that we can use to help answer these questions for agile projects. Let's examine four of these metrics that are commonly used.

» **Rate of progress.** How many features or user stories are getting completed and accepted by the product owner per week or month? Since some features are quick and easy to build, while others are more involved and time-consuming, agile teams usually estimate their work items in story points or some other relative metric (as explained in chapter 5). So a simple piece of work might be sized as 1 story point, a large chunk of work might be sized as 8 story points, and the project's rate of progress KPI will be expressed in story points per unit of time, such as 20 points per week.

» **Remaining work.** How much work is left in the backlog? This KPI attempts to quantify the remaining work. We've seen that the backlog items are usually estimated by the development team in story points. And in addition to those estimates, usually some additional effort will be required for unanticipated work, fixes, and ongoing evolution of the product. So to use an example, let's say that the first half of our project was estimated as 400 points—but once developed, it actually turned out to be 500 points. In this example, if our remaining work is also estimated at 400 points, then this too will likely end up being 500 points by the time we have teased out all the true requirements.

» **Likely completion date.** We look at how much work there is left to do and divide it by our current rate of progress. This only requires some simple math. If we seem to be getting 20 points of work done per week, and there are 400 points remaining that may well expand to 500 points by the time we fully understand the work, then our likely completion date will be 500/20 = 25 weeks, assuming no changes in scope or breaks in the schedule (such as vacations).

» **Likely costs remaining.** For simple projects, this will be the salary burn rate for the team multiplied by the remaining weeks left. There may also be other fixed costs that we need to take into account, such as licenses, equipment, deployment and training costs, and so on.

Exam tip

KPIs aren't that important for understanding agile—so although this topic is included in the exam content outline, it might not appear on your exam. Remember that the exam focuses on the agile mindset and the most important agile practices.

Managing Risk

To maximize value, agile teams need to consider risks and technical dependencies. A choice that could lead to future problems or rework probably won't deliver the most value to the customer, even if it gives the illusion of early progress. Therefore, we need to consider risks and nonfunctional requirements and let the customer know how those elements will impact the project.

In fact, the concept of risk is closely related to value, so much so that we can think of negative project risks (threats) as *anti-value*—factors that have the potential to erode, remove, or reduce value if they occur. If value is the "heads" side of the coin, then risk is the "tails" side. To maximize value, we must also minimize risk, since risks can reduce value. This is why the value-driven delivery domain also encompasses the concept of risk reduction.

Here's another simple way to grasp the relationship between value and risk. Think of value-driven delivery as creating credits or deposits that are paid into your bank account. Risks—or at least the threats that actually occur and become issues on your project—are like withdrawals or charges that are taken out of your account. To create the most value, we need to maximize the inflow and minimize the outflow of value from the account.

The link between risk and value highlights the importance of risk management on agile projects. When a risk event occurs, it takes time and resources away from efforts that deliver value, threatening the project benefits. Therefore, agile teams not only plan to deliver high-value features early, they also plan to implement risk avoidance and risk mitigation activities early.

Risk management may seem like a traditional, process-driven project management effort that would not work well in an agile environment. However, agile practices are actually very well suited for rapidly identifying and reducing risks. We'll explore risk management in more detail in chapter 6; here we'll conclude by discussing what we mean by "risk management" on agile projects.

For agile purposes, a risk is a potential event or circumstance that could have a negative impact on the project. Although the *PMBOK® Guide* definition of a risk also includes "good risks" that can pose opportunities for the project, in agile most discussions of risk focus on events that can negatively impact or threaten the project.[6]

As for any project, agile teams need to engage the development team, sponsors, customers, and other relevant stakeholders in the process of risk identification. Their ideas, along with the lessons learned from previous projects, risk logs, and industry risk profiles, should all be considered to come up with a complete list of the known and likely risks for the project. It's especially important to engage the members of the development team in risk analysis. For one thing, they have unique insights into the risks since they are closer to the technical details. Involving them will also generate increased buy-in for the risk management plan and response actions. If the team has no involvement, they have no commitment.

The figure below shows the risk management process as outlined in the sixth edition of the *PMBOK® Guide*: [7]

Figure 2.11: *PMBOK® Guide* Risk Management Process

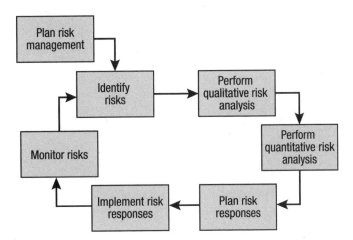

As depicted here, the sixth edition of the *PMBOK® Guide* introduced a new step in the risk management process called Implement Risk Responses. Previously, implementing risk responses was only implied by the Plan Risk Responses step that created actions for the project plan—so it is good to see it now called out explicitly. Implement Risk Responses is where we act on the risks identified. For an agile project, this will entail creating new tasks for the product backlog.

With an agile approach, there are many opportunities to actively attack the risks on a project before they can become tomorrow's problems. Iterative development allows high-risk work to be tackled early in the life cycle. Features or stories that carry high levels of risk can be tackled in early iterations to prove the technological approach and remove doubts.

So the iterative nature of agile projects helps us make sure we are driving the action needed to prevent risks. In planning each work cycle, we aim to tackle high-risk areas of the project sooner rather than later. This approach gets problems out in the open while there is still room in the schedule and budget to work on them. It also reduces the amount of effort invested in work that may end up being scrapped.

The primary tools that agile teams use to manage risk are the risk-adjusted backlog and risk burndown charts. These tools allow us to seamlessly integrate and prioritize our risk response actions into our backlog of development work. We won't delve into the use of those tools in this chapter because they are covered in domain VI (which addresses problem detection and resolution)—so you'll find that discussion in chapter 6.

T&T K&S Regulatory Compliance

Many agile projects operate in a highly regulated environment and have to deal with regulatory compliance issues. Although this concept isn't that important for the exam (you might see a few questions on it at most), it's worth taking a moment to consider how agile projects deal with compliance issues.

Regulations are typically designed to ensure safety; so while the mandated documentation and quality procedures may seem burdensome to the project team, they serve an important purpose. Personally, I have worked on military projects, food and drug projects, and energy transmission projects that have had regulatory compliance requirements. Usually these requirements are nonnegotiable—choosing to ignore, skip, or skimp on compliance isn't an option, since it would simply prevent the organization from being paid and the product being accepted and used.

Back in chapter 1, when we were reviewing the second Agile Manifesto value ("Working software over comprehensive documentation"), we discussed the idea of documentation being "just enough, just in time—and sometimes, just because." Regulatory compliance is one instance of where "just because" comes into play. Typically projects that are subject to regulatory compliance require special documentation to prove that the required practices were followed. In some industries, the regulators might want to see traceability from tests to requirements; in others they might want to see the use of regulated test harnesses and tools.

There are two simple approaches for incorporating regulatory compliance work into agile projects. The first is to weave it into the regular development work as the team progresses. The second is to allow time after creating the product to undertake the regulatory work and produce the required evidence and documentation. There are pros and cons to each approach:

» Doing the compliance work "as you go" keeps it linked and relevant to the current work. We can see if our work practices are making it easier or harder to pass the compliance tests, and adapt accordingly. However, the downside of this approach is that any subsequent changes to a product or its design may invalidate the work we have done before. In other words, the testing that has been done to date might need to be repeated.

» Doing the compliance work after product development doesn't allow us to tweak our practices on the go, but it does avoid the issue of rework. And assuming that the product passes, it can be sent for final approval, complete with its newly minted compliance documentation.

Most organizations adopt a hybrid approach; they will select a few architecturally significant components to test compliance on as they develop and refine their process, until they are comfortable that technology will satisfy the regulatory requirements. Then, once all the work is done, they will submit the entire product to final compliance testing and documentation again. This creates a duplication of effort in testing the components they did before, but it is the best way to balance and manage the risks of suboptimal processes and overall rework.

Prioritizing Value

Prioritization is a fundamental agile process. To understand why, recall that the second agile principle tells us that agile teams must "Welcome changing requirements, even late in development." Although that sounds great in theory, what actually happens when a new requirement is added to a fully loaded project? If the customer hasn't been prioritizing the work items, then the team has a dilemma. How will they squeeze in the new work? In effect, the customer is saying "I want it all!" However, if the customer has been continuously prioritizing the backlog, then that new requirement is simply inserted at the appropriate place in the backlog, and the lowest priority requirements drop off the team's work list to accommodate that change.

Agile teams also use prioritization to confirm that they are delivering value. At the end of each iteration, we sit down with the customer to review the backlog, asking "Has anything changed?" and "Do we still want to work on feature B next?" Any new priorities are captured in the backlog and revisited again at the next planning session. This helps ensure that we are continuing to make progress toward the desired target.

T&T Customer-Valued Prioritization*

Customer-valued prioritization refers to the agile practice of working on the items that yield the highest value to the customer first. The customer or product owner is responsible for keeping the items in the backlog (the list of work that needs to be done) prioritized by business value. As new work items are added, such as change requests and defect fixes, the customer prioritizes them so that their value can be compared to the existing items on the list. When the team is ready to plan a new iteration, they start working on the items at the top of the backlog, to make sure they are working on the highest-value items.

Different agile methodologies use different tools for customer-valued prioritization. Scrum has a "product backlog," FDD has a "feature list," and DSDM has a "prioritized requirements list"—but the idea is the same. The team works through a list of items that have identifiable value and have been prioritized by the customer.

By asking the customer what their top-priority features are, we learn about their motivations, risks, and acceptance criteria. A team that doesn't practice customer-valued prioritization is likely to miss out on identifying critical success factors.

I am often surprised by what the customer ranks highly when we first begin to prioritize the features and stories to work on. As we discuss these seemingly unlikely first choices, the conversation usually brings political concerns and residual risks to the surface. For example, "We are doing Premium Rates first because it is an objective for our quarter 1 milestones," or "We need to deliver the CIO dashboard to secure project funding for next year." While we might be tempted to argue these are not really "true" system business benefits, we are often wise to listen to these motivators, since the customer is our ally in the project, and the person who will declare what success looks like.

Customer-valued prioritization is an ongoing process throughout the project. Typically, the team will sit down with the customer at the end of each iteration to prioritize the remaining work items. These reprioritization sessions serve as important checkpoints that help ensure that the work is progressing toward the target of the project, which may itself be moving. In these sessions, we ask, "Have things changed?" and "Do we still want to move on to item X next?" After discussion, the new and evolving priorities are captured in the backlog, to be revisited again at the next planning session.

* Portions of the "Customer-Valued Prioritization" and "Relative Prioritization/Ranking" sections below were originally published in "Agile Prioritization" by Mike Griffiths on gantthead.com on April 26, 2001, copyright © 2011 gantthead.com. Reproduced by permission of gantthead.com.

K&S Prioritization Schemes

There is no specific prioritization scheme that is best for agile projects. Each team should choose their prioritization scheme based on the needs of the project and what works best for their organization. Let's look at some of the common prioritization schemes used on agile projects.

Simple Schemes

One of the simplest schemes is to label items as "Priority 1," "Priority 2," "Priority 3," etc. While this approach is straightforward, it can be problematic, since stakeholders have a tendency to designate everything a "Priority 1." If too many items are labeled "Priority 1," the scheme becomes ineffective. Business representatives rarely ask for a new feature and say it should be a Priority 2 or 3, since they know that low-priority items risk getting cut out of the project. For the same reasons, "high," "medium," and "low" prioritizations can also be problematic. Without a shared, defendable reason for what defines "high" priority, we end up with too many items in this category and a lack of true priority.

T&T MoSCoW

The MoSCoW prioritization scheme, which was popularized by DSDM, derives its name from the first letters of the following labels:

» "**M**ust have"
» "**S**hould have"
» "**C**ould have"
» "**W**ould like to have, but not this time"

The categories used in MoSCoW are easier to identify and defend than the "Priority 1" or "High Priority" labels of the simpler schemes. "Must-have" requirements or features are those that are fundamental to the system; without them, the system will not work or will have no value. "Should have" features are important—by definition, we should have them for the system to work correctly; if they are not there, then the workaround will likely be costly or cumbersome. "Could have" features are useful net additions that add tangible value, and "Would like" requirements are the "nice-to-have" requests that are duly noted—but unlikely to make the cut.

Monopoly Money

Another approach I have seen work well is to give the stakeholders Monopoly money equal to the amount of the project budget and ask them to distribute those funds amongst the system features. This approach is useful for identifying the general priority of system components, but it can be taken too far if the people distributing the money start to question activities, such as documentation, that they perceive as adding little value to the project. The Monopoly money technique is most effective when it's limited to prioritizing business features.

100-Point Method

The 100-Point Method, originally developed by Dean Leffingwell and Don Widrig for use cases, is another way to prioritize features. In this method, each stakeholder is given 100 points that he or she can use to vote for the most important requirements. The stakeholders can distribute the 100 points in any way: 30 points here, 15 points there, or all 100 points on a single requirement, if that is the stakeholder's only priority.

Dot Voting or Multi-Voting

This is another technique in which collective wisdom emerges through individual priorities. Here, each stakeholder gets a predetermined number of dots (or check marks, sticky stars, etc.) to distribute among the options presented.

To illustrate how this works, let's look at an example. Imagine we have completed a brainstorming session to come up with potential risks for the project. The session has resulted in a list of 40 unique risks that now need to be prioritized. Each person is given eight votes in the form of check marks that they can use to indicate which items they feel are most important. Each stakeholder is limited to a total of eight check marks to distribute as they wish. It could be one check mark on eight different items, four check marks on one item and two check marks on a couple of other items, or any other combination that reflects the stakeholder's priorities.

The facilitator then sums the votes for each item and creates a ranked list based on how many votes each item received. The voting can be public, or it can be private with someone tallying the totals off-line to prevent power struggles and strategic voting.

When deciding how many votes to give each person, a good rule of thumb is 20 percent of the total number of items. So if there are 40 items to be voted on, we would calculate $40 \times 0.2 = 8$, and everyone would get 8 votes to distribute.

T&T Kano Analysis

Kano analysis isn't really a prioritization scheme, but the exam content outline lists it as a prioritization method since it can be used to help determine priorities. This technique is used to classify customer preferences into four categories; although you may see different names for these categories, we will refer to them as Delighters/Exciters, Satisfiers, Dissatisfiers, and Indifferent. Project stakeholders can use these categories to understand how customer needs relate to customer satisfaction. Kano analysis can help us set the context for questions about features and build release plans that promote improved customer satisfaction.

Figure 2.12: Kano Analysis

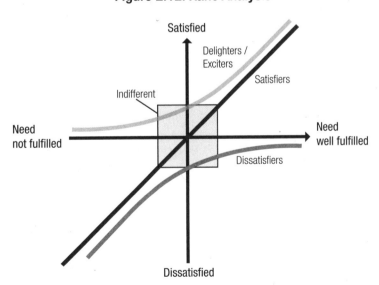

Let's look at each category in more detail:

» **Delighters/exciters**: These features deliver unexpected, novel, or new high-value benefits to the customer. For example, a delighter/exciter feature could be the visual mapping of data that was previously only available in tabular form. Delighters/exciters yield high levels of customer support.

» **Satisfiers**: For features that are categorized as satisfiers, the more the better. These features bring value to the customer. For example, a useful reporting function could be a satisfier.

» **Dissatisfiers**: These are things that will cause a user to dislike the product if they are not there, but will not necessarily raise satisfaction if they are present. For example, the ability to change a password within the system could be a dissatisfier.

» **Indifferent**: These features have no impact on customers one way or another. Since customers are indifferent to them, we should try to eliminate, minimize, or defer them.

Requirements Prioritization Model

The Requirements Prioritization Model created by Karl Wiegers is a more mathematically rigorous method of calculating priority than the other schemes we've discussed. With this approach, the benefit, penalty, cost, and risk of every proposed feature is rated on a relative scale of 1 (lowest) to 9 (highest). Customers rate both the benefit score for having the feature and the penalty score for not having it. Developers rate both the cost of producing the feature and the risk associated with producing it. The numbers for each feature are then entered into a weighted formula that is used to calculate the relative priority of all the features.

EXERCISE: MUST HAVE, SHOULD HAVE, COULD HAVE

Categorize the following features for a child's bicycle into "must-have," "should-have," and "could-have" features. The first one is done for you as an example.

Feature	Category
Two wheels and a frame	Must have
Ability to adjust the saddle to accommodate growth	should
Brakes for safe stopping	should
Bell or horn to alert others in proximity	could
Safety cover for the chain	should/could
Attractive color scheme	could
Stabilizers or the ability to fit them	should / could
Front suspension	could

Feature	Category
Valves for inflating tires	could
Pedals	should/could

ANSWER

Feature	Category
Two wheels and a frame	Must have
Ability to adjust the saddle to accommodate growth	Should have
Brakes for safe stopping	Should have
Bell or horn to alert others in proximity	Could have
Safety cover for the chain	Should have / Could have
Attractive color scheme	Could have
Stabilizers or the ability to fit them	Should have / Could have
Front suspension	Could have
Valves for inflating tires	Could have
Pedals	Should have / Could have

My answers may seem extreme—surely bikes need pedals, valves to put air in the tires, and brakes? However, a bicycle is a two-wheeled transportation device. By definition, it must have two wheels and a frame to link the wheels together, but beyond that, everything else is up for discussion and negotiation. My son learned to ride a bike on a "balance bike," which had solid tires and no pedals or brakes; instead, he just scooted it along with his feet.

Often there is a disconnect between expectations and requirements, and this disconnect is highlighted in this exercise. As sophisticated consumers of products today, we often have a high level of expectations, but high expectations are quite different from true must-have requirements. The qualifying question for a must-have requirement is closer to "Can it work without it?" than "Would I buy one that did not have it?"

T&T **Relative Prioritization/Ranking**

Regardless of the prioritization scheme that the team uses, their focus should be on the end goal—
understanding the priority of features. Sometimes the effort involved in refereeing the schemes can detract
from meaningful discussions about prioritization. For this reason, I am personally a fan of simply asking
the customer to list features in order of relative priority—no category 1, 2, or 3; no high, medium, or low;
no must-haves, etc. A simple list removes the categories that people tend to fixate on from the debate and
allows the focus of the discussion to be on priorities.

Let's look at an example of a simple relative priority list:

Figure 2.13: Relative Priority List

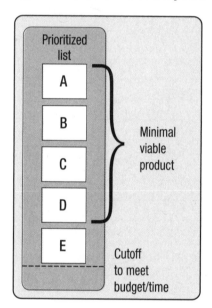

In the diagram above, each feature is listed in order of priority. The items at the top of the list—features A
through D—are part of the defined minimal viable product. If scope needs to be cut to meet the budget and
schedule objectives, it's clear by looking at this simple list that adjustments should be made to item E.

Relative prioritization also provides a framework for deciding if and when to incorporate changes. When a
change is requested, the team can ask the product owner, "What items are more important than this change?"
The new change can then be inserted into the prioritized work list at the appropriate point, as shown below.

Figure 2.14: Incorporating Changes into a Relative Priority List

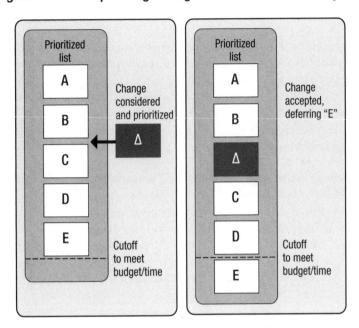

While agile methods provide tremendous flexibility in accepting late-breaking changes, they cannot bend the laws of time and space. So if our project time or budget was estimated to be fully consumed with the current scope, then adding a new change will inevitably force a lower-priority feature below the cutoff point of what we expect to deliver. So, yes, the team can accept late-breaking changes on projects—but only at the expense of lower-priority work items.

A single prioritized work list is useful for simplifying the view of all the remaining work. Rather than having separate "buckets" of work that represent change requests, defect fixes, and new features, a single prioritized list combining these items gives everyone a clear, complete view of everything that remains to be done on the project. When first adopting an agile approach, many teams miss this point and instead retain separate buckets for bug fixes and changes. However, this separation makes it difficult to gauge the team's velocity. A single prioritized list of work-to-be-done, regardless of its origin, offers better transparency and control over trade-offs, because the relative priority of the work is clearly shown by where each item is placed on the list.

Although I endorse the use of a simple list, there is no single best method to use in prioritizing features on every project. Whatever technique you use, try to diagnose any issues that arise in the prioritization process, be it "lack of involvement" or "too many Priority 1s." You can then add different approaches, such as using Monopoly money, applying the MoSCoW prioritization scheme, or simply listing the features in order of priority, to see if you can resolve the issues by framing the discussion in a new way. The goal of such efforts is to understand where features lie in relation to each other, rather than to simply assign a category label. We will then end up with a flexible list of prioritized requirements that we can revisit and reprioritize as needed. This allows us to maintain the agility to deliver the highest-value set of features within the available time and budget.

K&S Delivering Incrementally

Incremental delivery is another way that agile methods optimize the delivery of value. With incremental delivery, the team regularly deploys working increments of the product over the course of the project. In the case of software development projects, the working software is usually deployed to a test environment for evaluation, but if it makes sense for the business, the team could deliver functionality directly to production in increments. If we can deliver the "plain-vanilla" version of a product or service while working on the more complex elements, we have an opportunity to start realizing the benefits of the product and get an early return on investment.

Even if we are incrementally delivering software into a test environment, rather than into production, this approach can still help with the overall delivery of value. As shown in the cost of change curve below, issues that are found in software deployed in a test environment for evaluation (point 1) are much cheaper to fix than issues found during production (point 2). Incremental delivery reduces the amount of rework by finding issues earlier and thereby contributing to the delivery of value on the project.

Figure 2.15: Cost of Change

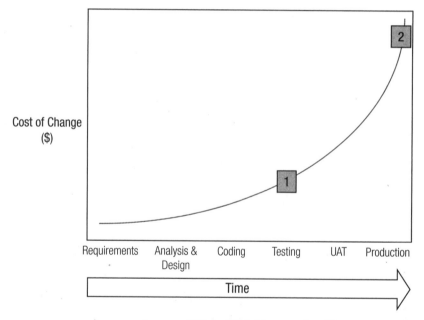

Image copyright © Scott W. Ambler, www.agilemodeling.com

T&T Minimal Viable Product (MVP)

When planning a release of features to customers, the release has to make sense, be useful, and deliver value. This applies to all types of agile projects, whether it's a release of software, a new electrical product, or an engineering increment. The term "minimal viable product," or MVP (also known as "minimal marketable feature," or MMF) refers to this package of functionality that is complete enough to be useful to the users or the market, yet still small enough that it does not represent the entire project.[8]

For a cell phone, for example, a minimal viable product could be a phone that can be used to make and receive calls, store contact names and numbers, and access voice mail, but that doesn't have a camera, Internet connectivity, or a music player in its first release. Instead, this functionality could be added in subsequent releases and evaluated independently. Keep in mind, however, that the functionality of the phone that is released in the MVP needs to be complete. So all the attributes related to making phone calls should be present as part of the MVP to allow the customer to comprehensively review this functionality.

In software development, it may be possible to transfer increments of the final product to the user community early so that the business can start getting some benefits from the application before the entire project is completed. For example, if the order entry and billing system is complete, but the management reporting and marketing links have not been built yet, the company might still gain benefits from deploying this early version of the system. This incremental release can allow for some return on investment while the team develops the remaining functionality. It also provides an opportunity to field-test the functionality of the order entry and billing system. These field tests may then result in change requests that can be rolled into the final product while the development team is still in place.

EXERCISE

Create a list of the functionality that should be included in the minimal viable product for each product listed here, and then also list other potential functionality that could be developed for subsequent releases.

Pencil

MVP	Additional Releases
Writing utensil	Eraser

Car

MVP	Additional Releases
Vehicle that can meet all legal aspects of operation (brakes, lights, wipers, etc)	Heat /AC Radio Upholstery Passenger seats

Automated teller machine (ATM)

MVP	Additional Releases
Machine that gives access to money + deducts from acct	Transfers Balance Checks Denomination Choices

ANSWER

There could be a wide range of correct answers for this exercise. The following are some possible options.

Pencil

MVP	Additional Releases
Makes a mark on paper	Eraser
Can be held in one hand	Visually attractive
	Self-sharpening or continuous lead
	Comfortable

Car

MVP	Additional Releases
Transport occupants from point A to point B	Air conditioner and heater
Road legal	Fuel efficient
Safe	Aesthetically pleasing
	Sporty performance
	Comfortable

Automated teller machine (ATM)

MVP	Additional Releases
Dispenses money	Accepts cash deposits
Displays balance	Accepts check deposits
Protects against attack	Remembers user's favorite withdrawal amounts
Keeps user information secure	

T&T Agile Tooling

Just as the Agile Manifesto values "Individuals and interactions over processes and tools," agile teams prefer low-tech, high-touch tools over sophisticated computerized models. It might seem ironic that the agile approach, so closely associated with software development, wouldn't take full advantage of the latest digital gizmos. However, high-tech tools actually present some disadvantages for agile teams. Although scheduling software can illustrate deep hierarchies of tasks, support task dependency integrity checks, and calculate interesting metrics such as slack, subassembly costs, and resource utilization, this technical sophistication isn't ideal for agile methods. The highly polished graphs, statistics, and quantitative reports that these tools can produce tend to disguise the volatile nature of what's being analyzed—project tasks and estimates. Sophisticated scheduling tools can also exclude team members from interacting with the tools and discourage whole-team collaboration.

When we use high-tech tools to perform scheduling calculations and forecasting, two problems arise: data accuracy perception increases, and barriers for stakeholder interaction are created. Let's look at these problems in more detail.

1. **Data accuracy perception increases**: Just because we can enter a developer's estimate into an expensive scheduling tool does not alter the fact that it may be a lousy estimate or, more likely, today's best estimate that will change as the project progresses and more information surfaces. Scheduling tools can create sophisticated models of the future that imply more credibility than their base data support.

2. **Barriers for stakeholder interaction are created**: Once tasks and estimates are codified into a Gantt chart, the number of project stakeholders who can readily enhance, improve, and update the plan is drastically reduced. With a sophisticated schedule, it is usually just the project manager or team leader who is responsible for updating the plan. The project manager or team leader might regularly ask the team how work is progressing and update the task durations based on what has already been done and the estimate to complete—but how often will the team members get to reschedule the tasks and insert new task groups?

These observations are reflected in Donald Reinertsen's book, *Managing the Design Factory*, in which he warns of sophisticated models:

> *There is a solid practical reason for not using a more sophisticated analysis: not everyone will understand it. The more complex we make a model, chasing after a bit more accuracy here and there, the more we create a formidable maze of calculations that will not be trusted by the team. It is far better to choose a simple modeling technique and have 100 percent of the team understand the model than it is to risk confusing half the team with an elaborate model that is 5 percent more accurate.* [9]

Low-Tech, High-Touch Tools

Instead of using high-tech tools, agile teams prefer to employ a "low-tech, high-touch" approach to planning and tracking. As implied by that term, these tools are simple, such as cards and charts, and therefore easy for all team stakeholders to manipulate by moving the cards, reordering the lists, etc. By adopting these deliberately primitive techniques, we avoid a tool-related perception of data accuracy and allow more people to update the plans as appropriate for the reality of the project.

One key reason agile teams value these types of tools is that low-tech, tangible objects promote communication and collaboration, which is where learning and knowledge transfer really occur on a project. In contrast, sophisticated tools can produce impressive-looking reports and graphs, but often alienate participants due to their complexity and the learning curve required to master them.

Low-Tech, High-Touch Military Tools

The preference for low-tech, high-touch tools isn't unique to agile. The coordination of military battles is often still performed by manipulating physical tokens for boats and troops, despite huge budgets for computer models, because moving tokens around better engages the participants and leads to less confusion.

In keeping with the low-tech, high-touch approach, agile teams make widespread use of tangible tools such as task boards, user stories written on 3-by-5-inch index cards, and numbered card decks for planning poker sessions (this technique is explained in chapter 5).

Here is a comparison of a task list generated by scheduling software and the same task list transferred to a Kanban board. Notice how much easier it is to grasp at a glance the status of the work.

Figure 2.16: Moving from Scheduling Software to a Task/Kanban Board

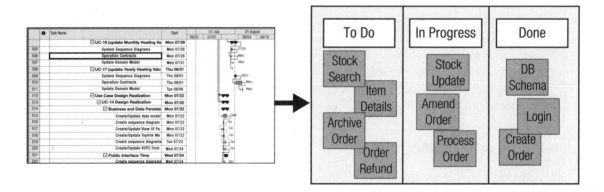

Agile teams use low-tech, high-touch tools such as task boards as their primary method of tracking and reporting value. These tools often take the form of information radiators, which are large charts displayed in prominent places. (We will explore information radiators further in the next chapter.)

© 2015 RMC Publications, Inc • 952.846.4484 • info@rmcls.com • www.rmcls.com

Implementing Low-Tech, High-Touch Tools

If your team isn't already using low-tech, high-touch tools, there are benefits to incorporating them into your work or implementing a visual system to plan iterations and track overall project status. For example, you could dedicate a wall in a team room or high-traffic corridor for big visible charts (large information charts that are easy for people to see and understand). Get the customer in front of the planning wall and show them the work that is done and the work that is planned for the upcoming iterations. When changes are requested, the team and the customer can work together at this wall to visually prioritize the new work against the backdrop of the existing features.

If this all sounds too loosey-goosey and lacking in professional best practices, you can back it up with a traditional planning approach. When I first started managing agile projects in the early 1990s, I was working at IBM in the United Kingdom, and the company had standardized project planning and reporting formats. My PMO was not flexible in its processes and demanded standard plans, so I maintained a set of plans for reporting purposes and let the team maintain the work plan. After several projects, I had the proof I needed. The team plans tracked what was really happening on the project and reacted to project changes and challenges faster than I could rebaseline and level a resource profile. I could make my plans and statistics indicate whatever I wanted, but the low-tech, high-touch team plans were a faithful reflection of the project status (good and bad).

If you operate in a highly regulated environment, you may also need to keep parallel traditional plans—but let your traditional plans be driven by the progress reported in the team's low-tech, high-touch plans. If you are concerned about losing the data in the low-tech plans, take digital photographs of your story cards and planning boards on a regular basis as a low-interference safeguard.

T&T Task/Kanban Boards

Let's look at some examples of one of the most fundamental agile tools, the task or Kanban board, which I've already mentioned a few times. "Task board" is the generic agile term for "Kanban board," and you may see either of these terms on the exam.

These boards originated in the Kanban methodology, where they are the primary tool for planning and monitoring the progress of the work. While there may be variations, a task or Kanban board is generally a whiteboard with columns that show the various stages of work. The tasks that are being worked on are represented by sticky notes that team members move through the columns to show their progress.

As explained in chapter 1, a Kanban team uses their board differently than an agile team uses a generic task board, since Kanban uses a pull system. However, in many contexts this difference isn't relevant, and as a result these terms are often used interchangeably to refer to the concept of a whiteboard with columns that show various stages of work.

Figure 2.17: Task Board

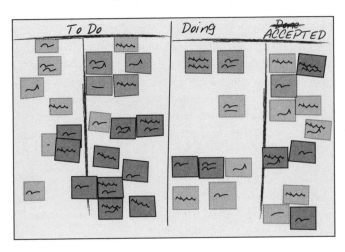

This next figure shows a task board that was actually in use on one of my projects:

Figure 2.18: Sample Task Board

Image copyright © Leading Answers, Inc. Reproduced with permission from Leading Answers, Inc. www.leadinganswers.com

In this example, we were tracking all the activities that needed to happen for a release to production. One of the benefits of meeting at the board and updating it daily is that the team will see a gradual left-to-right migration of cards across the board as work items move from "To Do" to "In Progress" to "Done." It is rewarding to the team—and reassuring to the sponsor—to see a growing collection of completed work as the deadline approaches. It also focuses everyone's attention on the few remaining issues, which really begin to stand out since they are the last ones to move.

While it may seem like a step backward to go from sophisticated software scheduling tools to cards on a wall, this versatile tool offers many benefits. For example, one of the primary advantages of using a task board is that it can help the team monitor and control their work in progress (WIP).

T&T **Work in Progress (WIP)**

Work in progress (WIP), also sometimes known as "work in process" or even "work in play," is the term given to work that has been started but has not yet been completed. Excessive levels of WIP are associated with a number of problems, including:

» WIP consumes investment capital and delivers no return on the investment until it is converted into an accepted product. It represents money spent with no return, which is something we want to limit.

» WIP hides bottlenecks in processes that slow overall workflow (or throughput) and masks efficiency issues.

» WIP represents risk in the form of potential rework, since there may still be changes to items until those items have been accepted. If there is a large inventory of WIP, there may in turn be a lot of scrap or expensive rework if a change is required.

T&T **WIP Limits**

Because of the problems listed above, agile approaches generally aim to limit WIP. A common way to apply WIP limits on agile projects is to use Kanban boards that restrict the amount of work in the system and help ensure that WIP limits are not exceeded. The boards can indicate WIP limits by displaying a preset number for how many tasks should be worked on at any given time. Another option is to restrict the amount of space designated for where the task cards can be placed on the board so only a select number of cards will fit into the space. Agile teams use tools such as Kanban boards with WIP limits to help identify and remove bottlenecks so they can keep the process running efficiently with optimal levels of WIP. As a result, these tools help reduce the risks of tied-up capital, rework, and waste on the project.

Without limits on WIP, a project team may be tempted to undertake too many different pieces of work all at once. This might be done with the best intentions, such as to fully utilize everyone's availability and to keep people busy working on the project. But the problem is, if we have a bottleneck in processing database requests or designing user interfaces, for example, then tasks may end up sitting for a while, and work accumulates in the system. When this occurs, it is difficult to identify where the bottleneck is, because everyone appears to be busy.

The figure below illustrates a Kanban board with no WIP limit. We can see that there is a lot of WIP, which means the team is busy. But we cannot see which tasks are idle and where the bottlenecks are that are contributing to a slow workflow.

Figure 2.19: Task/Kanban Board with No WIP Limit

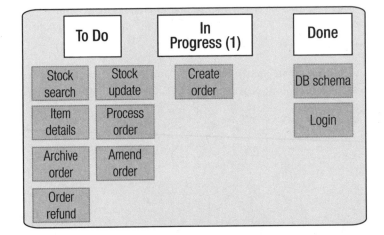

Limiting WIP is like draining the water out of a river; it shows us where all the boulders and obstacles are. We don't want to run the system with next-to-no WIP for too long—this would slow throughput because everyone but the bottleneck would be waiting for work. However, having a very low WIP limit for a short period of time is an effective method for identifying bottlenecks (or "constraints," if we use the terminology from the Theory of Constraints, which is the origin of this concept and will be discussed below).

The next figure shows a Kanban board with a WIP limit that is set too low. There is only one item in progress, which means some people are idle and there is a slow workflow. This low limit makes it very easy to identify which item is holding up the workflow, however.

Figure 2.20: Task/Kanban Board with the WIP Limit Too Low

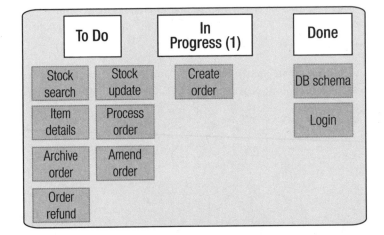

Once the bottleneck item is identified, we can elevate the constraint and remove the bottleneck by making that activity more efficient. For example, we may have other team members do some preprocessing for that activity, ask downstream consumers if they are willing to take a less-polished output, or increase resources for the constrained activity.

With the first constraint removed, we can run the process again and see if there is another item that is causing a slow workflow. We repeat the process until people are sometimes idle but tasks are being worked on most of the time and the throughput is fast.

Finally, we have a Kanban board with an appropriate WIP limit for the project. We can see that there is sufficient work occurring. People are sometimes idle, or have some slack, but the bottlenecks are cleared and there is a fast workflow.

Figure 2.21: Task/Kanban Board with the WIP Limit Just Right

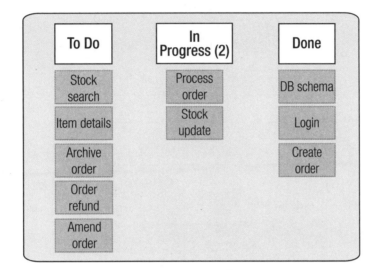

The aim of WIP limits is to optimize throughput of work, not to optimize resource utilization. This is often counterintuitive to people at first. We tend to think that all team members should be busy working at all times and that anything else is laziness or inefficiency. Consider a highway, however. When does it flow best— when it is fully utilized at rush hour (busy), or during off-peak hours when it has some slack (less busy)?

Figure 2.22: Fully Utilized Highway vs. Highway with Some Slack

Image copyright © Scott Ambler, www.agilemodeling.com

Limiting WIP helps identify bottlenecks and maximize throughput on a project, just like limiting the number of cars on a road helps traffic flow faster. On software projects, WIP limits equate to the number of features that are being worked on but are not yet accepted by the business.

T&T Cumulative Flow Diagrams (CFDs)

Cumulative flow diagrams (CFDs) are valuable tools for tracking and forecasting the delivery of value. They can help us gain insight into project issues, cycle times, and likely completion dates. Basically, CFDs are stacked area graphs that depict the features that are in progress, remaining, and completed over time. Here's an example of a CFD:

Figure 2.23: Sample Cumulative Flow Diagram

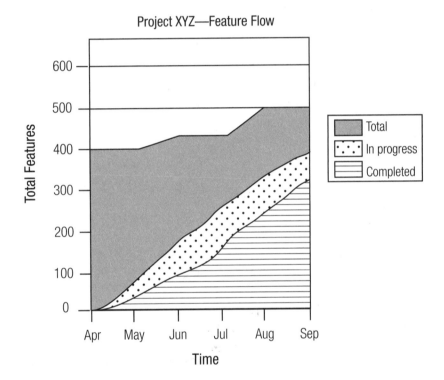

This figure shows the features completed versus the features remaining for a fictional project that is still in progress. The green area represents all the planned features to be built. This number rose from 400 to 450 in June and then to 500 in August as additional features were added to the project. The dotted section plots the work in progress, and the striped section shows the total number of features completed on the project.

Little's Law

Little's Law is a mathematical formula from queuing theory that can be used to analyze work queues (i.e., work in progress) on CFDs. As mentioned in chapter 1, this formula proves that the duration of a work queue is dependent on its size, which is why limiting WIP is such a key principle of the Kanban methodology. Now that we have talked about CFDs, let's look more closely at the illustration of Little's Law from chapter 1.

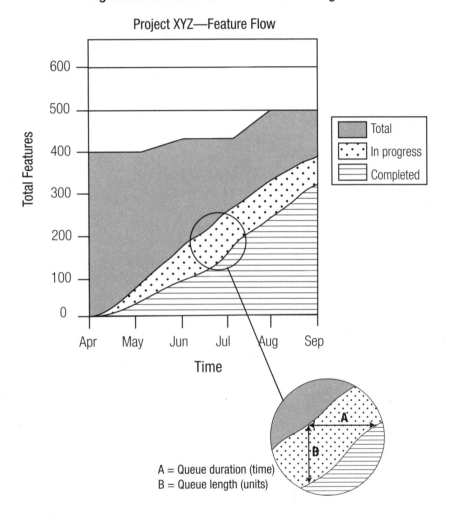

Figure 2.24: Little's Law and Work in Progress

On this CFD, the dotted section representing work in progress (WIP) is our work queue; it is the work we have started but not yet completed. The vertical (B) line tells us how many items are in the queue, and the horizontal line tells us the cycle time, or how long it will take to complete those items.

Cycle time is an important metric for lean systems. This data helps us predict when all the work currently in progress will be done. We should aim to keep WIP and cycle times as low as possible, since they represent sunk investment costs that have not yet resulted in business benefits. The more WIP and the longer cycle times we have on the project, the higher the amount of potential scrap we will have in the system if we encounter a problem.

Exam tip

Little's Law is named after the man who formulated it, John Little. However, as you proceed through this book, keep in mind that for the exam you don't need to know any of the research findings, statistics, dates, book titles, or names of companies or authors mentioned in our discussion. The exam isn't interested in those details; it focuses on the agile mindset, and your ability to apply agile practices and concepts in brief scenarios. So you only need to know names if they are the primary way of identifying important agile concepts—such as the Tuckman and Dreyfus models we'll discuss in chapter 4.

Bottlenecks and the Theory of Constraints

Eli Goldratt introduced the Theory of Constrains (TOC) as a tool for optimizing a production system. He observed that "changes to most of the variables in an organization usually have only small impacts on global performance. There are few variables (perhaps only one) for which a significant change in local performance will effect a significant change in global performance."[10] So to achieve the greatest benefits, we should find these constraints (or bottlenecks in the system) and focus on improving these issues.

One example of TOC thinking is question 3 from the daily scrum or stand-up meeting, which asks for any impediments (or blockers) to making progress. With this question, we are looking for roadblocks on the project so that they can be removed.

Constraints can also manifest themselves as restrictions on throughput capacity. For example, the database group might not be able to keep up with the changes coming from the development team, or perhaps the customer proxy cannot keep up with questions about validation rules for a new screen the team is designing. In both cases, the database group and the customer proxy are constraints, but these bottlenecks are not always easy to spot. We usually have a feeling where the bottlenecks are, but short of standing over people with a stopwatch and counting how many features they can process in a day, how can activity throughput and bottlenecks be objectively measured? This is where CFDs come into play.

We can use a CFD to find the bottlenecks if, instead of lumping all the work in progress as a single measure, we break it out by activity and plot the flow of this work. Here's an example of such a detailed CFD:

Figure 2.25: Detailed CFD

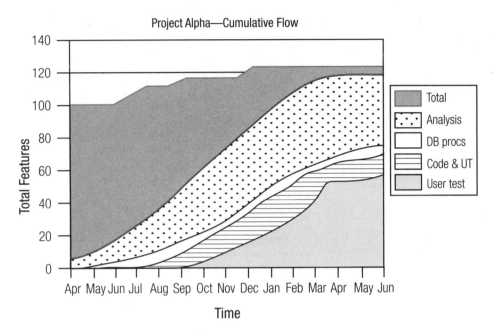

In this example, the work in progress has been broken out by activity, and the activities are stacked sequentially. Analysis, database procedures, coding and unit testing, and user testing work are being done on the project by different groups. When examining CFDs for bottlenecks, we look for areas that widen before the final activity is done. In this example, user testing is the final activity. The widening of an area indicates the growing completion of work. To explain this idea using mathematical concepts, a widening area is created when one line is followed by another line of shallower gradient. Since the line gradient indicates the rate of progress for an activity (features over time), a widening area is created above an activity that is progressing at a slower rate.

In the next figure, we can see that Analysis is a widening area activity, while creating the database stored procedures (DB procs) is a bottleneck activity.

Figure 2.26: Detailed CFD with Bottleneck Identified

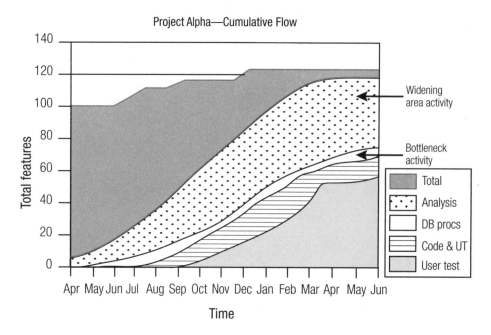

Regardless of whether the mathematical explanation makes sense to you, all you need to remember is that the bottleneck is the activity that lies *below* the widening band. The widening band is the feeding activity, not the problem activity. So in the example shown above, Analysis is going okay. However, the activity that lies below the widening band—DB procs—shows a slower rate of progress. This CFD warns us that creating the database stored procedures is currently a bottleneck in the system.

So we do not need to micromanage team throughput to determine where the bottlenecks are. Instead, by using CFDs to track progress, we can identify bottlenecks unobtrusively. Then, once we know where the problem is, we can start addressing the issue by applying the Five Focusing Steps of Goldratt's Theory of Constraints, starting with step 2:[11]

1. Identify the constraint.
2. Exploit the constraint.
3. Subordinate all other processes to exploit the constraint.
4. If after steps 2 and 3 are done, more capacity is needed to meet demand, elevate the constraint.
5. If the constraint has not moved, go back to step 1, but don't let inertia (complacency) become the system's constraint.

The use of agile methods inherently reduces the likelihood of activity-based bottlenecks because these methods promote cross-trained teams. By avoiding role specialization, people are able to move between roles more effectively and share the workload. However, while the goal is to have fully cross-trained teams, in practice, some role specialization and workflow management are the norm on most projects.

K&S **Agile Contracting**

Vendors are stakeholders who are external to the organization but are involved in the project because they are providing some kind of product or service. Agile teams should be careful in selecting vendors, and make sure they are prepared for working in an agile environment. If a vendor needs to use an agile approach to participate in the project, then this requirement should be outlined in the request for proposal (RFP).

Depending on their role in the project, however, some vendors may not need to use or understand agile practices. Before setting a requirement that all the vendors be familiar with agile, or developing an educational event for vendors, ask if it's worth it. For vendors who will play a minor role in the project, perhaps it isn't. But for vendors who will play a major or critical role, it will be important to educate them about agile. Like all decisions, in educating our vendors we need to consider the cost-benefit trade-off.

Since agile teams know that the requirements may change and the scope is negotiable, traditional vendor contracts based on formal specifications are problematic. Instead, agile projects typically use contracting models designed for an agile environment, such as the fixed-price work packages and graduated fixed-price contracts described below.

Agile Constraints and Contracts†

How can an agile project deliver value if some or all of the work is being done on a contract basis? While agile methods provide great flexibility and allow us to manage changing requirements and priorities, this adaptability and scope flexibility can create problems when defining acceptance criteria for contracts or outsourcing work.

These challenges have existed since the creation of agile methods, and a lot of effort has been put into resolving them. As you may recall from chapter 1, the first edition of the DSDM Manual presented the inverted triangle model shown below.

Figure 2.27: Inverted Triangle Model

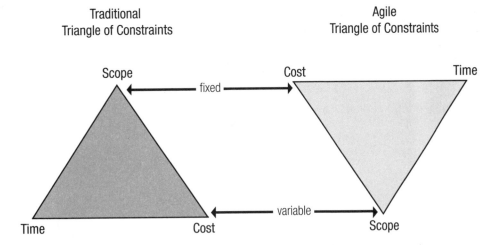

† *Portions of this section were originally published by gantthead.com in "Agile Contracts (Part 1)" by Mike Griffiths on February 8, 2011, and "Agile Contracts (Part 2)" by Mike Griffiths on March 15, 2011, copyright © 2011 gantthead.com. Reproduced by permission of gantthead.com.*

Unlike the traditional triangle of constraints, agile projects attempt to fix time and cost and adjust scope to achieve the highest-priority, best-quality product possible within the fixed constraints. If the product functionality (scope) is fixed, then there is a risk that the project will run out of money or time before completing it—or even worse, produce a poor-quality outcome.

This idea that the team might not deliver all the product functionality does not sit well with many people. They want an up-front estimate for delivering the whole product, not some subset of the easy parts; they also want a written contract so they can hold the vendor or project manager accountable. However, for a knowledge work project to achieve the complete functionality within the contracted cost and schedule, lots of time-consuming and costly analysis must be done and carefully crafted specifications must be produced. Also, substantial contingencies have to be incorporated into the contract to accommodate reasonable and unforeseen changes, technical issues, and potential setbacks. So agile customers *can* have fixed scope and firm estimates, but it will cost them—the estimates will be inflated as a buffer for uncertainty, and the customer will pay for activities that don't add value to the business after going live.

The goal of agile methods is closer cooperation between the project team and the customer, as expressed in the third Agile Manifesto value, which ranks "Customer collaboration over contract negotiation." This cooperation helps direct the team's efforts toward delivering value-adding features. An agile approach requires more trust between the parties than the traditional approach, in order to focus resources on what the team is trying to build, rather than bogging them down in debates about how changes will be negotiated or what the completion criteria really are.

The agile approach also requires the customer to be more involved in providing feedback on iteration deliverables, reprioritizing the backlog, and ranking the value of change requests against the remaining work items. For trusting, invested clients, agile contracts are great tools for extracting more value, and they give those clients a competitive advantage. For untrusting or hands-off clients, agile contracts will be a tough sell and may not be suitable.

There are different ways that agile contracts can be structured. Let's briefly look at some of them, starting with the DSDM contract.

Exam tip

On the exam, some questions may use the terms "buyer" for the customer who is buying services and "seller" for the vendor who is selling their services. Although this isn't agile terminology, these words are PMI-isms that could appear on the exam.

DSDM Contract

The DSDM contract was originally commissioned by the DSDM Consortium and continues to evolve. This contract focuses on work being "fit for business purpose" and passing tests, rather than matching a specification. The DSDM contract is used primarily in the United Kingdom and other areas of Europe.

Money for Nothing and Change for Free

Jeff Sutherland promotes a structure for an agile contract in his popular presentation, "Money for Nothing and Your Change for Free."[12] Sutherland suggests including early termination options and proposes a model that allows flexibility in making changes. Sutherland's structure starts with a standard fixed-price contract that includes time and materials for additional work, but then he inserts a "change for free" option clause.

The customer can only use this "change for free" clause if they work with the team on every iteration. Failure to be engaged like this voids the clause, and the contract reverts back to time and materials. Assuming the customer stays engaged, the product owner can reprioritize the backlog at the end of an iteration, and the changes will be free if the total amount of contracted work has not changed. This allows new features to be added if lower-priority items that require the same amount of time and effort are removed.

Figure 2.28: Change for Free

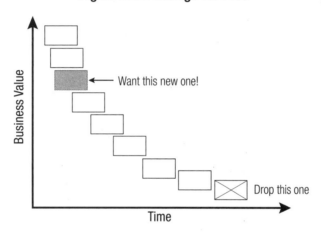

Like the "change for free" clause, the "money for nothing" concept is also only valid if the customer plays their part in the agile project. "Money for nothing" allows the customer to terminate the project early when they feel there is no longer sufficient ROI in the backlog to warrant further iterations. For example, the supplier might allow termination of the contract at any time for 20 percent of the remaining contract value, as shown below.

Figure 2.29: Money for Nothing

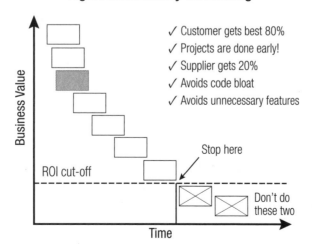

Asking the customer to pay 20 percent of the remaining contract value helps offset the supplier's risk that their team will be left idle before the anticipated contract time is up. And from the customer's standpoint, they get their top-priority business value, their project finishes early, and they are able to maintain good relations with the supplier. The product stays efficient without code bloat (for software projects) or unnecessary features. The arrangement remains open for additional releases if required at the contracted rates for time and materials.

Graduated Fixed-Price Contract

Graduated fixed-price contracts are promoted by Thorup and Jensen as another type of agile contract.[13] With this kind of contract, both parties share some of the risk and reward associated with schedule variance. Thorup and Jensen suggest using different hourly rates based on early, on-time, or late delivery. For example:

Project Completion	Graduated Rate	Total Fee
Finish early	$110 / hour	$92,000
Finish on time	$100 / hour	$100,000
Finish late	$90 / hour	$112,000

Under this arrangement, if the supplier delivers on time, they get paid for the hours worked at their standard rate. If they deliver early, they get paid for fewer hours—but at a higher rate. The customer is happy because the work is done early and they pay less overall. The supplier is happy because they make a higher margin. However, if the supplier delivers late, they will get paid for more hours, but at a lower rate. When this occurs, both parties are somewhat unhappy since they are both making less money, but at a gradual, sustainable rate that hopefully will not lead to the contract being terminated.

Fixed-Price Work Packages

Another approach to setting up an agile contract is to establish fixed-price work packages. Fixed-price work packages mitigate the risks of underestimating or overestimating a chunk of work by reducing the scope and costs involved in the work being estimated. For example, for one project Marriott International broke down their statements of work (SOW) into individual work packages, each with its own fixed price.[14] Then as the work progressed, the supplier was allowed to re-estimate the remaining work packages in the statements of work based on new information and new risks.

Using fixed-price work packages allows the customer to reprioritize the remaining work based on evolving costs. It also gives the supplier the ability to update their costs as new details emerge, removing the need for the supplier to build excess contingency funds into the project cost. The changes are then localized to small components (the work packages). If extra funding is required, it is easy to identify the need and justify it. The figure below illustrates the difference between a traditional SOW and fixed-price work packages.

Figure 2.30: Traditional Statements of Work versus Fixed-Price Work Packages

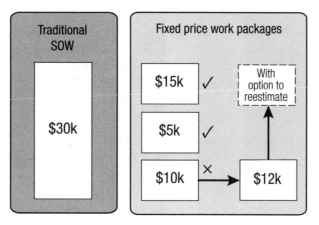

Customized Contracts

Different types of agile contracts can be pieced together to create a customized contract that best matches the needs of both the buyer and the seller. With such contracts, the customer retains flexibility to reprioritize work, and the seller is not penalized for sharing information about increased costs. They also remove the incentive for the seller to add large contingencies to the project price. By combining elements of a graduated fixed-price contract and fixed-price work packages and incorporating the concepts of early termination (money for nothing) and reprioritization (change for free), we can create a contract that protects both parties and encourages positive behavior.

Although agile contracts can be highly beneficial to both parties, we must remember that creating a contract between one party that wants to minimize the cost of a product or service and another party that is in business to maximize its revenue will always be a balancing act. On agile projects, procurement has always been particularly challenging, since the details of the scope can't be fully defined early in the project. In addition, the intangible nature of knowledge work products can make it difficult to evaluate and get acceptance for the work.

Any type of procurement—whether for agile or traditional contracts—works best when both parties want successful results that lead to future work. A project's success is ultimately determined by the level of ongoing collaboration between the customer and the seller. Agile contracts cannot generate or enforce that collaboration, but with some work and creativity, they can at least be better structured to support it.

Verifying and Validating Value

It is one thing to think we are building great products and services and another thing entirely to have the sponsors, users, or product owner confirm this. Agile methods are often used on projects that are intangible (such as designs, software, etc.). The intangible nature of these end products means it is all that much more important to validate that what we are building is, in fact, on the right track and seen as highly valuable by the business. This challenge can be illustrated in cartoon form, as shown below.

Figure 2.31: Gulf of Evaluation

| How the customer explained it | How the ScrumMaster understood it | How the developers built it | What the customer really needed |

Cartoons on this theme have been around for many years; they're popular because they highlight the communication failures that are likely to occur when people try to describe something intangible. What one person describes is often very different from how the listener interprets it. This semantic gap is called the "gulf of evaluation."

On manufacturing projects, the work is visible, tangible, and familiar; therefore, the gulf is small and quickly crossed. In contrast, on knowledge work projects, the work is often invisible, intangible, and new; this leads to a greater gulf of evaluation, and misunderstandings are more common.

Because having a gap between requested and delivered features may lead to rework, it is important to discover these differences early. Therefore, confirming value—meaning verifying that the team is building the right thing and that it works as desired rather than as first described, which could be different—is a key practice when using agile methods.

T&T Frequent Verification and Validation

Agile techniques are designed to resolve problems as soon as possible, before they can grow bigger and move up the cost of change curve. Like the old saying, "A stitch in time saves nine," agile uses regular testing, checkpoints, and reviews to address problems before they get bigger. This practice is referred to as frequent verification and validation.

Frequent verification and validation is practiced at many levels on agile projects. It is an effective antidote to both the fact that making mistakes is part of being human and the mismatch of expectations that arises when the team interprets the customer's end goal differently than intended. With frequent verification and validation, we are checking to make sure things are working and progressing as they should, as well as looking for any mismatches in expectations. In doing so, we enable the project to rapidly converge on the real solution, even while that solution is evolving from the originally stated requirements.

The figure below illustrates the cycles of verification and validation that take place in an XP software development project. We'll first examine how these feedback loops work in XP, and then extrapolate the principle of frequent verification and validation to other kinds of agile projects.

Figure 2.32: Frequent Verification and Validation in XP

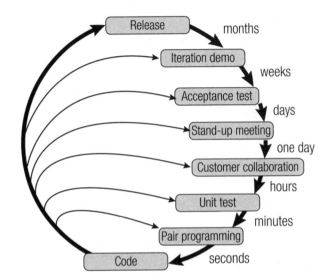

As depicted here, an XP project includes many overlapping testing and verification cycles that can range from a few seconds to several months. Let's see how some of these cycles work:

» The shortest feedback cycles take place during the XP practice of pair programming. When two developers are working together and the "reviewer" spots an error as the "coder" is writing, the issue can be resolved almost immediately.

» At the next level, XP developers run unit tests every few minutes to verify that the code is achieving the desired outcome and validate that the existing functionality is working properly.

» At the next level, customer collaboration should occur frequently. This provides feedback on whether the team's designs are correct and the customer's objectives are being met.

» In their daily stand-up meetings, the team will validate who is working on what and when the code should be done and ready to integrate.

» Every few weeks the team holds an iteration review meeting to demo the new work they have completed to the customer and verify that they are building the right product so far.

» Every few months (typically), at the end of each release, the customer reviews the deliverable to verify that it is "done" and ready for release.

As you can see, XP practices create cycles within cycles that are all focused on providing frequent verification and validation as the product is being built. Agile methods rely on multiple levels of deliberate checks, tests, and confirmation activities like this to catch issues as soon as possible, since they know that building new (and often intangible) products will bring communication challenges.

Although the above examples are derived from a software development environment, the same principles apply to other kinds of agile projects; only the specifics differ. For example, let's take the example of a group of instructional designers who are creating the lesson plan for a new course. They start by collaborating on the course goals, exchanging ideas in real time, like pair programmers. While writing the modules, they meet frequently to review the modules and coordinate their efforts. Once they have a first draft of the modules, they conduct a small sample test to get feedback from a volunteer study group. Even after the course goes live, they continue to ask for student feedback via anonymous surveys. In short, they use multiple feedback loops, ranging from very short to quite long, to verify and validate their course and make sure it works for the intended audience.

T&T Testing and Verification in Software Development

In this section, we'll examine the testing and verification techniques used in agile software development projects. You have probably noticed that we have emphasized the point that agile isn't only applicable to software development work. And it's true that the exam will mostly focus on the use of agile in generic knowledge work projects. However, agile was originally developed in a software development setting, so some of the terms and concepts used in testing software projects might appear on the exam, if only as wrong answers that you need to rule out.

Exam tip

In this section we'll be explaining how testing is done in software development projects to provide a solid grounding for those who aren't familiar with this field and would like to understand these agile concepts. However, this material is unlikely to be featured on the exam. So if you find this discussion to be too technical, you can skip this section and still do well. In terms of exam preparation, the most important concept in this discussion is the idea of refactoring, which we'll be revisiting in chapter 6.

Agile software projects automate as many of their tests as possible, which removes the human element from their execution. Automation also allows the tests to be run more frequently at a lower cost. The goal of frequent verification and validation is to find issues as soon as possible and keep them low on the cost of change curve. (As you may recall, this curve tells us that the sooner we catch an error, the less it will cost to fix.)

Exploratory and Usability Testing

The exam content outline mentions two specialized kinds of testing—exploratory testing and usability testing. These types of tests are commonly used in software development work and are easiest to understand in that context, although they aren't necessarily limited to IT projects. Let's see what is meant by these terms.

» *Exploratory testing* differs from scripted testing that attempts to exercise all the functional components of a system; instead, it relies on the tester's autonomy, skill, and creativity in trying to discover issues and unexpected behavior. This is a complement to scripted testing, and not in any way a replacement for it. A team will interweave exploratory testing with scripted testing to help find edge cases (problems that only occur in extreme conditions), system boundaries, and unanticipated behavior outside of the regular functions that can be tested with scripts. Using a combination of scripted testing (whether manual or automated) along with unscripted exploratory testing increases test coverage and reduces the risk that a defect won't be detected.

» *Usability testing* attempts to answer the question, "How will an end user respond to the system under realistic conditions?" The goal of this kind of testing is to diagnose how easy it is to use the system, and help uncover where there are problems that might need redesign or changes. This typically involves observing users as they interact with the system for the first time. For example, where do they pause and have to think about how to do something? Where do they get stuck and ask for help? Are there common tasks that require too many repetitive actions? Data might be gathered by videotaping the users, using eye-tracking tools, or performing post-test interviews.

T&T Continuous Integration

Continuous integration is a practice used by software developers to frequently incorporate new and changed code into their project code repository. This helps minimize the integration problems that result from multiple people making incompatible changes to the same code base. Typically, the more frequently we make these code commits, the smaller the amount of code that needs to be changed to allow a new build (or version) of the software to compile successfully. So continuous integration is one of the tools we can use to find and resolve problems as early as possible.

Continuous integration uses automated tools to start the integration process automatically whenever new code is checked in, or at timed intervals. In addition to checking whether any new and changed code compiles correctly, the team will use automated unit tests to ensure that the system still performs as intended after the new code is integrated.

The continuous integration process can be illustrated like this:

Figure 2.33: Continuous Integration

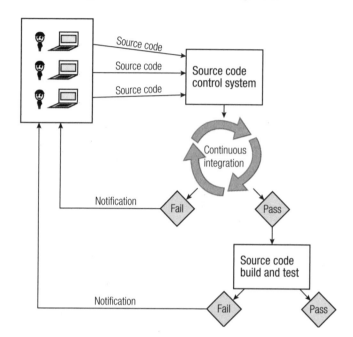

Let's take a closer look at the components of a continuous integration system.

» **Source code control system**: This is the software that performs version control on all the files that represent the product being developed.

» **Build tools**: The source code needs to be compiled before tests can be run. Most integrated development environments (IDEs) serve as a build tool to compile the code.

» **Test tools**: As part of the build process, unit tests are run to ensure that the basic functionality operates as planned. Unit test tools execute the small, atomic tests that are written in these tools to check the code for unanticipated changes in behavior.

» **Scheduler or trigger**: Builds might be launched on a regular schedule (such as every hour) or every time the system detects a change to the source code.

» **Notifications**: If a build fails, the team needs to be notified so they can correct the build as soon as possible. These notifications may be sent via e-mail or instant messaging.

Although continuous integration is a software development practice, the same concept can be applied to other knowledge work projects where multiple people are developing separate parts of the project and then bringing their pieces together to create a functional, valuable product.

Exam tip

Although the exam might mention continuous integration, it won't expect you to understand the details of how such a system works. It's helpful to understand this topic and the other software concepts discussed in this section, because they provide examples of the agile concept of frequent verification and validation that can be extrapolated to other types of agile projects. However, if you find all these technical terms to be too confusing, just think about what these practices are trying to accomplish and how they reflect the agile mindset and approach to projects.

Pros and Cons of Continuous Integration

Continuous integration is important for agile projects because it provides the following benefits:

» The team receives an early warning of broken, conflicting, or incompatible code.
» Integration problems are fixed as they occur, rather than as the release date approaches. This moves any related changes down the cost of change curve and avoids last-minute work before releases.
» The team receives immediate feedback on the system-wide impacts of the code they are writing.
» This practice ensures frequent unit testing of the code, alerting the team to issues sooner rather than later.
» If a problem is found, the code can be reverted back to the last known bug-free state for testing, demo, or release purposes.

The disadvantages or costs of using continuous integration are:

» The setup time required to establish a build server machine and configure the continuous integration software; this is typically done as an iteration 0 activity (before the development work begins on the project)

» The cost of procuring a machine to act as the build server, since this is usually a dedicated machine

» The time required to build a suite of automated, comprehensive tests that run whenever code is checked in

Despite the costs, continuous integration is established as a standard good practice for software development projects, and most multiperson teams use this approach. The benefits of this technology over older practices, such as the "daily build and smoke test," are significant; with continuous integration, far more tests are run per day, and less time passes before a problem is identified. This practice is a prime example of how agile teams can shorten the time between the occurrence of a defect, the detection of the defect, and its subsequent resolution.

Test-Driven Development (TDD)

Test-driven development (TDD) and test-first development (TFD) encourage teams to bring a different mindset to software development, and they employ short test and feedback cycles. There are differences between TDD and TFD, but for the exam, you simply need to understand at a high level the behaviors these techniques foster and why they are beneficial. For that reason, we will examine these techniques together under the single heading of test-driven development, or TDD.

The philosophy behind TDD is that tests should be written before the code is written. In other words, developers should first think about how the functionality should be tested and then write tests in a unit testing language (such as NUnit or JUnit) before they actually begin developing the code. Initially the tests will fail, since developers have not yet written the code to deliver the required functionality.

So with TDD, developers begin a cycle of writing code and running the tests until the code passes all the tests. Then, if necessary, they clean up the design to make it easier to understand and maintain without changing the code's behavior. This last process is called "refactoring."

Figure 2.34: Test-Driven Development Process

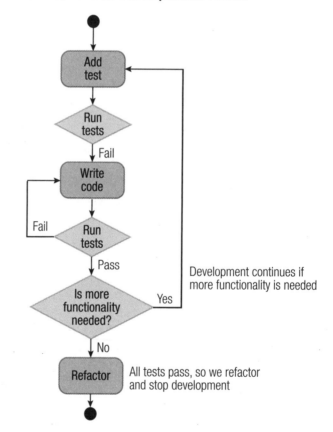

Red, Green, Refactor / Red, Green, Clean

As we've seen, most agile software developers use continuous integration tools to run builds. These tools tell them whether the code passed or failed by communicating a green or red status. The process of writing a test that initially fails, adding code until the test passes, and then refactoring the code is known as "Red, Green, Refactor" or sometimes "Red, Green, Clean."

The following are some key benefits of closely linking code to tests in this way:

» By focusing on the tests first, we must think about how the functionality will be used by the customer. This puts us in the mindset of being concerned with the outcome before the implementation, which studies have found leads to better designs and higher levels of client satisfaction.

» Writing the tests before the code ensures that we have at least some tests in place. Better test coverage of the code enhances systems quality and allows the development team and customer to be more confident in the code.

» Early and frequent testing helps us catch defects early in the development cycle, which prevents the defects from becoming widespread and costly to fix. Eliminating defects early in the development cycle while the code is still fresh in the developers' minds, rather than later in the project, also reduces the amount of time spent finding the cause of the defects.

» The approach of writing systems in small, tested units leads to a more modular, flexible, and extendable system.

There are also disadvantages or costs to this approach, including:

» The unit tests are usually written by the same developer who will implement the code; therefore, we will likely see the same misinterpretations of requirements in both the test and the code.

» Some types of functionality, such as user interfaces, are difficult or time-consuming to reliably test via unit tests.

» The tests themselves also need to be maintained, and as the project grows and changes, the sustainment load for test scripts goes up.

» As people see higher numbers of passing tests, they may get a false sense of security about the code quality.

Acceptance Test–Driven Development (ATDD)

The technique of acceptance test–driven development (ATDD) moves the testing focus from the code to the business requirement. As with TDD, the tests are created before work starts on the code, and these tests represent how the functionality is expected to behave at an acceptance test level.

Typically we capture these tests when we pull the user story from the backlog and discuss its desired behavior with the business representatives. The acceptance tests may be captured in a functional test framework, such as "FIT" (Framework for Integrated Testing) or "FitNesse."[15] The overall process goes through the four stages of discuss, distill, develop, and demo, as illustrated below.

Figure 2.35: Acceptance Test–Driven Development (ATDD) Cycle

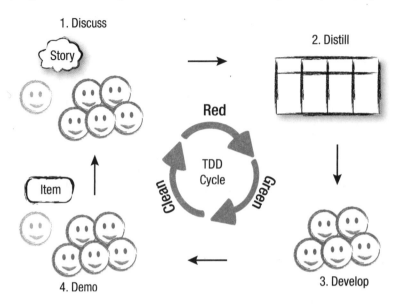

ATDD model developed by Elisabeth Hendrickson, Grigori Melnick, Brian Marick, and Jim Shore, based on a model by Jim Shore. Copyright © 2012 Quality Tree Software, Inc. Licensed Creative Commons Attribution.

Let's look at these four stages in a little more detail:

» **Discuss the requirements**: During the planning meeting, we ask the product owner or customer questions that are designed to gather acceptance criteria.

» **Distill tests in a framework-friendly format**: In this next phase, we get the tests ready to be entered into our acceptance test tool. This usually involves structuring the tests in a table format.

» **Develop the code and hook up the tests**: During development, the tests are hooked up to the code and the acceptance tests are run. Initially the tests fail because they cannot find the necessary code. Once the code is written and the tests are hooked to it, the tests validate the code. They may again fail, so the process of writing code and testing continues until the code passes the tests.

» **Demo**: The team does exploratory testing using the automated acceptance testing scripts, and demos the software.

When we combine the tasks of defining acceptance criteria and discussing requirements, we are forced to come to concrete agreement about the exact behavior the software should exhibit. In a way, this approach enforces the discussion of the "definition of done" at a very granular level for each requirement.

Exam tip

For all of these techniques—TFD, TDD, and ATDD—the main point you need to understand is that it is a good practice to think about how the system (or product, if we expand this concept to nonsoftware projects) will be tested *before* we start developing the product. If we can design acceptance tests before we develop the product, we are more likely to end up with tests for the majority of the work, which in turn means that the product will be better tested.

EXERCISE: TECHNICAL TERM FISHING

Review the following list of terms and circle any that are NOT associated with agile verification and validation:

TDD	Velocity	ATDD	SME
Automated build	WIP	Build server	Refactor
TED	Pair programming	UXP	ROI
TFD	Relative sizing	EI	Fast failure

ANSWER

TDD	Velocity	ATDD	SME
Automated build	WIP	Build server	Refactor
TED	Pair programming	UXP	ROI
TFD	Relative sizing	EI	Fast failure

Chapter Review

1. What isn't one of the practices your team can use to ensure they get feedback about the product being built?

   A. Pair programming
   B. Releases
   C. Wideband Delphi
   D. Stand-up meetings

2. In comparison to traditional projects, the agile approach to contracts doesn't require:

   A. More detailed specifications
   B. More customer involvement
   C. More trust
   D. More feedback

3. To test how people will respond to the system under real-world conditions, we would perform _____.

   A. User factoring
   B. Continuous testing
   C. Test-driven integration
   D. Usability testing

4. The contractor on your project doesn't understand the team's agile approach. As team coach, what should you do?

   A. Request that the PMO provide a week-long agile workshop for all potential contractors.
   B. Ask the product owner whether educating this contractor is a project requirement.
   C. Gather the team to discuss the contractor's role and decide how much they need to know about agile.
   D. You don't need to do anything, since the contractor isn't on the delivery team.

5. Your team is developing an online game that will have three beta (test) releases before launch. Which feature would not be included in the first beta release?

   A. Beta player sign-up and log in interface
   B. Password authentication
   C. Refer-a-friend marketing campaign
   D. Options for giving feedback to developers

6. The sponsor wants to use earned value metrics to measure the team's progress. You remind her that:

   A. The team prefers to use information radiators.
   B. The project plan isn't finalized yet.
   C. We need to establish a stable velocity first.
   D. Earned value metrics won't reveal whether the product is meeting the users' needs.

7. What would the developers on your team need to code first?

    A. Release 2.2 user stories
    B. Release 2.1 acceptance tests
    C. Release 2.3 unit tests
    D. Release 2.1 user stories

8. What is the main purpose of imposing limits on work in progress?

    A. To optimize throughput
    B. To minimize resource allocation
    C. To visualize lead time
    D. To balance workflow

9. Two team members have different opinions about what needs to be built to meet the customer's requirements. This is probably an example of _____.

    A. The definition of done
    B. Divergence
    C. The gulf of evaluation
    D. Pair programming

10. Your team has high levels of work in progress, and you are explaining to them why that is a problem. What isn't one of the issues you mention?

    A. It hides efficiency and throughput issues.
    B. It confuses the team members' roles.
    C. It doesn't deliver any return.
    D. It increases risk and potential rework.

11. Which of the following is not a form of frequent verification and validation?

    A. Pair programming
    B. Unit testing
    C. Iteration demos
    D. Iteration planning

12. A sponsor wants to evaluate a proposed three-year project against two proposed one-year projects. Which financial metric would be most helpful?

    A. NPV
    B. ROI
    C. MMF
    D. Velocity

13. You are working on a software development team that follows a test-driven development process. The sequence of activities you would undertake is:

    A. Write code, write test, refactor
    B. Write test, refactor, write code
    C. Write test, write code, refactor
    D. Write code, refactor, write test

© 2015 RMC Publications, Inc • 952.846.4484 • info@rmcls.com • www.rmcls.com

14. Which of the following is most likely to be a minimal viable product?

    A. A bicycle that has no handlebars
    B. A pen that has a reservoir for glow-in-the-dark ink as well as regular ink
    C. An order entry system that has no user interface
    D. A chair that has no back

15. The sponsor is trying to determine which project has the greatest business value—one that returns $4 million in three years or one that returns $5 million in four years. The interest rate is 5 percent for borrowing capital to develop the projects. The best approach to determine the highest-value project is:

    A. Select the "$5 million in four years" project, since it returns the highest amount.
    B. Select the "$4 million in three years" project, since it has the shorter payback period.
    C. Calculate the NPV of the projects, and choose the project with the highest value.
    D. Calculate the NPV of the projects, and choose the project with the lowest cost.

16. You have been asked to outline the basics of agile contracting for your steering committee. Which of the following statements best describes the recommended approach to contracting on agile projects?

    A. The contract is worded to allow for early completion of scope, and acceptance is based on items matching the original specification.
    B. The contract is worded to allow for reprioritization of scope, and acceptance is based on items matching the original specification.
    C. The contract is worded to allow for early completion of scope, and acceptance is based on items being fit for business purpose.
    D. The contract is worded to allow for reprioritization of scope, and acceptance is based on items being fit for business purpose.

17. What is the primary benefit of involving business representatives in the prioritization of work?

    A. To proactively engage the business representative between acceptance testing cycles
    B. To answer business-related questions about the requirements
    C. To promote communication about when features will be delivered
    D. To better understand the business needs for the project

18. Project X has an IRR of 12 percent, and project Y has an IRR of 10 percent. Which project represents the better rate of return?

    A. It depends on the payback period.
    B. Project Y
    C. Project X
    D. Project X or Y, depending on NPV

19. Little's Law demonstrates that:

    A. The duration of a work queue is dependent on its size.
    B. The duration of a work queue allows us to predict how long the project will take.
    C. WIP is equal to the duration of the work queue.
    D. Cycle time is equal to the size of the work queue.

20. How is risk related to value delivery?

   A. Risk is an inherent part of delivering value.
   B. Risk increases as value decreases.
   C. Risk is factored into value.
   D. Risk is anti-value.

Answers

1. **Answer**: C

   Explanation: This question will be fairly easy if you know that wideband Delphi is an estimating tool. But even if you don't, you should be able to think through the other three options to see that they all provide opportunities for getting feedback. Pair programming allows two programmers to give each other immediate feedback. Releases allow the end users to try out the product and provide their feedback. Stand-up meetings allow the team members to share information about what they are doing, when their tasks will be done, and whether they are having a problem. This provides daily feedback about how well the team's plan is going.

2. **Answer**: A

   Explanation: Like many questions you will encounter on the PMI-ACP exam, this question is written in a confusing way. You have to read it carefully to understand that you're being asked to identify the option that is more characteristic of traditional contracts than agile contracts. Although there are many kinds of agile contracts, in general they typically require less detailed specifications than traditional contracts; therefore, "more detailed specifications" is the correct answer.

3. **Answer**: D

   Explanation: As implied by its name, usability testing involves testing how end users will respond to the system under real-world conditions. All the other terms are made up. This question provides an example of how the PMI-ACP exam will use made-up terms that sound real; if you don't know the real agile terms, you won't be able to rule out other terms that sound reasonably similar to them.

4. **Answer**: C

   Explanation: Although vendors should be selected carefully for an agile project, they don't necessarily need to be fully educated about the agile methods that the team is using. How much education a given vendor will need typically depends on their role in the project—and it is the team members who are most likely to have the technical knowledge required to make that decision. Although educating the vendors could be a project requirement for some projects, that isn't the BEST answer.

5. **Answer**: C

   Explanation: This question tests your understanding of the agile concept of a minimal viable product. We would want the beta test players to be able to sign up, log in, use their passwords, and give feedback to developers about any bugs they find. What we don't need at this point is a program to convince more people to try the game, since it is still under development.

6. **Answer**: D

   Explanation: This question requires you to understand the drawbacks of using earned value metrics on an agile project. Three of the questions are distractors that aren't related to EVM. EVM charts can be displayed as an information radiator, so that is irrelevant. Agile plans aren't "finalized" until the project is done, so that can't be right, either. A stable velocity isn't necessary for EVM to work well. The correct answer is that earned value metrics don't reveal whether the deliverable being built is a good product that will meet the users' needs.

7. **Answer**: B

   Explanation: This question checks your understanding of test-driven development. Based on the release numbers, you should be able to tell that the correct choice must be either the acceptance tests or the user stories for release 2.1, since that release will be built before releases 2.2 and 2.3. The best answer would be the acceptance tests since in the test-driven development process the tests are written BEFORE the user stories they are designed to test. Although the question doesn't state that this team is using TDD, if you don't make that assumption there is no logical reason to choose either answer. The questions on the real exam may require you to "read between the lines" and make assumptions like this.

8. **Answer**: A

   Explanation: For the exam, it's important to understand that the purpose of using WIP limits is to optimize throughput (speed of workflow). Although WIP limits don't focus on maximizing resource allocation (keeping everyone busy), they certainly don't aim to minimize it, either. Finally, although a Kanban board can help a team visualize lead time and balance workflow, that isn't the purpose of implementing WIP limits.

9. **Answer**: C

   Explanation: If team members have different ideas about what needs to be built, it means they have different interpretations of the customer's description of the product—which is the definition of a gulf of evaluation. Although the team members might be disagreeing about the definition of done, their disagreement itself is not an example of that concept. If the question had stated that the team members had different opinions about HOW to build the product, then that could be an example of divergence (healthy debate that allows the team to converge on the best approach)—but that isn't what the question says. Pair programming involves writing and reviewing the code, not debating what to code.

10. **Answer**: B

    Explanation: To answer this question, we have to understand how work in progress impacts an agile project. If we have a lot of partially done work, how can we effectively measure efficiency and throughput? Also, all that work "on hold" is just sitting there, not delivering any return (value). Incomplete work also increases risk, since we might have to redo it if something changes before it is finished and accepted by the customer. The only problem listed here that isn't a result of high levels of WIP is confusing the team's roles, so that is the correct answer.

11. **Answer**: D

    Explanation: Iteration planning is not a way of verifying or validating a product. Instead, it is concerned with planning, estimating, and scheduling capacity.

12. **Answer**: A

    Explanation: The correct choice is NPV (net present value) because it converts multiyear returns on investment to a value in today's terms. It is the best option from the choices presented for evaluating projects with different durations. Also, MMF and velocity are not methods for assessing the value of a proposed project.

13. **Answer**: C

    Explanation: The correct sequence is to write a test (which will initially fail), write code until it passes all the tests, and then refactor the design to clean things up before moving on to the next item.

14. **Answer**: D

    Explanation: A bicycle without handlebars and an order entry system without a user interface are both unusable products. Since we have no use for them, they aren't minimal viable products (MVPs). A pen that contains two types of ink is unlikely to be an MVP since glow-in-the-dark ink isn't an essential feature for a writing implement. That leaves the chair as the correct answer. We can sit on a chair that has no back, and if the back is added later it will be a value-added feature. A minimal viable chair just needs to provide a surface to sit on.

15. **Answer**: C

    Explanation: To evaluate the value of two projects that are completed at different times, we can use net present value (NPV) to translate the amounts into today's values. Therefore, the approach of calculating the NPV and choosing the project with the highest value is the way to go. Although there may be valid arguments for one of the other answers, the question is asking for the *best* approach, which is to calculate NPV.

16. **Answer**: D

    Explanation: Two components common to agile contracts are an ability to reprioritize work and the goal of satisfying the business, rather than conforming to a spec. The closest match to these characteristic is the option "The contract is worded to allow for reprioritization of scope, and acceptance is based on items being fit for business purpose."

17. **Answer**: D

    Explanation: While involving business representatives in prioritizing the work does help us promote communication and answer questions, the primary reason for that practice is to help the team better understand the business needs for the project. "Proactively engaging" the customer between testing cycles really just means keeping them busy, which is not a concern of the team.

18. **Answer**: C

    Explanation: The answer is project X, simply because it has the higher rate of return (IRR). We don't need to consider the payback period or the NPV because the question asks which project has the better rate of return, which is its IRR.

19. **Answer**: A

    Explanation: Little's Law is a mathematical formula that demonstrates that the duration of a work queue is dependent on its size. WIP is the size of the work queue, and cycle time is the duration of the work queue. Although Little's Law provides information about how long it will take to complete work items, that doesn't allow us to predict how long the entire project will take, since it doesn't tell us about variations in the team's velocity or future distractions, roadblocks, or scope changes.

20. **Answer**: D

    Explanation: Risk can be considered the opposite of value, or anti-value, since risks or threats to the project have the potential to erode, remove, or reduce value if they occur. The other options are made-up ideas.

CHAPTER 3

Stakeholder Engagement

Domain III Summary

This chapter discusses domain III in the exam content outline, which is 17 percent of the exam, or about 20 exam questions. This domain focuses on working with the project stakeholders, including establishing a shared vision, collaboration, communication, and interpersonal skills.

Key Topics

» Active listening
» Agile chartering
» Agile modeling
» Assessing and incorporating community and stakeholder values
» Brainstorming
» Collaboration
  – Collaboration games
» Communication management
  – Face-to-face (F2F)
  – Social media
  – Two-way (trustworthy, conversation-driven)

» Conflict resolution
  – Levels of conflict
» Definition of "done"
» Emotional intelligence
» Facilitation
» Information radiators
» Knowledge sharing/written communication
» Negotiation

» Participatory decision models (convergent, shared collaboration)
  – Decision spectrum
  – Fist-of-five voting
  – Simple voting
  – Thumbs up/down/ sideways
» Personas
» Stakeholder management (stewardship)
» Wireframes
» Workshops

Tasks

1. Engage empowered business stakeholders.
2. Share information frequently with all stakeholders.
3. Form working agreements for participation.
4. Assess organizational changes to maintain stakeholder engagement.
5. Use collaborative decision making and conflict resolution.
6. Establish a shared vision to align stakeholders.
7. Maintain a shared understanding of success.
8. Provide transparency for better decisions.
9. Balance certainty and adaptability for better planning.

Stakeholders, as defined in the *PMBOK® Guide*, are any people or groups who will be impacted by or have an impact on the project.[1] This is a pretty broad term that includes the customers, sponsors, and business representatives, as well as the project leaders, development team, vendors, and other people inside or outside of the organization who will be affected by the project or its results, including the product's end users. Getting stakeholders involved—in other words, engaging them in the project—is absolutely essential for the success of any project.

This chapter covers a variety of practices that agile teams use to successfully work with project stakeholders. To organize this discussion, we'll group these practices into five themes—taking care of stakeholders, establishing a shared vision, communicating with stakeholders, working collaboratively, and using critical interpersonal skills.

K&S Taking Care of Stakeholders

The PMI-ACP exam content outline includes the term "stakeholder management." However, that phrase is a bit misleading, since agile approaches don't really consider stakeholders as something to be "managed" in the traditional sense. So a more appropriate term for this effort would be "stakeholder stewardship." The idea of stewardship basically involves comprehensively looking after something that is worth nurturing, safeguarding, and preserving—in this case, the project stakeholders.

Stakeholder Stewardship versus Stakeholder Management

In traditional project management, "stakeholder management" is the process of managing the people who are involved in the project. This starts with identifying the project stakeholders and continues with controlling and directing their participation in the project plan, until they are released from the project. While this makes logical sense, it seems to be a rather inhuman process that treats people as resources, which is not the best starting point for fostering support and collaboration.

When we talk about "managing" people, it implies that we are telling them what to do and controlling their activities. Of course, this isn't always the case—most good project managers know that we get the most out of people by acknowledging their skills and getting them engaged in planning their work. However, in some agile communities the term "management" has come to mean top-down, command-and-control direction.

This is why I prefer the term "stewardship." The idea of safeguarding and looking after people is better aligned with agile's servant leadership model. Stakeholder stewardship means looking after everyone on the project and making sure they have everything they need to succeed. This isn't a passive role—sometimes "serving and safeguarding" may mean directly addressing a problem of poor performance or inappropriate behavior. However, even this is done with a mindset of serving the team rather than trying to tell people how to do their jobs.

Like stakeholder management, stakeholder stewardship includes the initial efforts to identify who the stakeholders are. Some stakeholders will be obvious—for example, even at the start of the project we typically know the sponsors, users, and functional managers who will be involved. However, some of the other roles may be trickier to identify, such as auditors, upstream providers of project inputs, or downstream consumers of project outputs. It's important to identify all the stakeholders and carefully follow their involvement in the project, because excluding or alienating any of them may put the successful execution of the project at risk.

Educating Stakeholders about Agile

If the project stakeholders are new to agile methods, they may need some basic education about how agile projects operate to help them understand the approach that will be used, address any myths about agile, and guide their expectations. This education should include the goals, values, practices, and benefits of the agile approach to help them understand why the project will be executed in this manner.

It is normal for any change to be met with some degree of skepticism and caution. This isn't necessarily a bad thing. After all, the world would be pretty chaotic if we adopted every new idea, operating system, or systemic change, such as which side of the road we drive on. Some resistance to change is healthy because it helps ensure that only worthwhile changes prevail. So when we are initially engaging stakeholders in an agile project, we need to recognize some of the concerns that the stakeholders may have, and address them directly. Here are some examples of common concerns about agile:

» **Executives and project sponsors**: Executives and sponsors are often concerned about the risk of failure due to the use of unprecedented practices and counterintuitive planning approaches, even though an agile approach actually provides better feedback loops for preventing failure.

» **Managers**: Managers may fear a loss of control or erosion of their role when projects assume an agile approach.

» **The development team**: The development team may resist agile methods if they feel that this new approach is being forced upon them by management.

» **The user community**: The user community is often worried they will not get all features they want or require, or that the early iterations will result in poor quality deliverables.

» **Supporting groups**: Other groups may be concerned about an apparent lack of control, continual requests for their involvement, or the lack of a clear end point.

Keeping Stakeholders Engaged

In addition to the initial efforts to identify stakeholders, educate them, and address their concerns, stakeholder management involves continuing to engage them in the project as it progresses. One benefit of short iterations is that they prevent stakeholders from losing interest in the process. During an iteration, we meet with stakeholders, agree on what should be done, do some work, and then one to four weeks later, we show them what we have developed. Since brief iterations are good for maintaining stakeholder engagement, they are preferable to long development phases where visibility is low and stakeholder interest tends to wane as a result. And as you may recall from the previous chapter, this increased visibility throughout the project is one element of the agile value proposition, as shown in the upper left part of this diagram.

Figure 3.1: Agile Value Proposition

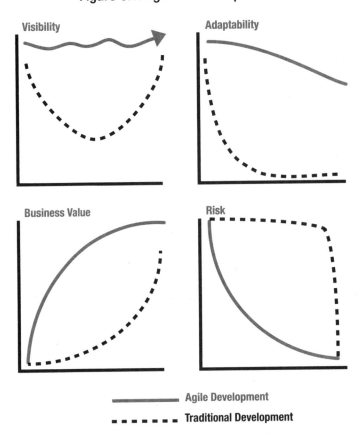

One key reason to keep stakeholders engaged is to ensure we will hear about change requests as soon as possible. An ongoing dialogue with our stakeholders will also help us identify potential risks, defects, and issues. Agile methods provide multiple touch points with stakeholders and regular communication events designed to facilitate early and continuous feedback. For example, in Scrum, sprint planning and sprint review meetings act like the bookends to an iteration, during which the team initiates and incorporates stakeholder feedback. Other agile methods include similar practices.

One point to keep in mind when engaging stakeholders in project events is that not all stakeholders can be handled in the same way. Some people may cause problems and actually be impediments to project progress. In such cases, the ScrumMaster, project manager, or other designated person needs to use their emotional intelligence and interpersonal skills to try to understand these stakeholders' concerns and find a positive way to engage them with the project. If this is not possible, it may be necessary to try and shield the team from their disruptive or corrosive influence.

Another aspect of stakeholder stewardship is establishing a process for escalating issues that need a high level of authority to resolve. Agreement on such a process is essential to keep the project operating smoothly. Then, if the team encounters a problem or issue that they don't have the authority or influence to resolve, they will be able to follow the established procedure to quickly escalate it to the appropriate stakeholder for resolution.

K&S Incorporating Stakeholder Values

As we've seen, agile methods emphasize incorporating the sponsors' and users' priorities into the project's priorities and execution. In other words, they focus on bringing project priorities into alignment with stakeholder priorities. Incorporating stakeholder values means making sure we do not plan or initiate work that the stakeholders don't support or value at this time.

The most important way that agile incorporates stakeholder values into a project is by engaging the product owner in the prioritization of the backlog. Executing the work according to the customer's priorities (taking into account the necessary dependencies and risk mitigation work) allows the team to deliver the highest-priority items and deliver early value to the business.

Another way to incorporate stakeholder values and interests into the project activities is to invite stakeholders to planning meetings and retrospectives. At the end of the day, it is the customers and sponsors who will determine whether our project is a success or a failure, so it is smart to plan our work and actions according to their values.

K&S Incorporating Community Values

In addition to reflecting the priorities of the project stakeholders, for any team to be effective its members must also share the values of their broader community—in this case, the community of agile practitioners. In addition to the fundamental values and principles set forth in the Agile Manifesto, the two most popular agile methods, Scrum and XP, each define their own set of core values. So let's revisit the two values that are shared by both Scrum and XP—respect and courage—from the standpoint of community values.

Respect

Agile approaches seek consensus. They rely on group participation in estimating, decision making, and uncovering ideas for improvements during retrospectives. Respect is necessary to make these practices work; they need to be centered on the guiding principles "Don't judge suggestions" and "There are no stupid ideas." Agile methods depend upon shared respect for differing points of view, suggestions, and opinions.

Courage

Agile methods encourage early evaluations based on prototypes—and asking people to present uncompleted work for review requires them to exhibit courage against the fear of criticism. Pair programming also takes courage, as does asking a product owner to prioritize the company's requirements. (It would be far easier—but much less useful—to just insist that everything is important.) Other examples of courage in agile include focusing on transparency by openly posting our velocity data and defect rates and addressing the question "What did not go well?" during a retrospective.

Why the Big Focus on Stakeholders?

Projects are undertaken for people and by people. Knowledge work projects often have no tangible product; therefore, effective communication and stakeholder engagement are critical to ensure that team members know what they are building and what the customer is asking for. For example, if a team is developing a large business system that will operate in a frequently changing environment, it is vital for the team members to engage the product owner throughout the project, to make sure they understand the correct details of each component and are able to incorporate late-breaking changes. Only through such ongoing stakeholder engagement can the team be confident they will deliver an optimum product or service.

Without continuous stakeholder engagement, the project is likely to fall into the "gulf of evaluation" shown in the cartoon we saw in chapter 2:

Figure 3.2: Gulf of Evaluation

| How the customer explained it | How the ScrumMaster understood it | How the developers built it | What the customer really needed |

The gulf of evaluation is so common because it is so easy for mismatches to arise between what one person envisions and tries to describe and how another person hears and interprets that description. If the mismatch is left unchecked for too long, then costly rework or project failure can (and often does) result.

Agile methods use many tools and practices to try to bridge this gulf—for example, we talked about the importance of frequent verification and validation in chapter 2. In this chapter we'll cover some other techniques, including user story workshops, participatory decision making, collaboration games, and frequent discussions of the definition of "done." But that list is just a sampling—as you read this book, notice how many of the agile practices we discuss are designed to provide feedback to ensure that the team is on the right track to meet the stakeholders' goals.

Principles of Stakeholder Engagement

The exam will assume that you understand the following principles of stakeholder engagement. As you read through this chapter, keep this summary in mind, and think about how each concept we discuss relates back to these principles.

» **Get the right stakeholders**. Projects won't be successful without the right stakeholders, so we should push hard to get the people we need. Of course, we want to have the right people on the development team, but we also need to have access to people in the organization who can most effectively help us understand the project requirements and make the necessary decisions to move things forward. Often the best people are the busiest people. Therefore, it may be beneficial for the project to pay to backfill these resources (bring additional people in) to help them complete their other duties so they have more time to contribute to the project.

» **Cement stakeholder involvement**. It is essential for stakeholders to stay engaged with the project, so we need to do all we can to make stakeholder involvement "stick." For example, we could document in the project charter how important the various stakeholders' involvement is on the project. We could also report on the benefits or issues resulting from the stakeholders' involvement (or the lack thereof) in the project's status reports and other reports to steering committees. Whatever methods we choose, we need to make sure stakeholders' involvement is visible and monitored on the project.

» **Actively manage stakeholder interest**. To keep stakeholders engaged, we should add some carrots to the mix. This means taking actions to recognize and reward stakeholder involvement, such as celebrating project accomplishments with the stakeholders, talking to their managers about how to recognize their contributions, or making sure project-related feedback becomes part of their performance reviews.

» **Frequently discuss what "done" looks like**. Since knowledge work projects often create intangible products and services, there is a lot of potential for a gap to arise between what the customer wants and describes and what the development team hears and interprets. It is critical to bridge that gap early and often, with each new idea presented on the project. Frequent discussions of what "done" looks like are essential if the team hopes to avoid nasty surprises, mismatched expectations, or poorly accepted products.

» **Show progress and capabilities.** Projects take a long time, and people generally want things as soon as possible after they've described them. For these reasons, it is important to frequently show stakeholders what has been built on the project. These demos or presentations not only allow us to check that we are building the right thing before we get too far into the project, they also show progress to our customer and sponsor. In turn, stakeholders stay engaged and informed about when things will be completed.

» **Candidly discuss estimates and projections**. One advantage of engaging the customers and showing them pieces of the project as those pieces are built is that the true rate of project progress is highly visible. The negative side of this is that the true rate of progress is often not what the sponsor or customer wants to hear. However, if bad news is delivered early enough, it can be valuable. By tracking and discussing the features delivered versus the features remaining, the team and the customer can make important trade-off decisions based upon solid data.

Exam tip

Don't try to memorize this list of principles, since they probably won't be tested directly. It is more important to thoroughly understand these ideas and their implications for agile projects. So think through these key concepts and try to incorporate them into your mindset for approaching situational questions about agile stakeholders.

Establishing a Shared Vision

As we've seen, agile methods recognize the importance and challenge of overcoming the semantic gap between what the customer asks for and how the team interprets that description. A variety of techniques have been created to help facilitate discussions, quickly bring mismatches to the surface, and work toward consensus and agreement. On agile projects, failing is okay as long as the failure is fast and cheap and it's still possible to recover the project. We ideally want to uncover disconnects in understanding at the lowest possible cost. By doing so, we save the organization money that it can then invest in other projects.

In this section, we'll discuss some key tools that agile teams use to establish a shared vision among all the stakeholders—agile charters, the definition of done, workshops, modeling, wireframes, and personas.

T&T K&S **Agile Chartering**

The project charter is one of the first documents produced for a project. It describes the project's goal, purpose, composition, and approach, and it provides authorization from the sponsor for the project to proceed. Agile charters can range from very lightweight worksheets and barely expanded vision statements to fairly detailed documents.

Agile charters acknowledge that scope may change and that initially some aspects of a project may be unknown. Therefore, rather than trying to fully specify the scope, agile charters focus on the goals envisioned for the project. They also describe the processes and approaches that the team will use to iterate toward the final product, as well as the acceptance criteria that will be used to verify the project outcomes.

Agile versus Non-Agile Charters

The technique of chartering in agile projects has the same general goal as the Develop Project Charter process defined in the *PMBOK® Guide*,[2] but the level of detail and the assumptions are different. As with a non-agile charter, the goal of an agile charter is to describe the project at a high level, gain agreement about the project's W5H attributes (Who, What, Where, When, Why, and How), and obtain the authority to proceed. However, since agile methods are often used on projects where the technology or requirements are uncertain or a high level of change is expected, they typically have less detail than non-agile charters, are shorter documents, and focus more on *how* the project will be run than on exactly *what* will be built.

When we are aiming at a static target (with unchanging requirements or technology), it is appropriate to plan, plan some more, and then execute. But on a dynamic project with a moving target, a high degree of planning may be inappropriate because key elements of the project are likely to change—and to do so quickly. For such projects, we need to allow for mid-flight adjustments and make sure we have the processes (such as prioritization, demos, retrospectives, etc.) in place to allow us to make such adjustments effectively and efficiently.

Any elements of the agile methodology that may be different from an organization's normal processes, such as the way changes are approved and prioritized into the backlog after approval, should be clearly outlined in the charter. This is especially important if agile is new to the organization or if the approach includes a departure from how certain aspects of a project, such as change control, have been handled in the past.

The bottom line is that chartering in an agile environment results in a flexible document that allows the team to respond to changing needs and technology and deliver high-value components that the organization can begin using quickly.

Developing an Agile Charter

An agile charter will typically answer a subset (or all) of the following W5H questions:

» **Who will be engaged?**—A list of the project participants and involved stakeholders
» **What is this project about?**—A high-level description of the project's vision, mission, goals, and objectives
» **Where will it occur?**—Details of work sites, deployment requirements, etc.
» **When will it start and end?**—The project start and target end dates
» **Why is it being undertaken?**—The business rationale for the project
» **How will it be undertaken?**—A description of the approach (This is particularly important if agile methods are new to the organization or if changes from the standard approach, such as the increased involvement of the customer, need to be explained.)

The process of chartering helps align stakeholders with the project. One way to help stakeholders explore and cement the basics of the project is to have them jointly develop a project elevator statement. Elevator statements are short descriptions of the project goals, benefits, and decision attributes that quickly describe the project or product. The following is a popular format for elevator statements:

For:	Target customers
Who:	Need (opportunity or problem)
The:	Product/service name
Is a:	Product category
That:	Key benefits/reason to buy
Unlike:	Primary competitive alternative(s)
We:	Primary differentiation

Let's build an example of a project elevator statement, using the scenario of describing a new course to prospective students:

For:	Project managers
Who:	Want to become agile project leaders
The:	"Learning to Lead Agile Teams" class
Is a:	Three-day course
That:	Takes participants through a comprehensive agile development life cycle, incorporating real case studies and hands-on exercises
Unlike:	Agile courses from generic training organizations
We:	Only use instructors with hands-on agile project experience to ensure they can answer all your questions, and our supplementary materials include valuable tools, case studies, and cheat sheets.

The final text of this elevator statement would read:

For project managers who want to become agile project leaders, the "Learning to Lead Agile Teams" class is a three-day course that takes participants through a comprehensive agile development life cycle, incorporating real case studies and hands-on exercises. Unlike agile courses from generic training organizations, we only use instructors with hands-on agile project experience to ensure they can answer all your questions, and our supplementary materials include valuable tools, case studies, and cheat sheets.

Another quick exercise that can help align stakeholders around the project is to create a project Tweet. This exercise requires stakeholders to describe the goal of the project in 140 characters or less. The intent of this activity is not to create an all-encompassing description, but to gauge stakeholders' high-level understanding of the project and their priorities.

Exercises like creating an elevator statement or a project Tweet ultimately help in developing the charter. Then, once the charter is complete and approved, it serves as a device to launch the project, giving authority for the team to proceed, following the approach described or referenced in the charter.

Business Case Development

Business case development for agile projects isn't that different than it is for traditional projects. This topic isn't listed in the exam content outline, and it is unlikely to be covered on the exam, so we're including it here as background information. For the most part, the differences lie in a lighter level of documentation on agile projects. In addition, the business case for agile projects may include a discussion of opportunities for realizing early benefits, which is not as typical for traditional projects.

The business case describes why the sponsoring organization should undertake the project. Since agile projects are business-value-driven, any calculations of value, such as return on investment (ROI), internal rate of return (IRR), and net present value (NPV), will be captured in the business case. When there are multiple projects competing for a limited supply of funding, the organization will assess the business case for each project to determine which one to undertake.

The amount of detail in an agile business case varies from project to project. For a small agile project, there may not be a separate business case document. Instead, the business case and business benefits may be contained in the project charter, the project vision document, or the project elevator statement. For large agile projects or agile projects undertaken in an organization that requires a business case document, the following sections may be included:

- » **Project overview**: A brief description of the project background and objectives

- » **Anticipated costs**: How much we estimate the project will cost, including reasonable contingency funds

- » **Anticipated benefits**: The benefits, monetary or otherwise, that we estimate the project will bring, such as new revenue, cost savings, better service, improved quality, compliance, improved working conditions, increased employee satisfaction, etc.

- » **Business models and indexes**: Calculations of ROI, IRR, and NPV, as required by the organization

- » **ROI assumptions and risks associated with the project**: The assumptions that were made when calculating the business models and indexes, such as inflation rates and customer adoption rates

- » **Risks of not undertaking the project**: The risks (if any) that the organization faces if it chooses not to undertake the project

- » **SWOT/PEST analysis**: The relevant results of SWOT or PEST analysis. SWOT (which stands for Strengths, Weaknesses, Opportunities, and Threats) and PEST (which stands for Political, Economic, Social, and Technical) analyses are graphical tools for examining different attributes of the project.

- » **Recommendations**: The project sponsor's recommendation for the project, such as Mandatory (regulatory compliance) or Critical (the sponsor's highest priority), or "High," "Medium," or "Nice-to-have" priority

© 2015 RMC Publications, Inc • 952.846.4484 • info@rmcls.com • www.rmcls.com

T&T Definition of "Done"

Next, we'll examine another essential agile concept related to establishing a shared vision—the definition of done. Creating a shared definition of done is crucial for satisfying our stakeholders' expectations. Therefore, a shared definition of done is necessary at every level of an agile project, including:

» **User stories**: "For this story, done will mean developed, documented, and user acceptance tested."

» **Releases**: "The first release will be deemed done when system Alpha is replaced and there are no Priority 1 defects or change requests."

» **Final project deliverables**: "Done for the project will mean all high- and medium-priority features are implemented, there are two months of trouble-free operation, and the project receives satisfaction scores of greater than 70 percent from the user community."

Background information

What "Done" Might Look Like for an IT Project

For software projects, James Shore provides the following list of elements that should be discussed and checked before we declare that anything is "done":[3]

» **Tested**: Are all unit, integration, and customer tests finished?
» **Coded**: Has all code been written?
» **Designed**: Has the code been refactored to the team's satisfaction?
» **Integrated**: Does the user story work from end to end (typically user interface to database) and fit into the rest of the software?
» **Builds**: Does the build script include any new modules?
» **Installs**: Does the build script include the user story in the automated installer?
» **Migrates**: Does the build script update the database schema if necessary? Does the installer migrate data when appropriate?
» **Reviewed**: Have customers reviewed the user story and confirmed that it meets their expectations?
» **Fixed**: Have all known bugs been fixed or scheduled as their own user stories?
» **Accepted**: Do customers agree that the user story is finished?

For a non-IT perspective on the definition of done, let's look at another example—the acceptance criteria for a new car. Since there are many stakeholder groups involved that we need to satisfy, the final list of attributes describing a car's suitability for sale and use would be very long. In addition to ensuring that its component and build costs meet its financial targets, the car itself needs to:

» Start, drive, and meet the performance envelope specifications
» Meet the review criteria defined by the design and styling department
» Include the interior space and cargo requirements specified by the marketing department
» Pass regional crash test safety standards
» Pass federal emissions test standards
» Pass all target market lights, indicator, and identification standards

By the way, although car manufacturing is a classic industrial age process, there are actually some car manufacturers who use agile approaches. For more information, do a web search for "agile car manufacturing."

Agile works best when we make a little progress on every aspect of our goal every day, rather than reserving the last few days of the sprint for getting the user stories "done-done." So as we develop the stories and features, we need to involve the customer and check that all aspects of the product work together—otherwise we could end up with a lot of fixes and feedback that need to be done at the last minute.

T&T Agile Modeling

The term "agile modeling" refers to the various modeling techniques that are commonly used on agile projects. While models are important in agile methods, their main value often lies in the discussion and creation of the model, rather than the final output. As a reflection of this, agile models are often sketched on whiteboards and then photographed as a means of recording them. The value is in the creation, not the preservation of the model in a specialized modeling tool.

So agile models are typically lightweight, or "barely sufficient," capturing the design without a need for further polish. We can create a model for predictive purposes to help clarify a design and identify the issues, risks, and elements that we need to test. We can also use modeling for reflective purposes to investigate problems and help find solutions.

Scott Ambler, an agile modeling expert, asserts that the value of modeling peaks earlier than traditional theory has led us to believe, as shown below. He recommends limiting modeling to the point where the model is "barely good enough," and then moving on to the next task.

Figure 3.3: Value of Modeling

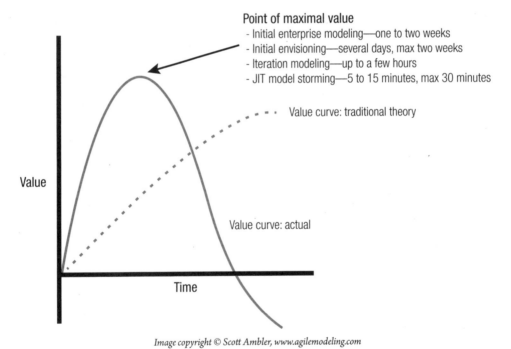

Image copyright © Scott Ambler, www.agilemodeling.com

The types of agile models that can be created during agile modeling include:

» Use case diagrams (see figure 3.4)
» Data models (see figure 3.5)
» Screen designs (see figure 3.6)

Regardless of the type of model you create, remember that the goal is still to deliver valuable but not extraneous documentation. So keep it light, move on quickly, and remain adaptable for changes.

Figure 3.4: Sample Use Case Diagram	Figure 3.5: Sample Data Model

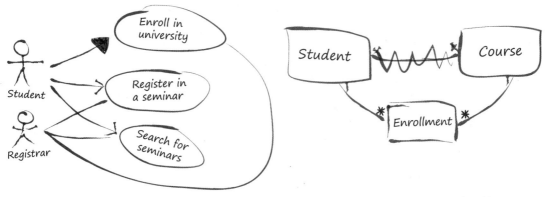

Image copyright © Scott Ambler, www.agilemodeling.com	*Image copyright © Scott Ambler, www.agilemodeling.com*

Figure 3.6: Sample Screen Design

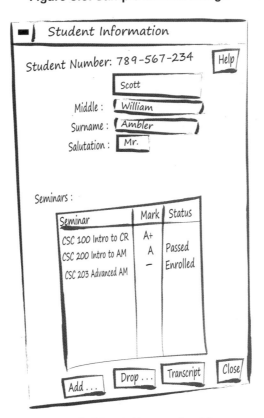

Image copyright © Scott Ambler, www.agilemodeling.com

T&T Wireframes

Wireframes are a popular way of creating a quick mock-up of a product. For example, in software development, a wireframe depicts individual screens and the flows between the screens, as shown below. This diagram helps confirm that everyone has the same understanding of the product. If there are discrepancies in understanding, the wireframe serves as a useful visual tool for stakeholders to refer to and adjust until they achieve consensus.

Figure 3.7: Sample Wireframe

MoviesOnline Basic Ordering Workflow

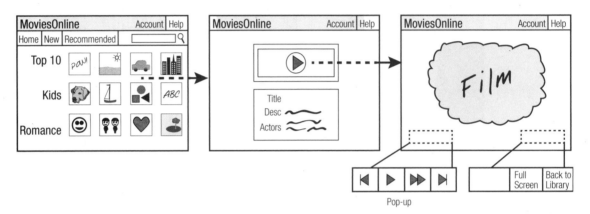

Wireframes are a form of "low-fidelity prototyping." In other words, they are quick and cheap ways to get feedback on something. Agile teams may use wireframe models created in tools like Visio®, Balsamiq® Mockups, or PowerPoint®, or they may draw the models freehand on whiteboards or sheets of paper stuck to a wall that allow for easy repositioning to change workflows.[4] The purpose of these tools is to help clarify what "done" looks like and validate the approach the team plans to take before they commit large amounts of time to building (potentially wrong) increments of the product.

T&T Personas

Personas are quick guides or reminders of the key stakeholders on the project and their interests. Software projects, for example, commonly create personas for the different types of people who will use the system that is being built. Personas may be based on profiles of real people or composites of multiple users. When they are used as a project tool, personas should:

» Provide an archetypal description of users
» Be grounded in reality
» Be goal-oriented, specific, and relevant
» Be tangible and actionable
» Generate focus

Personas are not a replacement for requirements, but instead augment them. Personas help the team prioritize their work, stay focused on the users, and gain insight into who the users will be. These tools help team members empathize with users of the product or solution.

Here is a sample persona for an online movie service based on a common template. The components of this template are a name that is easy to remember, an image or picture, a description of how the stakeholder will interact with the solution, and an explanation of what that person values about it. You may also see (or want to use) other formats, but this covers the basics of most personas.

Figure 3.8: Sample Persona

Name: Bob the Movie Buff	
  Description: Bob loves movies. On average, he rents 5 movies a week from his local rental store. His two children also like to watch children's TV shows. They often like to watch the same shows more than once, which means that Bob sometimes has to pay late fees. Bob's wife has different movie tastes than Bob and often spends a lot of time choosing a movie.	**Values:** Bob would like to be able to order movies from the comfort of his home. He would like to be able to search for movies by title, actor, genre, and director. He would also be interested in knowing how other viewers rated the movie. He is looking forward to unlimited movies so his children can watch shows multiple times without having to pay additional fees. He would also appreciate a "recommended" feature to help him and his wife choose movies.

Personas can help keep a team focused on delivering the features that users will find valuable, and this leads to better decision making on the project. They can also be used as shorthand in stakeholder discussions by referring to a persona that is familiar to the team. A question like "Does Samantha need it?" or a declaration like "Samantha doesn't need that!" can serve as a shortcut to help stakeholders make decisions more quickly and keep them connected to the project vision.

EXERCISE

Referring to figure 3.8 for guidance, create the beginnings of a persona for the typical reader of this book—someone who is preparing for the PMI-ACP exam. Think about their goals, motivation, and definitions of success when completing the personal and professional goals sections of the table. (This format is an alternative to the template described above.)

Personal Profile	Jen the pm Jen is a project manager who works with various client personalities. Jen wants to learn agile methodology for use with specific like minded clients

Project Experience	
Personal Goals	
Professional Goals	

K&S **Communicating with Stakeholders**

Since knowledge work is often invisible, we can't easily tell the state of a project by looking around the office. It's critical for agile stakeholders to communicate frequently to ensure that everyone is on the same page and kept up to date, since many project failures can be traced back to a failure of communication. Somewhere along the way, an issue or overrun occurred that wasn't communicated in a timely manner. Because that issue wasn't corrected, the project went from bad to worse. To help combat this problem, agile projects focus on making their internal communications as intuitive and efficient as possible.

In this section, we'll explore several fundamental concepts of agile communication: face-to-face and two-way communication, knowledge sharing, information radiators, and social media. In chapter 4 we'll cover more topics related to communication within an agile team, including tacit knowledge, co-location, and osmotic communication.

Face-to-Face Communication

The preferred way for agile stakeholders to communicate is through face-to-face (F2F) communications. Face-to-face encounters have the highest bandwidth of all forms of communication—in other words, they transfer the most information in a given period of time. And unlike static methods of communication, such as written reports, they provide opportunities for interactivity in real time.

Alistair Cockburn has developed a communication effectiveness model that compares common communication methods in terms of their effectiveness and their richness, or "temperature."[5] The figure below, which is based on Cockburn's model, maps two key factors—interactivity and information density—for several ways of communicating. This concept is important for agile since the interactivity and information density of a communication method indicate its power to transfer complex information efficiently.

Figure 3.9: Effectiveness of Different Communication Channels

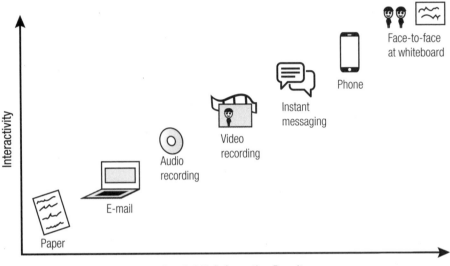

In this diagram, paper-based communications are the lowest in terms of interactivity and information density. Written documents take a long time to create and, because they aren't interactive, the documents have to be written in such a way that all recipients can understand the information, regardless of their level of knowledge or expertise. Paper documents are also low in bandwidth, so they typically do not convey the writer's emotional tone, feelings about the information, or implicit assumptions.

At the other end of the scale, face-to-face communication at a whiteboard has the highest efficiency. Here, the participants can both converse and draw their ideas on the whiteboard. They can use shortcuts for well-understood concepts to speed the exchange of information, and they can ask each other questions and get immediate feedback. Such conversations convey a lot of emotional bandwidth through nonverbal communication such as gestures, facial expressions, and tone of voice. This allows us to quickly tell if the person we are talking to is puzzled, angry, or passionate about what we are saying.

Although agile recommends face-to-face communication because of its efficiency and high bandwidth, other communication formats may still be used for specific purposes on an agile project. For example, there are some times when information does need to be recorded on paper; agile teams just need to bear in mind the inefficiency of that medium, and use it only when necessary.

T&T Two-Way Communication

Unlike industrial workers, who typically perform a well-understood task, knowledge workers tend to know more about their work than the people who lead and steward these projects. As such, it is always wise to ensure that the information flows between stakeholders are bidirectional—that is, they consist of two-way transfers of information. For example, when discussing the goals for a sprint, the product owner or business representative should not only ask the development team for their confirmation, ideas, risks, and concerns, but actually listen to what they have to say and take it into account. There's a good chance that the team members have valuable information that can't be accessed in any other way.

The figure below compares this two-way model of collaborative communication to the "dispatching" model traditionally used in a command-and-control approach to running a project.

Figure 3.10: Collaborative versus Dispatching Model of Project Communication

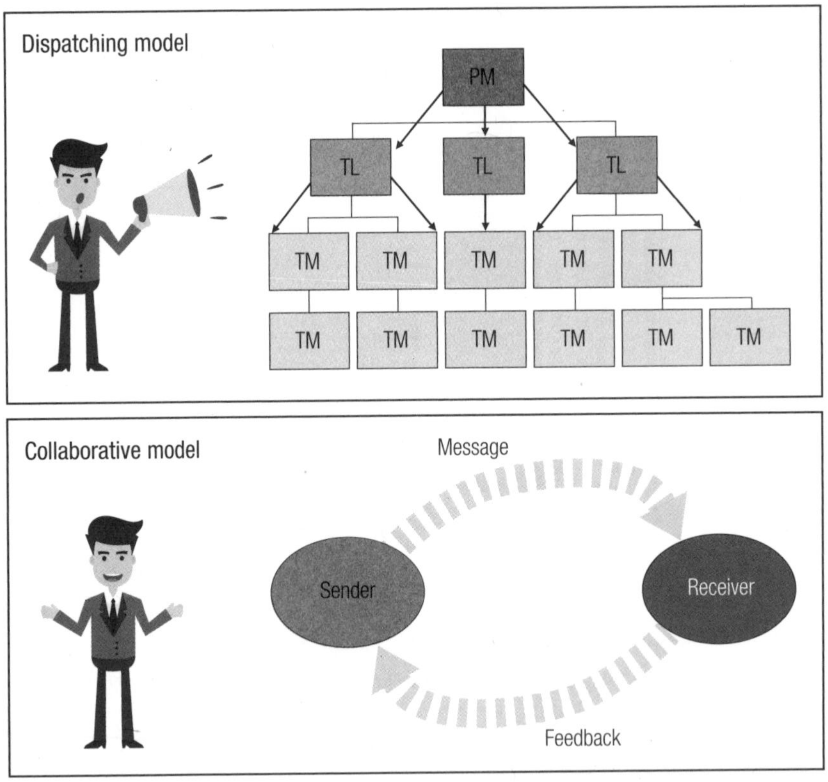

In the traditional top-down dispatching model, instructions for doing the work are communicated down the chain of command to those who actually perform the job. The project manager (PM) tells the team leads (TL) what to do, and the team leads pass those instructions on to the team members (TM). There is little, if any, upward flow of information (and when it does occur, it typically isn't welcomed or encouraged).

By comparison, in the collaborative, two-way communication style used on agile projects, hierarchies are flatter and feedback from the receiver is expected—in fact, it is necessary for the project to reach its goals.

T&T **Knowledge Sharing**

Knowledge sharing is a key component of agile methods. This should come as no surprise, since agile methods are designed for knowledge work projects, which are characterized by subject matter experts collaborating to create or enhance a product or service. Given that information (i.e., knowledge) is the basic commodity of agile projects, it is only right that we share it. Therefore, agile projects are encouraged to take an abundance-based—rather than scarcity-based—attitude toward sharing knowledge. This means we aim to share information and make it available to everyone who might want to consume it, rather than hoarding it to secure our jobs or increase our project stature.

It's better to share information than to hoard it because the more people who know about something, the more people there will be who can help you when you get stuck. Sharing knowledge also helps balance the workload between team members. For example, the XP practice of collective code ownership means that any developer can maintain any portion of the system. If a developer didn't create that part of the code, it might take them longer to work on it—but we can avoid the absolute bottlenecks that occur when we have to wait for the person who has the "right" knowledge or skills to become available.

Also, when information is shared throughout the team, it greatly reduces the risk of taking a hit to team productivity if the one person with key knowledge leaves the team. Instead of relying on a single expert who has all the secrets, agile tries to spread this information amongst all the team members.

Some projects do present special complications for knowledge sharing. For example, I've worked on confidential projects and military projects where knowledge can't be shared as freely. Some companies use physical security solutions so that only people who have the appropriate security clearance can gain access to the team room. Passwords and encryption may be used on electronic information. However, even the most secure projects are able to find ways to share knowledge with the right people without compromising security.

Agile Practices Promote Knowledge Sharing

Agile knowledge sharing happens at many levels, in both obvious and subtle ways. Product demonstrations are an obvious example. The main purpose of such demonstrations is not to show off the product, since the team knows very well what works and does not work; instead, demos are done because they are high-ceremony ways to share knowledge, through the following kind of dialogue:

Team to customer: Here is what we think you asked for and what we have been able to build. Please tell us if we are on the right track.

Customer to team: I like these bits, and this is okay, but you got this piece wrong. Oh, and that reminds me—we really need something over here to do X.

Other great examples of agile information sharing include Kanban boards, information radiators, personas, and wireframes. These tools all support knowledge sharing by ensuring that the project information is out in the open, and none of it is the sole domain of any one person—the project manager, ScrumMaster, testers, or developers. Instead, that knowledge is publicly available for any and all interested stakeholders to consume. In a similar way, the agile emphasis on collaborative agile planning, estimating, and retrospectives allows everyone on the team to be exposed to key project information, instead of just funneling it through one or two people.

Agile methods also emphasize knowledge sharing by using low-tech, high-touch tools like cards on a wall to plan and schedule the project. These simple approaches help more team members get involved and require less skill to update than software programs like Microsoft Project® or Primavera.[6] Simple tools also help build a common understanding of the plan.

Here's another example—since software developers are a tech-savvy bunch, why aren't our daily stand-up meetings done via e-mail or some other electronic medium? The answer is simple. It's because the real goal of the stand-up meeting is to share knowledge amongst the team members, not just generate lists of work done, work planned, and issues that have arisen.

Another, less obvious way to share information is team co-location. This practice is not done to save space or ease management overhead; instead, it is done to speed the sharing of information that occurs in face-to-face environments through osmotic communication and tacit (unwritten) knowledge.

In fact, just about every agile practice is structured for maximum knowledge sharing. We want knowledge to be shared throughout the team, not just reside in one person. This concept applies to most agile practices; when we examine how they are structured through the lens of maximizing knowledge transfer, all the peculiarities suddenly make sense. For example, think about XP's core practices. Most of these practices have a knowledge transfer component, as shown below.

Figure 3.11: Knowledge Transfer in XP's Core Practices

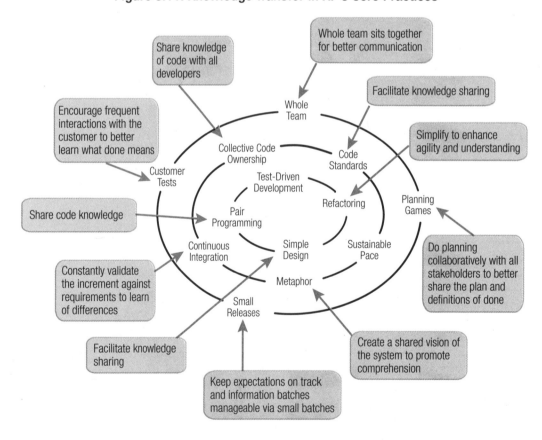

© 2015 RMC Publications, Inc • 952.846.4484 • info@rmcls.com • www.rmcls.com

Encouraging Knowledge Sharing

If we understand that knowledge sharing is a good thing, why is it sometimes so difficult to achieve and sustain? In the book *Knowledge Management in Theory and Practice*, Kimiz Dalkir notes that "individuals are most commonly rewarded for what they know, not what they share."[7] This reward system actually discourages knowledge sharing. To promote the idea of sharing knowledge, the organizational culture should instead encourage and reward the discovery, innovation, and transfer of information.

Background information

Measuring Up to Encourage Desired Behavior

In an article on The Lean Mindset website, Mary Poppendieck discusses the problem of encouraging knowledge sharing, quoting measurement expert Robert Austin:

> *In the book "Measuring and Managing Performance in Organizations," . . . Robert Austin discusses the dangers of performance measurements. The beauty of performance measurements is that "You get what you measure." The problem with performance measurements is that "You get only what you measure, nothing else." You tend to lose the things that you can't measure: insight, collaboration, creativity, dedication to customer satisfaction.[8]*

Although this is indeed a quandary, fortunately people have found ways to measure and reward knowledge sharing so that we don't lose its benefits—to show that this can be done, Poppendieck describes the case of Nucor Steel.

Nucor Steel has been one of the most successful steel companies in the United States from the 1980s through the present day. Where other companies have folded in the face of fierce foreign competition and poor labor relations, Nucor has remained competitive and profitable and has maintained good worker relations. They attribute a lot of this success to their "pay for performance" scheme, which is based on team productivity.

At Nucor, unlike companies that use traditional productivity-based bonus schemes, if you are a steel plant manager, you do not get paid based on how well your plant does, but on how well all the steel plants perform. At the next level down, department managers do not get a bonus based on the performance of their department, but on the performance of all the departments. This approach continues all the way through the organization, including team leads who get paid based on the productivity of all teams, not just their own, and individuals who get paid based on their team's performance, rather than their own.

This approach strongly encourages knowledge sharing. If a plant manager invents an enhancement to a process or discovers a money-saving idea, he or she is motivated to share the idea with the other plant managers. Likewise, teams that streamline their processes are rewarded for sharing that improvement with other teams, and team members who help out their teammates are also rewarded.

The approach described in the Nucor case explained in the box is called "measuring up," which refers to measuring something at one level above the normal span of control (e.g., the team level, rather than the individual level) to encourage cooperation and knowledge sharing. This happens a lot on agile projects, even though it does not get much publicity. Mary Poppendieck advises, "Instead of making sure that people are measured within their span of control, it is more effective to measure people one level above their span of control. This is the best way to encourage teamwork, collaboration, and global, rather than local, optimization."[9]

We see examples of measuring up in the way we track velocity. We could quite easily trace velocity to individual team members and determine who is the most productive, but this approach would be likely to encourage negative behaviors and a lack of cooperation. So instead we measure velocity at a team level; as a result, team members are motivated to help each other. The same concept applies to knowledge sharing. For knowledge sharing to occur, we need to base our tracking and rewards on team accomplishments, so that there are no benefits to hoarding information or being the guru of subject X.

In sharing knowledge, we follow the mindset "Let us show you what we've done, and then tell us if we are right or wrong." This is a great way to surface and resolve misunderstandings, but it can be an alien approach to organizations that are more reserved and cautious in their communications. The mindset is not about boasting or showing off, however. It is about managing risks, confirming that the way we are doing things is appropriate and valuable, and focusing on knowledge transfer. These are all key components when undertaking knowledge work projects, which transform information rather than concrete and steel.

EXERCISE: KNOWLEDGE SHARING

Test yourself! Summarize the key points about knowledge sharing. Write down why it is a valued practice and how it can be achieved on both co-located and geographically dispersed projects.

ANSWER

The main point to understand can be stated very simply—the more knowledge is shared, the better. Agile methods have a lot of knowledge-sharing events, including retrospectives, demos, and planning meetings. They also have a lot of knowledge-sharing practices, such as pair programming and physical co-location of the team, which allows for the osmotic communication of knowledge. If team members are geographically separated, electronic tools like instant messaging and Skype headsets can be used to allow information sharing activities to continue, despite the physical distance.

T&T **Information Radiators**

"Information radiator" is agile's umbrella term for highly visible displays of information, including large charts, graphs, and summaries of project data. These tools, sometimes referred to as "visual controls," are usually displayed in high-traffic areas to maximize exposure, where they can quickly inform stakeholders about the project's status. The term "information radiator" was coined by Alistair Cockburn, who was contrasting agile's approach to the practice of locking project information away in an "information refrigerator," where nobody knows what is going on. Instead, information radiators use highly visible charts or graphs that "radiate" information about the project quickly to anyone who is interested.

The sort of data that might be displayed on an information radiator includes:

» The features delivered to date versus the features remaining to be delivered
» Who is working on what
» The features selected for the current iteration
» Velocity and defect metrics
» Retrospective findings
» List of threats and issues
» Story maps
» Burn charts

Below is an example of an information radiator that shows a team's defects work in progress. For this project, we were interested in keeping on top of the defects—in other words, keeping the volume of open defects low—to minimize the cost of changes and the amount of code that might be written on top of faulty code.

Figure 3.12: Information Radiator Tracking Defects Work in Progress

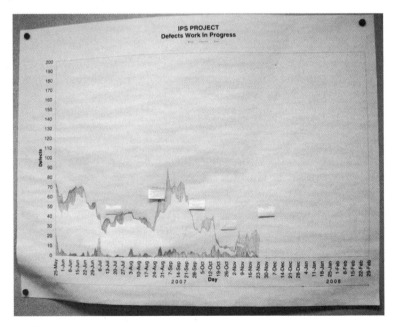

For this project, we also tracked our defect cycle time—the total time between defect injection (when a defect occurred or was introduced into the code) and defect correction (when it was fixed). Our goal was to fix each defect within one business day. To keep everyone informed about our progress, we posted a graph of defect cycle time like the one shown below as another information radiator.

Figure 3.13: Information Radiator Tracking Defect Cycle Time

T&T Social Media

The way we communicate has changed dramatically over the last several years. The updates from relatives that we used to receive in a yearly Christmas card are now instantly available on Facebook. Young workers who "grew up digital" are very familiar with social media and tend to believe that e-mail is a formal and stuffy medium you might use to send a thank you note to a friend's parents. Instead, they rely on instant messaging and social media tools to stay in touch and communicate.

It's no surprise then that these tech savvy workers want to leverage the power and convenience of social media to keep themselves informed about the projects they are working on. Since they carry their devices—phones, tablets, smart watches—with them at all times, it's easier for them to use social media to stay informed than corporate e-mail or a project website. Also, since many people work remotely or travel a lot, devices and tools that can connect via wi-fi, cellular, or satellite data provide them with a wider range of connection options than having to access a single corporate source for project information.

Project tools are able to output information in a variety of social media formats, including Twitter, SMS message (text), Chatter, Instagram, and so on. These tools aren't necessarily better than other methods of sharing project information—this is just how many of today's workers prefer to consume it. Social media tools enable more team members to work remotely while still keeping up with project changes and developments. Near-instantaneous push notifications allow team members to collaborate and stay informed nearly as well as working in a common team room.

Although social media tools are changing how agile teams work, they do have some downsides—for example, hacking and information theft are becoming common. While most of our projects are so dull that we struggle to keep stakeholders interested in them, some business domains are too sensitive or confidential for social media tools to be used. It's true that there are some commercial tools that provide military-grade encryption for social media collaboration and information sharing. However, it is the ease of access and general popularity of the readily available social media tools that make them desirable in the first place. Updates via Facebook or Twitter are great if the stakeholders are using these tools anyway—but not so good if you have to install special software and then input two-factor authentication details every time you want to know what's going on.

T&T Working Collaboratively

The Agile Manifesto highlights stakeholder collaboration in both its third value ("Customer collaboration over contract negotiation") and its fourth principle ("Business people and developers must work together daily throughout the project"). In this section we'll look at the benefits and requirements for working together effectively, as well as some of the collaborative techniques that are commonly used by agile teams.

The benefits of collaboration are widely acknowledged; a study by Steven Yaffee from the University of Michigan found the following benefits:[10]

1. **Generates wiser decisions** through the understanding of complex, cross-boundary problems via shared information
2. **Promotes problem solving** rather than procedural decision making
3. **Fosters action** by mobilizing shared resources to get work done
4. **Builds social capital** by building relationships and understanding
5. **Fosters ownership of collective problems** by valuing participation and shifting power downwards

There are some powerful concepts here that are worth a second look. It's pretty obvious that engaging a larger group of stakeholders would lead to better decisions; however, the real benefits of collaboration come from the changes that happen *within* the group. By engaging people, not only do we get better input and ideas, we also encourage problem solving over following orders, action instead of passivity, social understanding instead of functional silos, and the collective ownership of ideas over information hoarding. The project can gain more benefits from having a motivated, energized, and empowered team of stakeholders who are proactively contributing than by relying on a single leader or project manager for direction and decisions.

Another point about Yaffee's list that is worth mentioning is that "valuing participation and shifting power downwards" fits extremely well with the ideas of empowered teams and servant leadership that agile methods rely on. As we'll see, many agile practices are based on a collaborative approach, such as reporting progress via daily stand-ups, estimating via planning poker, participatory decision making, and team-based problem solving.

Green Zone/Red Zone

Collaboration can only flourish in a safe, supportive environment—so let's examine what that context looks like and how agile leaders can help nurture it. For collaboration to be most effective, we have to assess the group's openness to working together. Lyssa Adkins recommends using the "Green Zone, Red Zone" model to diagnose how much support there is for collaboration.[11]

A Person in the Green Zone...	A Person in the Red Zone...
Takes responsibility for the circumstances of his or her life	Blames others for the circumstances of his or her life
Seeks to respond nondefensively	Responds defensively
Is not easily threatened psychologically	Feels threatened or wronged
Attempts to build mutual success	Triggers defensiveness in others
Seeks solutions rather than blame	Does not let go or forgive
Uses persuasion rather than force	Uses shame, blame, and accusations
Can be firm, but not rigid, about his or her interests	Is black/white, right/wrong in thinking
Thinks both short term and long term	Focuses on short-term advantage and gain
Is interested in other points of view	Feels victimized by different points of view
Welcomes feedback	Does not seek or value feedback
Sees conflict as a natural part of the human condition	Sees conflict as a battle and seeks to win at any cost
Talks calmly and directly about difficult issues	Is rigid, reactive, and righteous
Accepts responsibility for consequences of his or her actions	Is unaware of the climate of antagonism he or she creates
Continuously seeks deeper levels of understanding	Has low awareness of blind spots
Communicates a caring attitude	Communicates high levels of disapproval and contempt
Seeks excellence rather than victory	Sees others as the problem or enemy
Listens well	Does not listen effectively

It is only when people spend most of their time in a Green Zone state of mind that they can be truly effective at working as a collective. Of course, it is acceptable—and human—to occasionally have Red Zone thoughts, but as soon as we notice such thoughts, we want to try to redirect them and move back into the Green Zone. To promote effective collaboration, agile leaders, coaches, and facilitators aim to stay in the Green Zone and continuously model Green Zone behavior. This strengthens and supports Green Zone thinking and behavior in the other project stakeholders, so that the group can direct their energy toward working together effectively.

T&T Workshops

Workshops are meetings in which the participants get work done. I use this definition to distinguish them from meetings where people are not engaged in work. Retrospectives are a form of workshop, as are estimating sessions and planning sessions. Workshops should have clear goals and a schedule that is visible to everyone. There should be no confusion as to why we are here, what we are trying to do, or what steps we plan to follow. It should also be clear that input is both expected and encouraged from every participant.

Here are some tips that can make workshops more effective:

» Diverse groups reflect a wider range of viewpoints than just a few experts, and therefore are likely to generate a wider range of options. Adding more diverse voices to a group can lead to valuable new ideas and solutions.

» To prevent dominant individuals and extroverts from monopolizing the discussion, the facilitator can use techniques such as going round-robin style around the group or generating ideas on sticky notes.

» Another useful tip is to start with an activity that gets everyone participating within the first five minutes. This signals that this is a working session, and that everyone is expected to contribute, rather than simply be a passive audience for a tangential dialogue that may not concern them. This first engagement doesn't have to be a complex task; it could simply be asking people to define what done will look like for the session or what problems they are facing. The point is to force some action and independent thinking to reinforce the active nature of the workshop and what is expected from the participants. We'll describe some specific ways to kick off a session when we talk about retrospectives in chapter 7.

User Story Workshops

User story workshops (also known as "story writing workshops") are the preferred mechanism for gathering candidate user stories and starting the process of refining them into well-formed stories. Although we could just interview the stakeholders individually about what they would like the solution to do, people are more likely to overstate their wishes and gold-plate their requirements if their peers aren't present. Also, in large organizations people often don't know what happens to information outside of their group. By bringing all the major stakeholders together in one room, the downstream consumers of the flows can fill in the gaps so that the process is better understood.

One benefit of these workshops is that it often turns out that workflows can be optimized. For example, the sales department might be passing their customer, order, and part reports to the finance department—but all finance really wants is the customer and order total fields on a single report. If we have representatives from both groups in the workshop, we'll be able to discover this issue and simplify the process.

Another important reason for conducting user story workshops is to engage the key stakeholders in the design process. As they work with the development team and discuss the trade-offs and priorities of the work, two things will happen. First, the team will get a better understanding of the stakeholders' needs, without jumping directly to possible solutions; and second, the business will get a better sense of the costs and options of the various approaches before they commit to one of them.

T&T **Brainstorming**

Brainstorming is a collaborative technique in which a group tries to rapidly generate a lot of ideas about a problem or issue. Depending on the method of brainstorming being used, the participants are asked to write down or spontaneously call out potential suggestions to help solve the problem. Since the goal of a brainstorming session is to maximize the number of suggestions generated, the working rules are "There are no stupid ideas" and "We'll sort through the ideas later." To encourage a creative flow of ideas, it's important to record all the suggestions and not judge or make fun of any of the solutions proposed.

Brainstorming can be very useful on knowledge work projects. Agile teams can use this approach to help identify options, solve issues, and find ways to improve processes. For example, the team might brainstorm:

» Product roles to feature in personas
» Features to include in the minimal viable product for a release
» Potential risks that could impact the project
» Solutions to a problem raised in a retrospective

Of course, brainstorming is no silver bullet. Like all good approaches, it has its downsides—in fact, it can actually stifle innovation and lead to groupthink. In a 2012 article in *New Yorker* magazine, Jonah Lehrer describes numerous studies that found brainstorming groups think of fewer and lower-quality ideas than the same number of people who generate ideas alone and then get together to pool their ideas.[12] So when running workshops it's best to not rely solely on brainstorming techniques, but to combine them with other techniques for generating ideas, as well.

Brainstorming Methods

There are many different methods that can be used for brainstorming. To give an idea of the range of these methods, we'll examine three of them—Quiet Writing, Round-Robin, and Free-for-All.

» **Quiet Writing**: With this method, the participants are given five to seven minutes to generate a list of ideas individually before the group gathers to share their ideas. This approach minimizes the effects of peer influence because the stakeholders generate their ideas in isolation before they share them.

» **Round-Robin**: In this format, people take turns by passing a token around the group. When a participant receives the token, he or she will suggest an idea and then pass the token to the next person. This method has the advantage of allowing ideas to build on each other; however, people have to be comfortable sharing their ideas in front of each other for it to work.

» **Free-for-All**: Another approach is the free-for-all format. With this method, people just shout out their ideas spontaneously. This method can be collaborative, as people build on each other's suggestions through discussion—but it can only work in a supportive environment. And even with a supportive environment, the quieter team members may not be heard, or may not feel as if they have an equal opportunity to participate.

Once ideas have been captured in the brainstorming sessions, the next steps are to sort them, prioritize them, and then implement the best ideas. Sorting the ideas is often done by putting them on a board, consolidating similar ideas, and removing duplicates until a list of distinct suggestions emerges. The group can then use any of the techniques described in chapter 2 to prioritize that list; some of the most common approaches are simple prioritization ("High, Medium, Low"), MoSCoW, and dot voting.

T&T Collaboration Games

Collaboration games, also known as innovation games, are facilitated workshop techniques that agile stakeholders can use to get a better understanding of complex or ambiguous issues and reach consensus on options and solutions.

Here are some examples of the collaborative games used on agile projects:

» **Remember the Future**: This is a vision-setting and requirements-elicitation exercise.
» **Prune the Product Tree**: This exercise helps stakeholders gather and shape requirements.

» **Speedboat (aka Sailboat):** The goal of this exercise is to identify threats and opportunities (risks) for the project.
» **Buy a Feature:** This is a prioritization exercise.
» **Bang-for-the-Buck:** This exercise looks at value versus cost rankings.

To get a feel for the games that agile teams use, we'll examine the first three of these games in more detail, looking at how they work and the theory behind them.

Remember the Future

This is a facilitated exercise that asks project stakeholders to imagine that an upcoming release or iteration has been successfully completed. They are asked to "look back" and describe what happened to allow the iteration or release to be successful.

How It Works

For this explanation, we'll use the example of planning a release six months out. We gather together the project stakeholders, including the development team, users, and sponsors, and ask them to imagine that it is now six months plus two weeks from the current date. (The reason we ask them to imagine two weeks after the release date is because that's often how long it takes for the acceptance and implementation "dust" to settle.)

Explain to the group that in the first part of the exercise they will work individually to write a report for their boss about how the release went, in which they will list everything that was completed and delivered to make it successful. So for the first 20 minutes, the participants work alone to come up with their lists, recording each item on a sticky note.

Once the 20 minutes are up, everyone transfers their sticky notes to a wall. Then, as a team, they work together to group the sticky notes into associated clusters and remove any duplicates. This process can take another 20 minutes as people clarify the meaning of their sticky notes and create headings to identify each group of items. Here's an example of what this might end up looking like:

Figure 3.14: Remember the Future

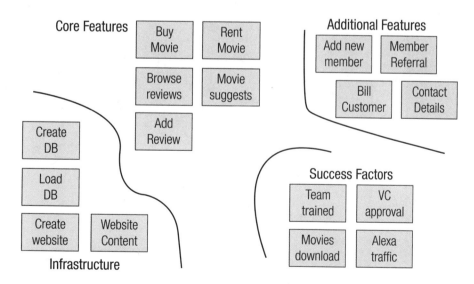

Image originally published in "The Melin Exercise" by Mike Griffiths on gantthead.com on April 20, 2009, copyright © 2009 gantthead.com. Reproduced by permission of gantthead.com.

Theory Behind the Game

The game is based on the findings of numerous studies in cognitive psychology. When asked the open-ended question, "What should a system or product do?" people struggle to generate a complete list of features and interim steps. However, if we vary the question slightly by asking people to imagine it is now some point after the delivery date and then ask them to "remember" all the things the system or project did to be successful, we will get significantly different results. Because the event is now "in the past," people have to mentally generate a sequence of events that led to this result. This leads to improved definitions and more detailed descriptions of the interim steps. Of course, predicting the future isn't really the purpose of this game. Instead, we are trying to better understand the stakeholders' definition of success and how we can achieve that successful outcome.

Prune the Product Tree

This group exercise engages the participants in brainstorming a product's features and functionality.

How It Works

For this game, we start by drawing an outline of a big tree with a trunk and branches on a whiteboard or flip chart. Artistic ability doesn't matter here—the tree is just a placeholder for features. We explain to the group that the tree is the product—its trunk represents what we already know or have built so far, and its outer branches represent new functionality that has yet to be designed. Next, we ask the participants to record each of the product's desired features on a sticky note, and place those notes on the tree. We ask them to group related features close together and place supporting features closer to the trunk. The features that are dependent upon those supporting features should be further out or higher up on the tree.

The result might end up looking something like this:

Figure 3.15: Prune the Product Tree

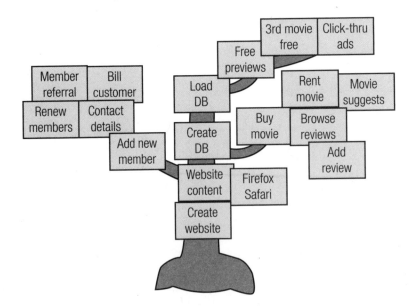

Once this is done, encourage people to add new sticky notes to the tree to identify additional features. Then ask them to expand into more details of the product as the features are split out into user stories and other supporting requirements emerge.

Theory Behind the Game

By thinking about how the features relate to each other and to the existing functionality, stakeholders can better understand the process of setting priorities and defining development sequences. This exercise assists participants in the process of progressive elaboration. If you use yellow sticky notes, the features end up looking like autumn leaves, and you can expect to see a lumpy tree with patchy leaves emerge as the participants continue to identify and add more features, or "leaves."

Speedboat (aka Sailboat)

Once the stakeholders have identified the features and user stories on their product tree, they may be tempted to jump right into prioritizing and scheduling those tasks; however, there is one very important step that should occur first. We need to identify and plan to either reduce or avoid potential project threats and take advantage of potential opportunities. This step has to happen before we prioritize our user stories because many of the risk response steps will need to be factored into the prioritization process. If we leave risk mitigation until after we've finished prioritizing our user stories, we will find ourselves trying to shoehorn important tasks into already-full iterations, and this approach simply will not work.

The Speedboat game (which is also called Sailboat) starts with the features and user stories identified in the Prune the Product Tree game. It focuses on gathering risks that pose threats to the project as well as potential opportunities. This exercise is very quick to set up and facilitate, and it typically results in a good list of project threats and opportunities.

How It Works

To start, place another whiteboard or flip chart to the left of the product tree diagram created in the previous exercise. On it, draw a waterline and a picture of a boat, with the boat moving in the direction of the product tree. Explain to the participants: "This boat represents the project heading toward the goal we just defined. What are the anchors (or threats) that could slow us down or even sink us? And what are the helpful winds (or opportunities) that could fill our sails and help propel us toward our goal?"

Next, ask the participants to work as a group to create "anchor" sticky notes for the threats to the project and "wind" sticky notes for the opportunities. Ask them to post the anchors below the waterline and the winds above the waterline. (To further distinguish the project threats from the opportunities, agile teams often use yellow or red sticky notes for the threats and blue or white sticky notes for the opportunities.)

Finally, ask the group to identify any "rocks" we'll need to steer around—these are threats that we just need to accept, since we can't mitigate or influence them. This might include such things as regulatory changes or market moves by major competitors. The final diagram might look something like this:

Figure 3.16: Speedboat or Sailboat

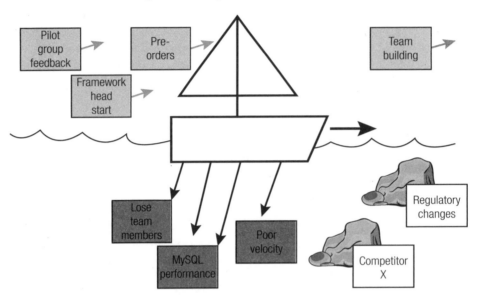

Theory Behind the Game

In addition to helping the stakeholders identify the threats and opportunities for the project, this game provides another important benefit. Some people need a way to articulate their concerns before they are comfortable committing to a goal. Once their worries have been openly recorded and recognized, they are less encumbered by these concerns and are happier to contribute. They have said their piece and done their duty in identifying the risk, and can now move on.

This is not to say that simply recording the risks will make the threats go away or the opportunities happen. This is really just where the hard work of threat reduction/avoidance and opportunity enhancement begins. The team still has to go through the effort of determining how to respond to the threats, build those response actions into the project, and assess their effectiveness. However, this game helps people remove their mental obstacles so they can move on more happily, knowing that their concerns are on the project's "radar" of things to watch for and manage.

Using Critical Interpersonal Skills

The adage "The soft stuff is the hard stuff, and the hard stuff is the easy stuff" speaks to the fact that interpersonal skills are often more difficult to master than technical skills. The good news is that while our intelligence quotient (IQ) peaks when we are in our 20s, our emotional quotient (EQ) continues to develop through our 40s and 50s. Interpersonal skills are covered on the PMI-ACP exam because they are critical success factors for agile projects. Poor interpersonal skills can quickly demoralize and disenchant a technically strong team, while a leader who effectively uses these skills can get amazing results from an average team.

We use the term "interpersonal *skills*" instead of "interpersonal *talents*" because unlike talents, which we are largely born with, skills can be acquired and improved through education and practice. In this section, we'll cover the following interpersonal skills in the agile toolkit:

» Emotional intelligence
» Active listening
» Facilitation
» Negotiation
» Conflict resolution
» Participatory decision making

T&T **Emotional Intelligence**

One of the best ways to stay flexible in leading or participating in unpredictable knowledge work projects is to continuously try to improve our emotional intelligence. Emotional intelligence is our ability to identify, assess, and influence the emotions of ourselves, other individuals, and groups. The model shown below organizes the different aspects of emotional intelligence into four quadrants. It represents the emotional skills related to "Self" on the left side and those related to "Others" on the right side. Each of those columns is also split into two skill areas, "Regulate" and "Recognize," to make four cells.

Figure 3.17: Quadrants of Emotional Intelligence

Self	Others	
Self-Management **Self-Control** Conscientiousness Adaptability Drive and motivation	**Social Skills** **Influence** Inspirational leadership Developing others Teamwork and collaboration	Regulate
Self-Awareness **Self-Confidence** Emotional self-awareness Accurate self-assessment	**Social Awareness** **Empathy** Organizational awareness Understanding the environment	Recognize

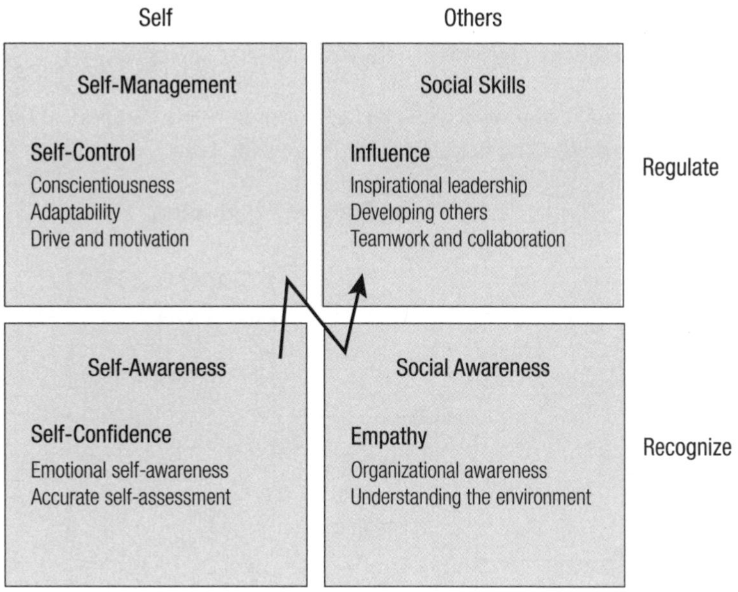

While everyone has some level of skill in all quadrants, it is usually easiest to start improving our emotional intelligence by recognizing our patterns in the "Self-Awareness" quadrant (bottom left). After that, we can learn to regulate ourselves through "Self-Management" (top left), then build our "Social Awareness" (bottom right), and finally, hone our "Social Skills" (top right).

In other words, we first need to recognize our own feelings. Once we understand our emotions, we can begin to control them. So as a start, we need to recognize what makes us angry, frustrated, happy, or thankful. We can then realize we have the power to choose how to respond to what we are feeling. We could simply follow the normal pattern of a stimulus leading to a response—but as humans, we have the unique ability to insert a decision between the stimulus and the response. We are able to choose if we want to continue to allow a situation to upset us or if we want to respond differently. Recognizing that we have a choice is a key part of becoming self-aware and eventually mastering self-management.

We should also keep in mind that how well we manage ourselves and our attitude has an impact on those around us, particularly if we are in a position of leadership. As emotional intelligence expert Daniel Goleman explains, "The leader's mood and behaviors drive the moods and behaviors of everyone else. A cranky and ruthless boss creates a toxic organization filled with negative underachievers who ignore opportunities."[13]

Once we've sorted ourselves out in the areas of self-awareness and self-management, we are able to work on developing social awareness and empathy for others. As an agile leader, coach, or manager, we need to be able to identify when team members are stuck, frustrated, or upset in order to help them. Once we're able to recognize when people need help, we can use social skills, such as the ability to influence, inspire, lead, and develop others, to help them get unstuck, do their work, and collaborate with each other.

T&T Active Listening

Active listening is hearing what someone is really trying to convey, rather than just the meaning of the words they are speaking. The expression "Do what I mean, not what I say" speaks to this concept. On agile projects, we need to listen for the message, not just the string of words being spoken.

Active listening is a skill that can be improved with practice. According to the authors of *Co-Active Coaching: Changing Business, Transforming Lives*, our listening skills progress through three levels, as shown below.[14]

Figure 3.18: Three Levels of Listening

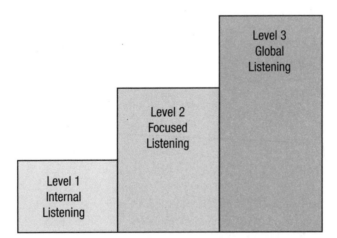

Level 1: Internal Listening

At this first level of listening, we hear the words being spoken, and although we may be very attentive, we interpret them through our own lens. When listening, we are thinking "How is this going to affect me?" and miss the speaker's real message. For example, if a team member starts telling us about some challenges she is having with a new team member, we might be so focused on our goal of quickly getting the new team member integrated and up to speed that we miss what she is really trying to say about the new person.

Level 2: Focused Listening

When listening at this level, we let go of our own thoughts and put ourselves in the mind of the speaker. We empathize with their thoughts, experiences, and emotions as they tell us about the situation. In our example of the person who is having challenges with a new team member, we are able to empathize with the speaker's feelings and recognize that the training of this new person has added to her workload at a stressful time. We look for emotional indicators in her words and pauses, voice and tone, and facial expressions for more information about how she feels about what she is describing.

Level 3: Global Listening

When listening at this level, we build on the approach taken in level 2, adding a higher level of awareness, like longer antennae, to pick up on subtle physical and environmental indicators. These indicators can include the speaker's movements or posture, energy level, and the atmosphere or "vibe" in the room. We notice factors like whether the person is speaking openly in front of others or privately, the mannerisms of any listeners within earshot (e.g., do they seem to agree, or are they averting their eyes or otherwise distancing themselves from the conversation?), and many other subtle clues to help us understand a fuller context of the information being shared. In our example, as the speaker is telling us about the new team member, we recognize the importance and sensitive nature of the conversation because she came to us privately to discuss the matter. We also notice that she is physically tense, clenching her hands into a fist, and pacing back and forth as she is speaking. Taking these clues into account, we can help her recognize and explore her feelings to better understand her reaction to working with the new team member.

Leading and coaching agile teams requires that we listen intently to people. And we cannot listen when we are talking ourselves, so it's important to talk less and listen more. Wait for others to speak. Try counting slowly to ten to give other people enough time to get comfortable and speak up. In general, while our mouths are flapping we are not learning.

K&S Facilitation

The questions about facilitation on the exam are designed to test whether you understand how to run effective meetings and workshops. When facilitating a meeting or session (or when you encounter such questions on the exam), keep the following in mind:

» **Goals:** People often feel that meetings are a waste of their time, especially if the meeting discusses a wide range of topics and the participants don't understand why they need to be there or what their contribution should be. Establishing a clear goal for each meeting or workshop session can help people get engaged in the discussion from the start. Plus, having a clear goal and keeping everyone focused on that goal, rather than allowing the session to be sidetracked, can shorten the session time, making the discussion feel more valuable to all involved.

» **Rules:** Establishing some basic ground rules is another important technique for holding effective sessions. For example, there might be rules regarding the use of cell phones, starting and ending the sessions on time, or respecting the views of all participants. It is not enough to simply set the rules, however. The rules must also be enforced during each session.

» **Timing**: Timing is always important when we are trying to get a group of people together, and it can be easy to lose track of the time once the session is going. Therefore, the duration of the session should be established ahead of time, and someone should be designated as the timekeeper. It is also useful to determine in advance when the session breaks will take place.

» **Assisting**: The session facilitator needs to make sure the meeting is productive and that everyone has a chance to contribute. In addition to keeping the group focused on the session goal and enforcing the ground rules, this may include making sure junior or quieter members have the opportunity to express their thoughts, coping with dominant or aggressive participants, and otherwise keeping the session flowing smoothly.

T&T Negotiation

Negotiation happens throughout an agile project, especially when discussing the requirements or priorities of features and what "done" should look like. For example, let's say that a customer asks for the ability to print out product codes. In negotiating this requirement, the team presents options to the customer. The low-cost option might be to use the Print Screen function on the product code screen, paste the screen shot into Microsoft Word®, and then print that document to get the product codes.[15] A more costly solution might be a fully formatted report with headers, footers, and page numbers. There would also be additional options that fall between these low-cost and high-cost solutions. The team and the customer will need to negotiate the trade-offs between functionality and cost to come up with a balanced solution.

Negotiating on agile projects does not have to be—and typically should not be—a zero-sum game with a winner and a loser. Instead, healthy negotiations allow each party to investigate the options and trade-offs and present alternative perspectives. There should be an opportunity for each viewpoint to be fully described, noting the pros and cons of the different options. Negotiations are most effective when the interactions between participants are positive and there is some room for give and take on each side.

T&T Conflict Resolution

Conflict is an inevitable part of project work. Whenever people come together to solve problems, there will be differences of opinion and competing interests. Some degree of conflict is healthy, to ensure that ideas are sufficiently tested before they are adopted. However, we need to make sure the conflict does not escalate beyond healthy skepticism and friendly teasing, or we will end up with a negative and repressive project environment.

Creating an environment in which people can use conflict constructively is a key part of successfully engaging stakeholders on a project. We must watch for instances when conflict moves beyond normal, healthy debate and becomes destructive and harmful to the relationships and the team. Conflict resolution expert Speed B. Leas offers a framework that helps us judge the seriousness of a conflict and better understand how conflicts may escalate from Level 1 (Problem to Solve) to Level 5 (World War).[16]

Level	Name	Characteristic	Language	Atmosphere/Environment
Level 1	Problem to Solve	Information sharing and collaboration	Open and fact-based	People have different opinions or misunderstandings, or there are conflicting goals or values. The atmosphere isn't comfortable, but it isn't emotionally charged either.
Level 2	Disagreement	Personal protection trumps resolving the conflict	Guarded and open to interpretation	Self-protection becomes important. Team members distance themselves from the debate. Discussions happen off-line (outside of the team environment). Good-natured joking moves to half-joking barbs.
Level 3	Contest	Winning trumps resolving the conflict	Includes personal attacks	The aim is to win. People take sides. Blaming flourishes.
Level 4	Crusade	Protecting one's own group becomes the focus	Ideological	Resolving the situation is not good enough. Team members believe that people "on the other side" will not change and need to be removed.
Level 5	World War	Destroy the other!	Little or nonexistent	"Destroy!" is the battle cry. The combatants must be separated. No constructive outcome can be had.

Understanding this framework of the stages of conflict can help us look at a situation more objectively, moving past our own judgments to see what is really happening. Identifying the stage of a conflict can also help us determine what actions we should take or what tools or techniques may work in the given situation.

Therefore, when a team is in conflict, we should first take some time to observe the situation and make sure we are seeing both sides of the dispute, not just jumping in with a knee-jerk reaction. We need to allow time for proper observation, conversation, and intuition about the issues before taking action. This means that at first we simply listen to the complaints, without immediately trying to solve them. We feel the energy of the group, and assess the level of conflict. We look for glances, eye rolling, and words that halt conversations to ascertain if the conflict is out in the open, or if it is playing out below the surface.

One way to determine the level of conflict is to focus on the language the team is using and compare it to Leas's description of the five levels. So let's examine the language used at each level in more detail:

» **Level 1 (Problem to Solve)**: The language is generally open-hearted and constructive, and people frequently use factual statements to justify their viewpoints. For example, team members may make statements such as, "Oh, I see what you are saying now. I still prefer the other approach, however, because in the past we've seen fewer bugs and less rework using that technique."

» **Level 2 (Disagreement)**: The language starts to include self-protection. For example, team members may make statements like, "I know you think my idea won't work as well, but we tried your approach last time, and there were a lot of problems."

» **Level 3 (Contest):** The team members start using distorted language, such as overgeneralizations, presumptions, and magnified positions. They may make statements like, "He always takes over the demo" and "If only she wasn't on the team . . . ."

» **Level 4 (Crusade):** The conflict becomes more ideological and polarized. The team members begin to make statements like, "They're just plain wrong" and "It's not worth even talking to them."

» **Level 5 (World War):** The language is fully combative. The opposing team members rarely speak directly to each other, instead speaking mostly to those "on their side," expressing sentiments like, "It's us or them" and "We have to beat them!"

After observing and diagnosing the level of conflict, we can decide what to do about it. If the conflict is at level 1 through 3, do not take any immediate action to resolve it. Instead, first give the team a chance to fix it themselves. If the team can overcome the conflict on their own, they will have developed and exercised their own skills for resolving conflicts. It is okay for them to have some discomfort during this process, because that will better equip them to manage similar conflicts in the future. However, if the situation doesn't improve and instead seems to be escalating, the following guidelines can be useful in resolving the conflict:

» **Level 1 (Problem to Solve):** For a conflict at this level, try constructing a collaborative scenario to illustrate the competing issues and use that scenario to help build consensus around a decision that everyone can support.

» **Level 2 (Disagreement):** At level 2, conflict resolution typically involves empowering the relevant team members to solve the problem. This approach builds the team members' support for the decision and restores a sense of safety to the group.

» **Level 3 (Contest):** At this level, the conflict has become accusatory. To help fix the issue, we need to accommodate people's differing views. Although this may involve compromising on the work to be done, we should not compromise the team's values.

» **Level 4 (Crusade):** Resolving a level 4 conflict requires diplomacy. Since the communications between opposing sides have largely broken down, the team may need a facilitator to convey messages between the different parties. Our focus should be on de-escalating the conflict in an effort to take it down a level or two.

» **Level 5 (World War):** If a conflict gets to level 5, it may actually be unresolvable. Instead of trying to fix it, we may need to figure out how we can give people ways to live with it. At this level we might separate the opposing individuals to prevent further harm to each other.

In summary, conflict is normal and inevitable when people work closely together. Agile leaders, coaches, and ScrumMasters often feel obliged to help resolve a conflict. However, before rushing in, it is best to observe the situation to get a better view of the issues. Leas's model can help us objectively assess the severity of a conflict. We should pay attention to the language being used and give the team an opportunity to resolve the conflict themselves. If we do need to intervene, we should focus on de-escalating the problem by separating facts from emotions and looking for ways to help people move forward, despite their differences.

EXERCISE

Review the following snippets of conversation, and determine the conflict level illustrated in each.

Snippet	Conflict Level
"They have no idea, yet again. We would be better off without them!"	Level 4
"Okay, I get that you will have extra work if we choose this option. But so will I if we go with your method. And I'll have to redo this piece each time we set up a new page."	Level 2
"That's it! I warned you before. You and me—outside, right now!"	Level 5
"I know you have told me before, but I must be losing it. How do I request a ticket again?"	Level 1
"You're just pushing for this option because it makes your job easier. You never care about how it impacts anyone else! I'm tired of it. I think we should try something else for once."	Level 3

ANSWER

Snippet	Conflict Level
"They have no idea, yet again. We would be better off without them!"	Level 4
"Okay, I get that you will have extra work if we choose this option. But so will I if we go with your method. And I'll have to redo this piece each time we set up a new page."	Level 2
"That's it! I warned you before. You and me—outside, right now!"	Level 5
"I know you have told me before, but I must be losing it. How do I request a ticket again?"	Level 1
"You're just pushing for this option because it makes your job easier. You never care about how it impacts anyone else! I'm tired of it. I think we should try something else for once."	Level 3

K&S **Participatory Decision Making**

The last topic we'll discuss in this chapter is participatory decision making, which means engaging the project stakeholders in the decision-making process. The speed at which we make decisions and the group's level of agreement with the decisions will impact both the performance of the project and the cohesion of the team. Also, since knowledge work projects have no tangible product moving down a production line, communication and decision-making processes are more critical for keeping everyone informed and engaged.

Although agile methods use many tools to promote effective communication among stakeholders (including co-location, daily stand-up meetings, planning workshops, retrospectives, etc.), less is written about decision-making tools. This does not mean that stakeholder participation in decisions isn't important, however. If we do not involve stakeholders when making decisions, we run the risk of alienating some of them. This in turn leads to reduced commitment and participation, and could result in missing an important perspective that would help avoid pitfalls later in the project. Also, as stakeholders' involvement in the process increases, so too does their commitment to the outcome, as illustrated below.

Figure 3.19: Commitment Increases as Involvement Increases

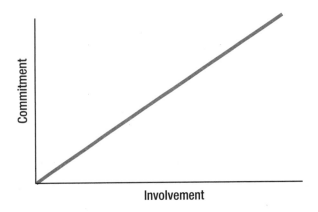

Convergent, Shared Collaboration

Agile methods favor more team empowerment and less command-and-control direction on projects. Although this higher level of team empowerment increases stakeholders' satisfaction and productivity, it also raises the need for an effective decision-making process. So without a project dictator, how do agile stakeholders make decisions and move forward? Below we'll describe several approaches that can be used to make collective decisions. As you read about these methods, notice that they share two common elements—convergence and shared collaboration. Let's examine what this means.

» **Convergence**: Participatory decision-making models aim for convergence, or collective agreement on the best answer. There may be outlying opinions that need to be clarified or perhaps removed from consideration, but on the whole the goal is to converge on a collectively agreed-upon best answer. This helps create buy-in and support for the decision since everyone has had an opportunity to voice their opinion, be heard, and actively participate in the evolution of a common view.

» **Shared Collaboration**: The second characteristic is that these approaches aim to share the decision-making process fairly. They are looking for group consensus rather than yielding to the will of a single influential individual. For example, in chapter 5 we'll examine two team-based estimating approaches, planning poker and wideband Delphi. These methods use anonymity (in wideband Delphi) and submitting estimates simultaneously (in planning poker) to ensure that the group's estimates aren't consciously or unconsciously pulled toward the opinions of the most senior or experienced person in the room. Agile teams try to avoid such bias by using tools that emphasize both convergence and shared collaboration.

Participatory Decision Models

The first thing to realize is that it's not realistic to expect the group to achieve total agreement on all issues and decisions. So we need some mechanisms for making tough decisions while still keeping everyone engaged in the project. These mechanisms are called participatory decision models. Let's look at some examples of these models, along with their advantages and disadvantages.

Simple Voting

One simple approach is to ask the team to vote "for" or "against" an idea by a show of hands. Although this is an easy technique, it limits our opportunities to refine the resulting decision. In striving for a quick result, this method can prevent the team from discovering and exploring better alternatives. What if someone has a suggestion for tweaking the options that are being voted on? A simple "for" or "against" vote omits refinement as an integral step. To help address this limitation, the team could discuss their thoughts before voting—but for most straightforward decisions, such a discussion would be a poor use of the team's time.

Personally, I would vote "No" on this approach.

Thumbs Up/Down/Sideways

Asking for a show of thumbs up, down, or sideways around the room is a more efficient way of achieving a simple vote while still allowing an opportunity to explore other options. With this technique, we ask those who are holding their thumb sideways why they cannot make up their mind. Sometimes these people are just neutral on the idea, but other times they have a conflict, concern, or question that needs further investigation. This approach is quicker than polling everyone in the group for input, since most people will have no concerns and will just want to move forward.

Fist-of-Five Voting

The "fist-of-five" approach has the advantage of speed (like the simple voting method), while still allowing people to indicate their degree of agreement (like the decision spectrum method discussed below). Using the fist-of-five approach, people vote by showing the number of fingers that indicates their degree of support.

A small problem with this approach is that two standards for using it have emerged. To counter this issue, the group simply needs to make it clear upfront which method they are using. In one version—which has been popularized by the American Youth Foundation—the participants show their level of support according to the number of fingers raised; a fist (no fingers) means no support, while five fingers indicates full support and a desire to lead the charge.

In the other popular version of this method, the fingers show the level of resistance or objection to the idea being voted on. With this method, the number of fingers raised indicates the following:

» **One finger**: "I totally support this option."
» **Two fingers**: "I support this option with some minor reservations that we probably don't need to discuss."
» **Three fingers**: "I have concerns that we need to discuss."
» **Four fingers**: "I object and want to discuss the issue."
» **Five fingers** (an extended palm like a stop sign): "Stop; I am against this decision."

Highsmith's Decision Spectrum

Jim Highsmith outlines a great decision-making tool in his book *Agile Project Management: Creating Innovative Products*. Using Highsmith's model, team members indicate how they feel about a decision by placing a check mark on a spectrum ranging from "fully in favor" to "mixed feelings" to "absolutely no, or veto." Highsmith's model is effective because it allows people to both indicate their support for a decision and express their reservations at the same time. It's important to give people an opportunity to voice their concerns if we hope to reach an agreement to go forward while still respecting dissenting views and keeping everyone engaged. This method invites those who are not entirely in favor of an option to share their concerns. Often, just giving people the opportunity to register their reservations is enough to help them commit to a new direction.

The following is a sample decision spectrum. The outline can be created on a whiteboard with tape and permanent markers and then reused for multiple decision-making sessions.

Figure 3.20: Highsmith's Decision Spectrum

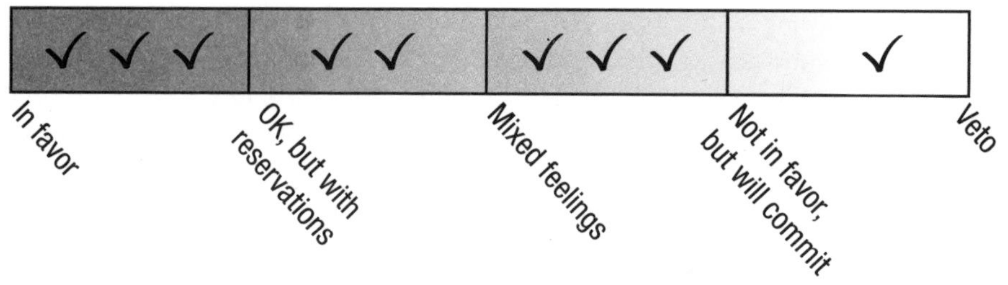

With participatory decision models, a key point to remember is, "not involved means not committed." We need to find ways to get our stakeholders involved in important project decisions, including iteration and release planning, estimation sessions, and retrospectives. If people are not involved, they will not be committed to the decision and, ultimately, will not be committed to the project.

Chapter Review

1. How will using short iterations help your team?

    A. Keep the team fully occupied.
    B. Keep stakeholders involved in the project.
    C. Keep stakeholder communications streamlined.
    D. Let the team relax and get acclimated at the start of the project.

2. What's the best way for your team to share their progress with the other project stakeholders?

    A. Information radiators
    B. Scrum of scrums
    C. Stand-up meetings
    D. Retrospectives

3. Everyone on the team seems to have a different opinion about how to build the next product increment. What could help them resolve this debate and move forward?

    A. Fishbone Analysis
    B. Shared communication
    C. Emotional intelligence
    D. Fist-of-five voting

4. Which tool wouldn't help your team share knowledge?

    A. Burnup charts
    B. Stand-up meetings
    C. Co-located team members
    D. Fibonacci diagrams

5. In interviewing candidates for an open position on your team, you're looking for someone who will be able to _____.

    A. Understand and influence the emotions of others.
    B. Work independently without asking others for help.
    C. Resolve stress and conflict between other people.
    D. Help others manage change and challenges.

6. The definition of done is _____.

    A. Provided by the product owner
    B. Determined by the team and the ScrumMaster
    C. Defined by the customer and the ScrumMaster
    D. Agreed upon by the team and the product owner

7. What is the most frequently used kind of workshop on a typical agile project?

    A. Team survey
    B. Iteration planning meeting
    C. Scrum inspection
    D. Project kick-off meeting

8. Which of the following statements best describes the role of an agile project charter?

    A. Forms the basis of the prioritized backlog and identifies the work to be completed by the project
    B. Describes the threats and issues the project may encounter and describes mitigation strategies for avoidance
    C. Defines the who, what, where, when, why, and how of the project and provides authority to proceed
    D. Outlines the roles and responsibilities of the project stakeholders and identifies any third-party contractors

9. Agile's emphasis on two-way communication means that:

    A. Distributed teams have to make an effort to communicate effectively.
    B. Agile teams need to share knowledge as widely as possible.
    C. The customer should solicit the delivery team's ideas, risks, and concerns and take them seriously.
    D. The stakeholders need to be kept in the loop about the team's progress.

10. Which of the circumstances outlined below would be a good fit for the use of personas?

    A. When the conversation is centered on the high-level flow of a process
    B. When we are trying to better understand stakeholder demographics and needs
    C. When we need to capture the high-level objective of a specific requirement
    D. When we want to communicate what features will be included in the next release

11. Your project team is in turmoil, and you're trying to diagnose the level of their conflict with another department. You notice that statements such as "Marketing always gets it mixed up!" are becoming commonplace. What level of conflict would you say the team is experiencing?

    A. Level 1
    B. Level 2
    C. Level 3
    D. Level 4

12. Which of the following isn't a form of agile workshop?

    A. The stakeholders gather to identify and define the user stories.
    B. The team gathers to reflect on their last iteration and identify issues and process changes.
    C. The delivery team gathers to estimate the user stories for the next iteration.
    D. The team gathers to share their progress yesterday, discuss what they will do today, and identify any impediments to progress.

13. What isn't part of emotional intelligence?

    A. The ability to control our own emotions
    B. The ability to identify, assess, and influence our own emotions
    C. The ability to identify, assess, and influence the emotions of other people
    D. The ability to control other people's emotions

14. The concept of knowledge sharing on an agile project is best characterized as:

    A. Encouraged where possible and where the team shows an interest
    B. Central to many agile practices
    C. Undertaken if there is time left at the end of an iteration
    D. Undertaken principally through stand-up meetings

© 2015 RMC Publications, Inc • 952.846.4484 • info@rmcls.com • www.rmcls.com

15. The gulf of evaluation refers to what?

    A. The gap between what the product owner knows and what the testers know
    B. The mismatch between the customer's vision of the solution and how the developers understand it
    C. The difference between what the ScrumMaster tells the team to do and what they actually produce
    D. The disparity between what the customer wants and what they really need

16. On an agile project, the definition of done is discussed frequently so that:

    A. Functionality can be negotiated until the last responsible moment.
    B. All stakeholders have a clear understanding of what completion means.
    C. Team members get to improve their negotiation skills.
    D. Active listening can reveal previously undiscussed requirements.

17. Agile modeling aims to:

    A. Capture the intent of the design in a barely sufficient way
    B. Capture the intent of the design in a comprehensive way
    C. Deliver actionable documentation for the project
    D. Leverage the fact that the value of modeling increases with time spent on it

18. Your team seems to unproductively debate even trivial decisions. To help them make collective decisions, you could try using:

    A. Fist-of-five voting
    B. Bare fist fighting
    C. Planning poker
    D. Brainstorming

19. Which of the circumstances outlined below would be a good fit for the use of wireframes?

    A. When the conversation is centered on the high-level flow of a process
    B. When we are trying to better understand stakeholder demographics and general needs
    C. When we need to capture the high-level objective of a specific requirement
    D. When we want to communicate what features will be included in the next release

20. Which of the following benefits isn't part of the agile value proposition?

    A. Reduced risk at the end of the project
    B. Increased visibility throughout the project
    C. Increased adaptability during planning
    D. Earlier delivery of business value

Answers

1. **Answer**: B

   Explanation: Short iterations help keep stakeholders actively involved in the project through frequent iteration planning and review meetings. Short iterations don't optimize resource allocation (i.e., keep the team fully occupied) or streamline communications. And since they mean that agile teams are always working toward a short-term target, they don't give team members any time to relax at the start of a project.

2. **Answer**: A

   Explanation: The most popular tools agile teams use to share their progress with other project stakeholders are the large visible displays of information known as information radiators. A scrum of scrums is a way of coordinating work between two teams. Stand-up meetings share information within the team. Retrospectives are primarily for the benefit of the team—and they are focused on improving the team's processes, not sharing progress.

3. **Answer**: D

   Explanation: The answer to this question can be found logically by thinking through the four options. Fishbone Analysis helps us get to the root cause of an issue; although this might be helpful, it isn't a tool for reaching consensus. Shared communication is a made-up term that doesn't really mean anything. Emotional intelligence is an interpersonal skill that helps us understand and influence emotions; again, although this might be helpful, it isn't the best answer. The correct answer is fist-of-five voting, which is a tool for collective decision making.

4. **Answer**: D

   Explanation: All the answer options are ways that agile teams share information except for "Fibonacci diagrams," which is a made-up term.

5. **Answer**: A

   Explanation: In weighing the options for this question, you need to apply your understanding of the agile mindset as well as common sense about how people work together. You should recognize that "understand and influence the emotions of others" is one aspect of emotional intelligence. Since emotional intelligence is an important skill on agile teams, then all you need to do is rule out the other answers to make sure this is the BEST answer—and as it turns out, it is. Agile puts a big emphasis on working collectively rather than independently. In agile, conflicts are ideally resolved by those directly involved, not by outside parties. And helping others "manage change and challenges" describes the role of a therapist, not an agile team member.

6. **Answer**: D

   Explanation: Although this question uses Scrum terms, it is equally applicable to generic agile teams. As a rule, the definition of done is the result of a conversation between the team and the product owner. If you think about it, this is logical; the team and the product owner each have information that the other party doesn't have. The product owner knows what is needed by the business, and the team knows what can realistically be built within the available constraints. Both of these perspectives need to be weighed to come up with a shared definition of done.

7. **Answer**: B

   Explanation: Two of these options aren't agile workshops—"team survey" and "Scrum inspection" aren't defined agile processes. To find the answer you just need to know how regularly the other two meetings are used on a typical agile project. By definition, a project kick-off meeting takes place once at the start of the project. Iteration planning meetings, on the other hand, are held at the start of each iteration, every one or two weeks—this is the correct answer.

8. **Answer**: C

   Explanation: An agile project charter defines the who, what, where, when, why, and how of the project and provides authority to proceed. Although this document may also list the scope that ends up in the backlog and the threats and issues the project might face, these are all just portions of the charter, and do not best describe its role.

9. **Answer**: C

   Explanation: Agile projects need to have bidirectional communication in which information flows both from the business to the team and from the team to the business. While all the options listed could be interpreted as correct statements, option C is the only one that refers to this kind of communication. If you thought option D was correct, notice that it simply refers to keeping the stakeholders informed, not to decision making based on that information.

10. **Answer**: B

    Explanation: Personas are quick guides or reminders of the key stakeholders on the project and their interests. So this tool would be a good fit when we are trying to better understand stakeholder demographics and general needs.

11. **Answer**: C

    Explanation: "Marketing always gets it mixed up" falls into the category of overgeneralizations, presumptions, and magnified positions, which indicates a level 3 type conflict. At level 2, they wouldn't be using overgeneralizations (such as "always"), and at level 4, their language would likely be more ideological and hostile.

12. **Answer**: D

    Explanation: The last option here isn't a workshop, because it describes the daily stand-up, which is a brief meeting, not a workshop. The other three options, in order, describe story-writing workshops, retrospectives, and estimating sessions, all of which are forms of agile workshops.

13. **Answer**: D

    Explanation: Emotional intelligence deals with our ability to identify, assess, and influence both our own emotions and those of other people, including our capacity for emotional self-control. However, it doesn't allow us to control the emotional responses of other people.

14. **Answer**: B

    Explanation: Knowledge sharing is central to many agile practices. It is true that stand-up meetings help the team members share information, but they are not the primary event for knowledge sharing. This practice is too important to be considered optional if the teams show an interest, or if there is time.

15. **Answer**: B

    Explanation: The gulf of evaluation refers to the gulf, or difference between, what one person envisions and tries to describe and how another person hears and interprets that description. On agile projects, there can be significant consequences if the developers don't have the same understanding of the project solution as the customer or product owner. The other options are made-up.

16. **Answer**: B

    Explanation: The reason we have frequent discussions about the definition of done is to prevent the mismatches that can occur when different people interpret the descriptions of new functionality in different ways. The definition of done is not intended to be used to negotiate functionality, improve negotiation skills, or surface new requirements (although that may occur). Instead, we have these discussions to make sure everyone has a common understanding of what completion or success will look like.

17. **Answer**: A

    Explanation: Agile models are lightweight, barely sufficient (just enough detail) models that aim to capture the high-value benefits of modeling without taking too much time to create very detailed or polished models. We want to focus on the product being developed, rather than on generating documentation. Finally, the value of modeling doesn't increase with time spent; there is a point of maximal value past which additional modeling isn't worth the effort.

18. **Answer**: A

    Explanation: When teams struggle to make collaborative decisions, it can be helpful to introduce participatory decision models such as fist-of-five voting. Planning poker is an estimation technique that would only help if they were having difficulty agreeing on estimates. Brainstorming is an idea-generating exercise that wouldn't help with decision making. And bare fist fighting is, sadly, in conflict with PMI's Code of Ethics and Professional Conduct.

19. **Answer**: A

    Explanation: Wireframes are a way of creating a quick mock-up of a product, including its process flows. So this tool would be a good fit when the conversation is centered on the high-level flow of a process.

20. **Answer**: A

    Explanation: Although the agile value proposition does include reduced risk, that effect is most pronounced at the start of the project, since agile teams aim to lower risk as quickly as possible. At the end of a project, no project risk remains since the project is over and we already know the outcome. The other three options are accurate descriptions of the other elements of the agile value proposition.

CHAPTER 4

Team Performance

Domain IV Summary

This chapter discusses domain IV in the exam content outline, which is 16 percent of the exam, or about 19 exam questions. This domain focuses on building high-performing teams, including adaptive leadership, empowering and coaching the team, collaborative team spaces, and performance tracking.

Key Topics

- » Adaptive leadership
- » Agile team roles
- » Building agile teams
  - Self-directing
  - Self-organizing
- » Burndown/burnup charts
- » Caves and common
- » Co-location (physical and virtual)
- » Developmental mastery models
  - Dreyfus (skill acquisition)
  - Shu-Ha-Ri (mastery)
  - Tuckman (team formation)
- » Global, cultural, and team diversity
- » Osmotic communication
  - Co-located teams (proximity)
  - Distributed teams (digital tools)
- » Tacit knowledge
- » Team motivation
- » Team space
- » Training, coaching, and mentoring
  - Individual vs. team coaching
- » Velocity

Tasks

1. Develop team rules and processes to foster buy-in.
2. Help grow team interpersonal and technical skills.
3. Use generalizing specialists to maximize flow.
4. Empower and encourage emergent leadership.
5. Learn team motivators and demotivators.
6. Encourage communication via co-location and collaboration tools.
7. Shield team from distractions.
8. Align team by sharing project vision.
9. Encourage team to measure velocity for capacity and forecasts.

This chapter is all about agile delivery teams. For the exam—as well as to be successful on real-world agile projects—you need to understand how to develop and support self-organizing, self-empowered teams and help them flourish and succeed. Since the members of the team are project stakeholders, many of the ideas we covered in chapter 3—such as communication, collaboration, facilitation, and emotional intelligence—are also relevant to team performance, especially for team leaders, ScrumMasters, and agile coaches. In this chapter we'll focus on some other topics that help build and maintain high-performing teams, organized around three themes: building agile teams, creating collaborative team spaces, and tracking team performance.

Why People Over Processes?

Agile methodologies seem to put a lot of weight on process, with their focus on practices such as iterations, backlogs, and reviews. This is ironic, given that the first value in the Agile Manifesto is "Individuals and interactions over processes and tools." So why do agile processes get so much attention? One major reason is that processes and tools are simply easier to describe, explain, and classify than the trickier topics related to individuals and interactions. Also, it's more difficult to formalize practices related to individuals and interactions because people vary so much—in their skill sets, attitudes, experiences, perspectives, culture, and so on. So the right thing to do with one team might be exactly the wrong approach to take with another.

This is why there is a common saying: "The soft stuff is the hard stuff." We need to recognize that the "soft stuff" is not just more difficult (harder) than processes and tools—it's also more important. We should focus on the people side of projects, even if that focus isn't our area of expertise or where we are most comfortable. Good people who have few or no processes in place can succeed even on difficult projects, yet poorly skilled or poorly aligned teams often fail, even with the best processes. As leaders of teams, we need to focus our efforts on people factors to get the maximum return on performance. This is why it's so important to understand how to build and support healthy teams.

Although trying to quantify the significance of good processes versus good people is problematic, data from the COCOMO II® estimation model (see box) validates that the Agile Manifesto's idea of valuing individuals over processes is indeed the path to project success. The COCOMO II® weighting factors for having best-in-class people (compared to industry averages) are always larger than having best-in-class processes (compared to industry averages). This tells us, based on thousands of completed projects, that people are more significant than processes. In other words, if we want to lower costs and boost performance, we should focus more time on training, retaining, and engaging our team than on trying to improve our processes. Of course, ideally we will focus on improving both people and processes—but given the choice, there will always be a better return from investing in people.

Agile Team Roles

In discussing the agile methodologies in chapter 1, we defined the team roles and terminology used by Scrum and XP. Our focus in that chapter was on the separate methodologies. However, for the exam it's essential to have a good grasp of the agile team roles from a broader perspective, rather than thinking of them only in terms of specific methodologies. To help you do that, let's take a moment to review the key team roles from a methodology-agnostic perspective, showing the multiple terms that are used for each role. Although there are subtle differences in the way these terms are used in each methodology, for the purposes of the exam, these terms are basically interchangeable.

Background information

What's More Important, People or Processes?

Here's more information about the COCOMO II® data that quantifies the importance of people versus processes. First of all, what is COCOMO II®? Estimating software projects is notoriously difficult because of the inherent risk involved in developing solutions for new business problems that have high rates of change. Despite this difficulty, software companies still need a way to bid on projects and estimate their likely budgets. As a result, many smart people have studied the subject extensively to come up with solutions. One of the most popular solutions that was developed was a software estimation model called COCOMO® (from the term "COnstructive COst MOdel").

The COCOMO® model was created by reverse engineering the inputs from thousands of completed software projects that had a known exact cost. The idea behind the model was to assess a large number of projects to see if there was some correlation between the project input variables and the final cost, and then use this data as a basis for estimating future projects. This method has proven to be successful, and the updated model known as COCOMO II® is at the heart of many commercial estimation systems used today.

The figure below shows the COCOMO II® weighting factors for seven of the input variables, one of which is people. In looking at this chart, we can see that the "People" factors have the largest score of all the variables, 33, while the "Tools and Processes" factors only have a score of 3. This means that when calculating the final cost of a software project, the impact of people factors (such as a team with poor, average, or best-in-class ability and skills) is over ten times more significant than the tools and processes those people are using.

Figure 4.1: Weighting Factors for COCOMO II® Input Variables

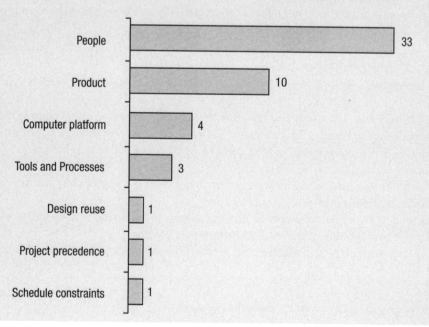

Development Team/Delivery Team

This group includes everyone needed to build and test a complete increment of the product, such as coders, writers, designers, analysts, and testers. Agile delivery teams rely on generalizing specialists—people who can perform multiple jobs and switch from role to role as the demand arises. These team members:

» Build the product increments, using agile practices and processes.
» Regularly update information radiators to share their progress with stakeholders.
» Self-organize and self-direct their working process within an iteration.
» Share their progress with each other in daily stand-up meetings.
» Write acceptance tests for the product increments.
» Test and revise the product increments until they pass the acceptance tests.
» Demonstrate the completed product increment to the customer in the iteration review meeting.
» Hold iteration retrospectives to reflect on their process and continually improve it.
» Perform release and iteration planning, including estimating the stories and tasks.

Product Owner/Customer/Proxy Customer/Value Management Team/Business Representative

» Maximizes the value of the product by choosing and prioritizing the product features.
» Manages the product backlog, making sure that it is accurate, up to date, and prioritized by business value.
» Makes sure the team has a shared understanding of the backlog items and the value they are supposed to deliver.
» Provides the acceptance criteria that the delivery team will use to prepare acceptance tests.
» Determines whether each completed product increment is working as intended, and either accepts it or requests changes (in the iteration review meeting).
» May change the product features and their priority at any time.
» Facilitates the engagement of external project stakeholders and manages their expectations.
» Provides the due dates for the project and/or its releases.
» Attends planning meetings, reviews, and retrospectives. (If this role is performed by a group of people, typically only one or two of them will attend these meetings.)

ScrumMaster/Coach/Team Leader

» Acts as a servant leader to the delivery team, helping them improve and removing barriers to their progress.
» Helps the delivery team self-govern and self-organize, instead of governing and organizing them.
» Serves as a facilitator and conduit for communication within the delivery team and with other stakeholders.
» Makes sure the delivery team's plan is visible and its progress is radiated to stakeholders.
» Acts as a coach and mentor to the delivery team.
» Guides the team's agile process and makes sure their agile practices are being used properly.
» Helps the product owner manage the product backlog.
» Helps the product owner communicate the project vision, goals, and backlog items to the delivery team.
» Facilitates meetings (planning, reviews, and retrospectives).
» Follows up on issues raised in stand-up meetings to remove impediments so that the team can stay on track.

Project Sponsor

» Serves as the project's main advocate within the organization.
» Provides direction to the product owner role (the person or team representing the business) about the organization's overall goals for the project.
» Focuses on the big picture of whether the project will deliver the expected value on time and on budget.
» Is invited to the iteration review meetings to see the product increments as they are completed, but might not attend.

EXERCISE

Here's a chance to check your understanding of the responsibilities of the various agile team roles.
Read the task in the first column and record the role responsible for that task in the second column.
(If you want to practice all the terms that might appear on the exam, also try to include all the terms
for each role listed above.)

Task	Responsibility
Provides the due dates for the project and/or its releases.	Product Owner, Customer, Proxy Customer, value mgt team, business rep
Facilitates meetings (planning, reviews, and retrospectives).	Scrum Master, Coach, Team leader
Manages the product backlog, making sure that it is accurate, up to date, and prioritized by business value.	Product Owner, Customer, Proxy Customer, value mgt team, business rep
Helps the delivery team self-govern and self-organize, instead of governing and organizing them.	Scrum Master, Coach, Team leader
Writes acceptance tests for the product increments.	Development Team, Delivery team
Demonstrates the completed product increment to the customer in the iteration review meeting.	Development Team, Delivery Team
Focuses on the big picture of whether the project will deliver the expected value on time and on budget.	Project Sponsor
Determines whether each completed product increment is working as intended, and either accepts it or requests changes in the iteration review meeting.	Product Owner, Customer, Proxy Customer, value mgt Team, Business rep
Acts as a servant leader to the delivery team, helping them improve and removing barriers to their process.	Scrum Master, Coach, Team lead

ANSWER

Task	Responsibility
Provides the due dates for the project and/or its releases.	Product owner/Customer/Proxy customer/Value management team/Business representative
Facilitates meetings (planning, reviews, and retrospectives).	ScrumMaster/Coach/Team leader
Manages the product backlog, making sure that it is accurate, up to date, and prioritized by business value.	Product owner/Customer/Proxy customer/Value management team/Business representative
Helps the delivery team self-govern and self-organize, instead of governing and organizing them.	ScrumMaster/Coach/Team leader
Writes acceptance tests for the product increments.	Development team/Delivery team
Demonstrates the completed product increment to the customer in the iteration review meeting.	Development team/Delivery team
Focuses on the big picture of whether the project will deliver the expected value on time and on budget.	Project sponsor
Determines whether each completed product increment is working as intended, and either accepts it or requests changes in the iteration review meeting.	Product owner/Customer/Proxy customer/Value management team/Business representative
Acts as a servant leader to the delivery team, helping them improve and removing barriers to their process.	ScrumMaster/Coach/Team leader

K&S Building Agile Teams

In their book *The Wisdom of Teams*, Jon Katzenbach and Douglas Smith define a team as "a small number of people with complementary skills who are committed to a common purpose, performance goals and approach for which they hold themselves mutually accountable."[1] There are several valuable aspects of this definition that are worth highlighting—and we will be expanding upon most of these points in this chapter.

First, note that teams are described as "a small number of people." Agile methods recommend keeping the delivery team small (typically 12 or fewer members) since this allows the team members to develop better relationships and communicate more directly. If the project requires a larger group, it will usually need to be broken into smaller subteams that coordinate their work.

Exam tip

In the real world, agile development teams may be large. However, if an exam question asks about optimal team size, agile methods recommend a team of no more than 12 people.

Second, team members have "complementary skills." While individual team members may not possess all the skills required to complete a project on their own, the team will collectively have all the necessary skills. This could mean the team consists of specialists who all own their role in the project, but agile methods prefer to use *generalizing specialists*—individuals who have cross-functional skills and can readily move between roles. For example, an agile software project might have business analysts who can also perform quality assurance work or developers who also have business analysis skills.

Third, teams are defined as being "committed to a common purpose." This means the team members are aligned behind a project goal that supersedes their personal agendas. Teams also share common performance goals and a common approach. In other words, team members are in alignment (if not always in agreement) as to how the goals will be measured and how the team should go about the work.

Finally, there's the idea that team members "hold themselves mutually accountable." In other words, the team has shared ownership for the outcome of the project.

Benefits of Generalizing Specialists

Let's take a closer look at the advantages of using a team of generalizing specialists. Having team members who can perform different tasks helps the team minimize handoffs and avoid peaks and troughs in their workload. Let's see why this is important.

By definition, specialists have a narrow skillset, since they specialize in one role or function; so if a project relies on specialists, there are two kinds of problems that can arise. First, there will need to be multiple handoffs between people who have different skills to get the work done—and handoffs of knowledge work are slow and risky. Since we are rarely building the same product twice, each handoff involves a lot of unique knowledge and information. A large portion of the work usually consists of theory, ideas, or models, so there are few tangible, visible assets to hand over. Instead, one specialist has to try and explain the ideas, goals, and design constraints they used to another specialist—and generally something gets lost in the transition.

The second problem with passing work from one specialist to another is that this sequential process can lead to bottlenecks. For example, the diagram below shows the team for a software project, consisting of four specialists—Sandy (analyst), Jorge (user interface designer [UI]), Mara (coder and unit tester [UT]), and Tim (systems tester).

Figure 4.2: Team of Specialists on a Software Project

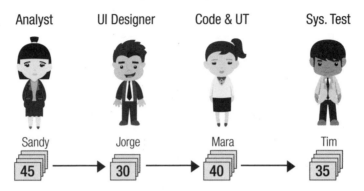

The numbers below each role represent the number of user stories the specialist can typically work through per sprint. Notice that each person is working through the stories at a different speed. In the first sprint at the start of the project, everyone will have enough work to do. But since Jorge, the UI designer, has the slowest throughput rate, eventually a bottleneck will build up before coding and unit testing. So all the specialists downstream from UI design will run out of work to do, as shown below.

Figure 4.3: Bottleneck at UI Design and No Work for Code & UT

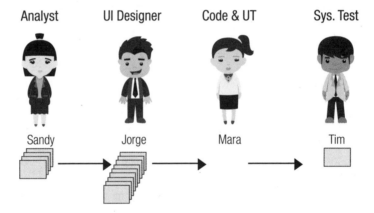

Using generalizing specialists instead of specialists can help solve both the handoff issue and the bottleneck issue. A generalizing specialist is skilled in more than one discipline, but not like a "jack of all trades and master of none"—more like a "king of several trades and master of each." Each additional role that a team member can play can eliminate a handoff. For example, if a coder is also skilled at interviewing subject matter experts to gather user stories, then we can avoid a handoff of requirements from a business analyst to that developer. This not only saves a lot of time, since handoffs are slow; it also means that no information will be lost in translation at this point in the project.

Generalizing specialists can also help solve the bottleneck problem by sharing the workload. Developers who are willing and able to do some testing when needed are far more valuable to a team than those who just want to code all day. Agile teams sometimes use the word "swarming" to describe collective problem solving—this can take the form of multiple people pitching in to help finish a task, remove a bottleneck, or move a deliverable across the line to production readiness.

Another way of understanding the difference between the roles of generalizing specialists and traditional specialists is by thinking of "T-shaped" people and "I-shaped" people. Like the letter "T," generalizing specialists spend most of their time deep in one role (the stalk of the "T"), but can and sometimes do spend time on activities that come before and after that work (the crosspiece on the "T"). I-shaped people, on the other hand, are deeply skilled in one role and spend all their time there, which is fine—but it means these people aren't as versatile, and therefore useful, to an agile team as T-shaped people.

Returning to our example, let's re-envision this team as four T-shaped generalizing specialists, as illustrated below. Now Sandy and Jorge have cross-functional skills, as do Mara and Tim. Since each of these two pairs can work as a unit, sharing tasks, their throughput is combined, which reduces handoffs and balances the workload of the two pairs so that there won't be any bottlenecks.

Figure 4.4: Reduced Handoffs and Bottlenecks through T-Shaped Generalizing Specialists

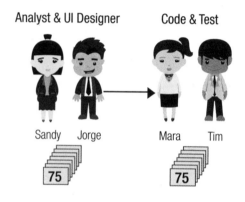

Exam tip

Although the exam questions won't go into the level of detail described in the above example, you should be familiar with the concept of generalizing specialists and why they are used on agile teams.

Characteristics of High-Performing Teams

Many people have researched how to build high-performance teams, including Carl Larson and Frank LaFasto, authors of the book *Teamwork*. Their work has influenced the following guidelines for managers:[2]

» **Create a shared vision for the team**. This enables the team to make faster decisions and builds trust.

» **Set realistic goals**. We should set people up to succeed, not to fail, so goals need to be achievable.

» **Limit team size to 12 or fewer members**. Small teams are able to communicate face-to-face and support tacit (unwritten) knowledge. If an agile project requires more resources, the preferred approach is to split the work among smaller subteams. Representatives of the subteams then need to come together every day to coordinate and synchronize work across the project (for example, in a scrum of scrums).

» **Build a sense of team identity**. A sense of team identity helps increase the team members' loyalty to the team and support for other team members.

» **Provide strong leadership**. Leaders should point out the way, then let the team own the mission.

Lyssa Adkins has also explored high-performance teams and has identified that they have the following eight characteristics:[3]

» They are **self-organizing**, rather than role- or title-based.
» They are **empowered** to make decisions.
» They truly believe that **as a team they can solve any problem**.
» They are committed to **team success** rather than success at any cost.
» The team **owns its decisions and commitments**.
» The members are motivated by **trust**, instead of fear or anger.
» They are **consensus-driven**, with full divergence and then convergence.
» They are in constant **constructive disagreement**.

Most of the characteristics listed by Adkins are self-explanatory, but let's look at some of them in more detail.

Exam tip

Remember that the exam will be testing your agile "mindset" as reflected by the concepts listed above, not the actual statements or specific details or sources. So aim to absorb these ideas into your global agile understanding rather than memorizing the lists.

Empowered Teams

Let's take a closer look at Adkin's first two points about teams—self-organization and empowerment. We can say that there are two ways in which agile teams are empowered—they are self-directing and self-organizing. Let's examine each of these characteristics in turn.

Self-Directing Teams

Being told what to do is never a recipe for runaway success. In such a scenario, the receivers of tasks end up second-guessing the direction or sequence of upcoming activities, and the providers of instructions get frustrated with unforeseen obstacles and technical issues. Pushing out instructions is like pushing rope— it's not very effective and it never really brings out the best in people.

In contrast, the members of empowered teams are freed from command-and-control management and can use their own knowledge to determine how to best do their job. This enables people to tap into their natural ability to manage complexity. We manage complexity every day, by juggling our work life, home life, e-mails, phone calls, and appointments. Organizations often fail to capitalize on this ability when it comes to executing project tasks, however. Instead of presenting team members with a number of items that have to be accomplished, they present a set of ordered tasks that, in reality, might best be done in a different way. Allowing teams to direct their own work enables the team members to draw upon and build their innate skills for managing complexity.

In addition, managers and schedulers do not have the same technical insight into task execution as do the people who are performing the work on a daily basis. So leaders of agile teams are better served by allowing team members to direct their own work. Instead of providing detailed task lists, the leader should describe the iteration goals at a high level and let the team determine how to best accomplish the work, within the ground rules of what is acceptable within the organization.

This acknowledgement that the team members are in the best position to direct the project work is liberating and motivating for them. People work harder and take more pride in their work when they are recognized as experts of their domain. When self-directing teams select work items from the queue of waiting work, they have the expertise to choose the items that are not blocked for any reason, that they are capable of doing, and that will bring them toward the iteration goal. This practice alleviates many of the technical blockages seen in push systems where a task list and sequence are imposed on the team.

So we need to delegate responsibility for success to the team and allow them to do what is necessary to achieve their goals. This is one aspect of agile's "servant leadership" model described in chapter 1. Instead of a command-and-control approach where instructions are passed from the project manager to the team leaders down to the team members, in agile projects the project manager and the team leaders serve the people who are doing the work by shielding them from interruptions, removing impediments, communicating the project vision, and providing support and encouragement.

Self-Organizing Teams

Self-organizing teams are empowered to work collectively to create their team norms and make their own local decisions. This means they not only figure out the best way to accomplish the work they have committed to do in an iteration, they also resolve many of the day-to-day issues that crop up along the way. Project managers and team leaders can support and reinforce this behavior by respecting the team's estimates and decisions and allowing them to make mistakes and learn from them.

This does not mean that managers and leaders abdicate their responsibility to the team. Instead it means the team is given freedom within the confines of an iteration. If the team's estimates are way off or if they make poor technical decisions, these issues will be identified and discussed at the iteration retrospective. During the next iteration these areas should improve, and within a more few iterations, the team's estimates and technical decisions are likely to be better than what the manager or leader could have produced.

Keep in mind that the self-directing and self-organizing attributes of agile teams are goals—we do not start there. The team members initially need support and guidance as they come to grips with the project scope and tools and learn to work together as a team. Once the team has stabilized (i.e., when it reaches the Norming phase, as explained below) we can introduce the goals of self-direction and self-organization as long-term objectives for the group.

Empowering and Encouraging Emergent Leadership

Empowered teams aim to create an emergent leadership model where different people step up to lead different initiatives. One identifying feature of high-performing teams is how seamlessly they swap leadership roles without an angry power struggle or a big debate. Instead, when a team member wants to take on a new initiative, the others are likely to respond, "You want to do that? Great, tell us what you need."

Like geese flying in formation, emergent leadership means that everyone takes a turn at the front. But unlike the geese, the team members don't switch in a linear way whenever someone gets tired. Instead, leadership happens organically on many levels, whenever someone sees a task that needs to be done or spearheaded and wants to take it on. The task itself could be something very simple—for example, Mara sees that the

3D printer is jammed again, so she asks the others not to use it while she cleans it out and resets it. Or it could be much larger and involve building consensus for a new approach, leading the exploration of that approach, and engaging other stakeholders for support.

The most important point is that these leadership roles are self-selected, not assigned, and that everyone is trying to help wherever they can best do so. This helps harness people's passions for trying new ideas and making improvements; it fosters and rewards an entrepreneurial spirit and makes the work more rewarding.

Create a Safe Place for Experiments

When I first started in project management I worked for a cranky old program manager. He was approaching retirement and cared little for being politically correct. I remember him telling me, "Mike, you have two hands. If you're using one to cover your rear, that only leaves one hand to work with!" That saying has stuck with me, and it illustrates the principle that we are much less effective when we are worried about making mistakes or getting into trouble.

People have built-in defense mechanisms that are triggered by mental as well as physical threats. When we sense a threat (or lack of safe environment), less oxygen and glucose are sent to the brain (to limit bleeding in the event of an attack). So when we feel threatened, it's harder to find smart answers because of our reduced cognitive resources for problem solving.

Conversely, when we are in a safe environment it fosters engagement, a state in which we are willing to tackle difficult tasks, take risks, think deeply about issues, and develop new solutions. This is just what we need for solving difficult knowledge work problems. The engaged state is linked to positive emotions of interest, happiness, joy, and increased dopamine levels. Research shows that when people are happy they explore more options when problem solving, solve more nonlinear problems that require insight, and collaborate better.[5]

So in leading agile teams we want to create a safe place where people can experiment, try new approaches, and make mistakes—and then get help, learn, and recover from their mistakes. How can we do that? First of all, we don't criticize people for experiments that fail. If the team tries a new approach for an iteration or two, and it results in worse rather than better results, we ask what can be learned from the experiment and then move on. (The lesson might simply be "Don't do that again!")

We should try to create an "engagement culture" that rewards people for problem solving, collaboration, and sharing ideas and inputs. This may be as simple as providing food for retrospectives and "lunch and learn" sessions, or it can be more formal, depending on what is acceptable in your organization.

Background information

The Cheese and the Owl

A scientific explanation of the importance of creating safe environments for people to experiment can be found in the research of Friedman and Förster, who measured the impact of threats on problem-solving ability and creativity.[4] Two groups of people were asked to solve a maze that had a picture of a mouse in the middle; their goal was to help the mouse get out. One group had a picture of cheese (a reward) at the outside, while the other group had a picture of an owl (a threat) hovering overhead, ready to snatch up the mouse if it couldn't get out. After completing the maze, both groups were given creativity tests. The group that was heading toward the cheese solved significantly more creative problems than those threatened by the owl. This study, which has been backed up by many similar studies, shows that even a small perceived threat can have a big impact on cognitive performance.

EXERCISE

For each behavior listed below, place a check mark in the appropriate column to indicate whether it is a command-and-control approach to managing projects or a servant leadership approach.

Behavior	Command-and-Control	Servant Leadership
Handing out detailed task lists	✓	
Doing administrative work for team members		✓
Creating the entire project's WBS one weekend so as not to disturb the team	✓	
Posting the project Gantt chart on the office wall	✓	
Posting a "suggestions" box on the office wall		✓

ANSWER

A command-and-control management approach does not accept that team members have the majority of the answers and excludes them from planning and scheduling activities.

Behavior	Command-and-Control	Servant Leadership
Handing out detailed task lists	✓ This behavior means the project manager does not recognize that team members are best placed to determine task-based work.	
Doing administrative work for team members		✓ By doing administrative work for the team, the leader is allowing the team to spend more time on value-added work.
Creating the entire project's WBS one weekend so as not to disturb the team	✓ By creating the project's WBS without the team, the leader is not incorporating the team's local knowledge.	
Posting the project Gantt chart on the office wall	✓ Gantt charts are typically controlled by one person and are not easy for team members to adjust. By relying on such a tool, the leader is not recognizing that the team may want to change things as the project progresses.	
Posting a "suggestions" box on the office wall		✓ This behavior means the leader is looking for input from the team members.

Encourage Constructive Disagreement

The second-to-last point on Adkins's characteristics of high-performing teams ("They are consensus-driven, with full divergence and then convergence") speaks to establishing a safe environment in which debating or arguing over issues is seen as healthy. Constructive conflict is encouraged because it ultimately leads to better decisions and stronger buy-in for those decisions once they are made. Divergence (argument and debate) and convergence (agreement about the best solution) increase the team's commitment. This practice also addresses Adkins's final point ("They are in constant constructive disagreement").

© 2015 RMC Publications, Inc • 952.846.4484 • info@rmcls.com • www.rmcls.com

Constructive disagreement is vital for really understanding and working out issues. Patrick Lencioni, author of *The Five Dysfunctions of a Team*, lists the following dysfunctions that damage and limit team performance:[6]

1. **Absence of trust**: Team members are unwilling to be vulnerable within the group.
2. **Fear of conflict**: The team seeks artificial harmony over constructive, passionate debate.
3. **Lack of commitment**: Team members don't commit to group decisions or simply feign agreement with them.
4. **Avoidance of accountability**: Team members duck the responsibility of calling peers on counterproductive behavior or low standards.
5. **Inattention to results**: Team members prioritize their individual needs, such as personal success, status, or ego, before team success.

These dysfunctions all stem from avoiding conflict (or constructive disagreement) and not having a safe environment in which it is okay to ask questions. Establishing a safe environment for disagreement is key to success; such an environment allows team members to build a strong commitment to decisions. If the team members have such a commitment when they encounter the inevitable obstacles on a project, rather than returning to management with a list of reasons why something cannot be done, they will instead push past the obstacles or find a way around them.

Models of Team Development

By this point it should be clear that to increase our chances for project success, we need to help our teams build mastery and develop into high-functioning units. There are three models of team development and skill acquisition that the PMI-ACP exam will expect you to be familiar with: Cockburn's Shu-Ha-Ri model, Dreyfus's model of skill acquisition, and Tuckman's model of team formation. We'll examine each of these in turn.

K&S Shu-Ha-Ri Model of Skill Mastery

Alistair Cockburn's Shu-Ha-Ri model originated in Japanese Noh theater. It describes a three-step process of increasing mastery that progresses as follows:[7]

1. Shu: Obeying the rules—*shu* means "to keep, protect, or maintain"
2. Ha: Consciously moving away from the rules—*ha* means "to detach or break free"
3. Ri: Unconsciously finding an individual path—*ri* means "to go beyond or transcend"

Cockburn's model tells us that when mastering a new skill or process, we have to move through three levels, or stages. As beginners at the *shu* level, we start by following the rules we've absorbed from our teachers, mentors, or learning experiences. Once we've mastered those guidelines through practice, we reach the *ha* level where we can let go of them because they've become second nature to us—we can break free of our training and work intuitively. In the final stage, *ri*, we reach full mastery; at this point we've integrated the rules so thoroughly that we can transcend them and strike out onto new paths for others to follow.

Let's look at some examples. One application of the Shu-Ha-Ri model addresses the appropriate time and place for agile process tailoring. When a team is new to agile, it's best to start by following a method that has already been tested, proven, and refined by others. We need to practice that system and develop some mastery over it before we can understand how and why all the pieces work together. To put it simply, we must really know how the "plain-vanilla" process works before we start removing things or inventing new flavors.

The Shu-Ha-Ri model can also be applied to the exam itself. Although the exam questions may appear to be tricky, they are actually focused at the *shu* level—following the rules. You are likely to encounter some ambiguous questions that make you think, "The answer depends on a whole bunch of other circumstances that the question isn't telling me!" When that happens, think back to the *shu* level—the agile basics outlined in this book. For example, ask yourself, "What would plain-vanilla Scrum, XP, or generic agile recommend?"

K&S Dreyfus Model of Adult Skill Acquisition

Another team development model that might come up on the exam is the "Dreyfus model of adult skill acquisition." This model by Stuart E. Dreyfus postulates that adults learn new skills over five stages—novice, advanced beginner, competent, proficient, and expert. As we move through each of these stages, in addition to improving our skills, our level of commitment, approach to decision making, and perspective on the task evolve as well.[8]

Let's examine each of the five stages of this model.

» **Novice**: Novices follow the rules they have been given and make analytical decisions. For example, if we are learning to drive a manual transmission, we might be told to change gears as the engine approaches 2,500 rpm. Although this rule is a good start, it could lead to problems if we try to drive on a hill, which will influence when we should switch gears.

» **Advanced beginner**: At this stage, we are still following rules and making analytical decisions, but now we have gained enough experience with real-world situations to begin to understand the context of the rules. To decide when to change gears, we might use a guideline such as "gear up when you hear the engine racing."

» **Competent**: As we gain competence, the number of rules and guidelines for different contexts becomes overwhelming. Since we can't apply them all, we begin to decide which rules are the best for each situation, and this makes us feel more personally responsible for the choices we are making.

» **Proficient**: At this level, our decision making is still analytical, but we are actively choosing the best strategy rather than relying on the rules. In the process, we become more emotionally involved in the task. For example, as a proficient driver, we will have a gut feeling if we are approaching a corner too fast on a rainy day.

» **Expert**: As we develop expertise, our decision making becomes intuitive—we are able to spontaneously assess the alternatives and select the best approach without having to first analytically examine all the possible strategies.

As reflected in the above descriptions, as we progress through the mastery of new skills, our perspective changes from neutral (no opinion) to choosing a perspective based on the circumstances, to an expert perspective rooted in our own experiences. Meanwhile our decision-making approach moves from analytical to intuitive, and our commitment from detached to involved. This transition is summarized in the diagram below:

Figure 4.5: Dreyfus Model of Skill Mastery

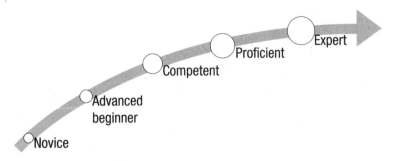

Commitment: Detached	Commitment: Detached	Commitment: Detached understanding and deciding; involved outcome	Commitment: Involved understanding; detached deciding	Commitment: Involved
Decisions: Analytic	Decisions: Analytic	Decisions: Analytic	Decisions: Analytic	Decisions: Intuitive
Perspective: None	Perspective: None	Perspective: Chosen	Perspective: Experienced	Perspective: Experienced

✓ Exam tip

The exam is unlikely to test the details of the Dreyfus model, so focus on grasping the big picture. For example, think of a learning curve you have experienced and try to identify how your skill level and point of view changed at each stage in this model. When did your decision making become intuitive? How did your level of commitment evolve over time from completely detached to fully involved? This will prepare you for any scenario-based questions about this model that you might encounter on the exam.

K&S Tuckman Model of Team Formation and Development

Agile methods use an approach called adaptive leadership in which leaders modify how they interact with team members based on the team's level of maturity, or stage of formation. We'll discuss adaptive leadership next, but before we can get to that, we need to examine the process of team formation and development that it is based on. The model of team formation we'll be describing originated with Bruce Tuckman. The four primary stages of his model are called Forming, Storming, Norming, and Performing. These are followed by a disengagement phase called Adjourning or Mourning, since people often miss being on a high-performing team after it's disbanded.[9]

Figure 4.6: Stages of Team Formation and Development

Forming
Storming
Norming
Performing
Adjourning

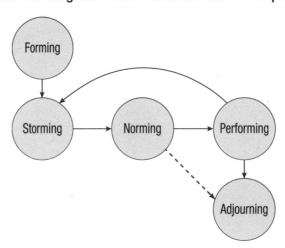

These stages are just what they seem—in Forming, people come together as a team. Then there is some turmoil, or Storming, as people learn to work together. In the Norming stage, the team normalizes, becoming comfortable in their roles and relationships. Eventually, they become a highly functional, or Performing, team that works effectively together.

As depicted in the above figure, a team may cycle through Storming, Norming, and Performing multiple times. This is likely to happen whenever there are changes to the project team, such as people leaving or joining the group. Each time, the team members have to reset and sort out their roles, relationships, and responsibilities within the new structure.

As a results-oriented team leader, I personally prefer another way to illustrate this model—from a performance perspective. As shown below, we need to get to the Performing phase to get the most out of the team.

Figure 4.7: Team Stages and Team Performance

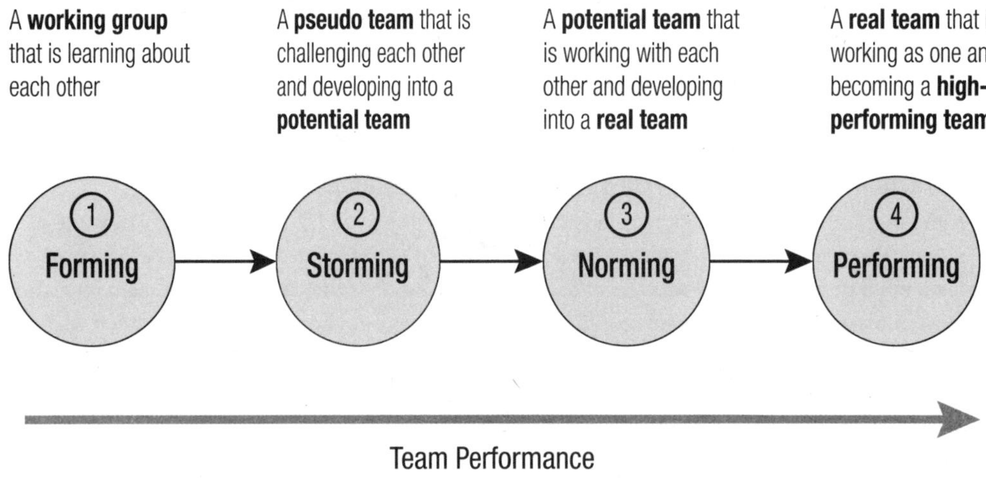

A working group that is learning about each other
A pseudo team that is challenging each other and developing into a potential team
A potential team that is working with each other and developing into a real team
A real team that is working as one and becoming a high-performing team
1 Forming
2 Storming
3 Norming
4 Performing
Team Performance

In this depiction of the model, a team starts in the lower right quadrant (Forming) as they learn about each other. Next, they move through Storming, where they challenge each other, and Norming, where they learn how to work together. Finally, they arrive at the Performing stage, where they can start working as one to reach their full potential. In other words, we typically start a project with a collection of people who are a "working group" rather than a real team. During Storming, we have a "pseudo team," which transforms into a "potential team" during Norming, and finally becomes a "real team" once they reach the Performing stage.

It seems so neat and clean to define the stages of team formation like this. Does that mean all teams go through these phases in a predictable way? No, each team is different. The people who make up the team and factors like whether any of the team members have worked together in the past will affect the way in which a team moves through the stages.

This brings us to another question: Do teams progress as a whole unit through the stages? Not really. People and teams are complex and messy. As leaders, the best we can do is be aware of these models, look for signs that the team is in a particular phase, and then act accordingly. These general pointers can be very useful, but we should never expect a team to proceed in an orderly fashion and follow the stereotypical stages.

T&T Adaptive Leadership

Now that we have discussed Tuckman's model, we can turn to the concept of adaptive leadership. As leaders, we serve the team best by adjusting our focus to meet team member's changing needs as they move through the stages of team development. One approach to this is the model of situational leadership developed by Ken Blanchard and Paul Hersey, who identified four leadership styles— Directing, Coaching, Supporting, and Delegating.[10] Tuckman's stages of development can be mapped to Blanchard and Hersey's styles of leadership as shown below.

Figure 4.8: Leadership Styles and the Stages of Team Formation

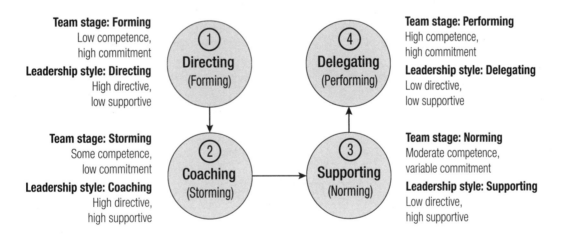

Team stage: Forming
Low competence, high commitment
Leadership style: Directing
High directive, low supportive

① **Directing** (Forming)

④ **Delegating** (Performing)

Team stage: Performing
High competence, high commitment
Leadership style: Delegating
Low directive, low supportive

Team stage: Storming
Some competence, low commitment
Leadership style: Coaching
High directive, high supportive

② **Coaching** (Storming)

③ **Supporting** (Norming)

Team stage: Norming
Moderate competence, variable commitment
Leadership style: Supporting
Low directive, high supportive

Starting in the lower right quadrant, we can see that Tuckman's Forming stage corresponds to Blanchard and Hersey's Directing style. Early in a team's formation, the role of the team leader is to directly help with project activities and present a clear, tactical picture of what needs to be done. The leader may also make a lot of requests like "Help me see it" or ask questions like "Where is the problem?" to assist team members in identifying and articulating the issues.

As the team passes into the Storming phase, there will generally be plenty of disagreement, open conflict, and harsh dialogue. During this stage, the leader needs to assume the Coaching style to help team members resolve conflicts without damaging relationships. Keep in mind that some conflict is good, so we don't want to mollycoddle the team too much. Let the disputes occur, but act as a referee or safety valve to ensure the conflict does not go too far.

When the team reaches the Norming phase, the members have successfully created rules (team norms) to help govern themselves. This doesn't mean the team leader can simply go into cruise control mode, however. Instead, the leader needs to play a Supporting role. The team will still need help with conflict resolution, as well as reminders to enforce the norms they have created. This is a good time for the leader to challenge the team with high-level goals such as "The team is responsible for tracking velocity on the project" or "Everyone owns testing." This is also a good time to tackle the issues raised at retrospectives.

The final Performing stage isn't a given—many, if not most, project teams never reach this phase because they go through too many changes (often initiated by the organization), which sends them back to repeat the Storming and Norming phases over and over again. Performing teams are autonomous, empowered, self-managing, and self-policing. They require little more than to be pointed in the right direction and given regular recognition and appreciation for their high performance. Blanchard and Hersey's Delegating leadership style for this phase means the leader brings work and challenges to the team for them to solve.

EXERCISE

In the columns below, match the Tuckman stage of team formation with the corresponding Blanchard and Hersey situational leadership style by drawing lines between the two columns.

Stages of Team Formation	Situational Leadership Styles
Forming	Supporting
Storming	Delegating
Norming	Directing
Performing	Coaching

ANSWER

The team formation stages and the adaptive leadership styles match up as follows:

Stages of Team Formation	Situational Leadership Styles
Forming	Supporting
Storming	Delegating
Norming	Directing
Performing	Coaching

(Forming → Directing; Storming → Coaching; Norming → Supporting; Performing → Delegating)

K&S **Team Motivation**

What motivates people goes far beyond paying them a salary. Salaries simply encourage people to show up to work every day. Once they are at work, a person's productivity level can vary from someone who undermines and is a net drain on the project to a critical contributor who brings passionate innovation to the organization, as shown below. Often, a person's position on this continuum is based on their level of motivation—so we can think of efforts to motivate people as the art of encouraging them toward the right-hand side of this range.

Figure 4.9: Continuum of Net Contribution

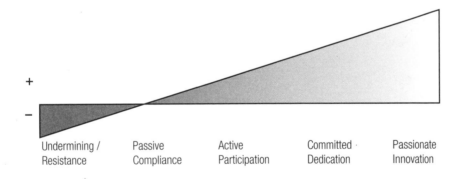

| + |
| − |

| Undermining / Resistance | Passive Compliance | Active Participation | Committed Dedication | Passionate Innovation |

In addition to motivating individuals, we can also approach motivation in terms of the entire team. Alistair Cockburn compares team motivation to the overall propulsion vector of team members in a raft. In other words, if the team members' individual motivations are personal and have no alignment toward the project goal, then the overall team vector (direction and speed) is likely to be small and not well directed toward the project goal, as shown below.

Figure 4.10: Team Members' Motivations
Aren't Aligned toward the Project Goal

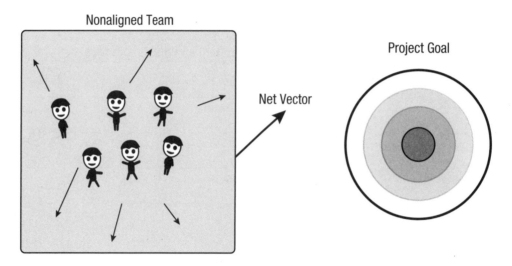

If, however, we can find a way to align the team members' personal goals with the project goal, we greatly magnify the project's net vector toward successful project completion, as illustrated below.

Figure 4.11: Team Members' Motivations
Are Aligned toward the Project Goal

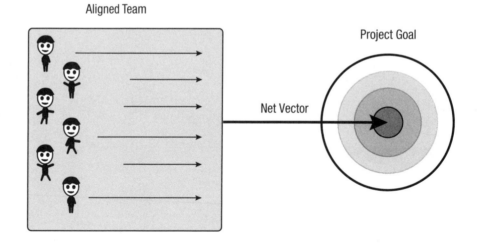

To achieve this alignment, we need to discover and fully understand the personal motivators of the team members and the motivators for the team as a group, using techniques like one-on-one interviews. Imagine, for example, that your team includes the following individuals:

» Bob is six months away from retirement and just wants his last months of work to go as easily as possible.
» Tim is mad because he wanted to use a particular technology for the website but the architecture group chose a different tool.
» Jane is looking for a new job because she was not given the team lead role that she wanted.

Once you've done the research and identified what motivates (and demotivates) these people, you can begin to determine if some elements of what they want can be worked into the project plan. For example, you might say, "Bob, I know you are retiring in six months, but wouldn't winning 'Project of the Year' be a great way to finish your career and be remembered? We have a good chance if we can pull this off." For Tim, the motivator may be, "I know we didn't select your preferred technology for the new site, but could we use it to build our help system and knowledge base? Can we put you in charge of that part of the project?" And to Jane you might say, "Jane, how about we make you team lead for the next two iterations with a review at the retrospectives? I know it is not the full-time role you wanted, but it will be good experience, and you can put it on your resume."

I've provided artificially simple examples here in the interest of keeping the descriptions short, but they help illustrate that most projects have opportunities for incorporating team members' personal goals into the project goals—even if it is only temporarily, for a couple of iterations. When we are able to align personal goals with project goals, people will see what's in it for them, and their motivation and productivity may increase substantially.

Candid conversations that explain why the project is important to the company can also help motivate the team. Try getting an executive or sponsor to outline for the team what success on the initiative means. Having a senior member of the organization communicate an inspiring vision of what the organization hopes the project will achieve is well worth the phone calls and e-mails needed to make it happen. Plus, having the team members know that someone who is significant to them cares if the project is a success is also a powerful motivator.

The final point I'd like to make regarding team motivation is that, while it is good to understand and incorporate individual motivators, we want to make sure we cement company and team objectives over individual objectives. It is important to do things like celebrating victories as a team and coming up with whole-team rewards, rather than giving individual rewards for discrete pieces of project work. We need to promote the "mutually accountable" aspect of teamwork.

K&S Training, Coaching, and Mentoring

The exam content outline includes the toolkit item "training, coaching, and mentoring." Since these terms are often used interchangeably, let's take a closer look and see what they mean.

» **Training** is the teaching of a skill or knowledge through practice and instruction. The agenda is usually created by the trainer, and the format is very structured (and typically prepared in advance). An example of training would be attending a one-day course on agile planning techniques.

» **Coaching** is a facilitated process that helps the person being coached to develop and improve their performance. Coaching often starts with an initial topic that is then expanded and developed in collaboration with the coach. Coaching sessions are usually scheduled in advance and have a defined structure. An example of coaching would be arranging a session with the ScrumMaster to get help with project reporting.

» **Mentoring** is more of a professional relationship than a specific activity. The mentor can be a sounding board for tackling issues on an as-needed basis. The mentee owns the agenda, and the format is free-flowing. An example of mentoring would be having a relationship with an experienced agile leader who you meet with from time to time, bounce ideas off, and ask for advice.

In the context of an agile team, coaching is the most common of these activities. The role of the ScrumMaster, team coach, or team leader includes serving as a coach to the members of the delivery team as needed to improve their effectiveness and keep the project on track. So let's take a closer look at coaching an agile project team.

Coaching

The goal of coaching is to help the team members stay on track, overcome issues, and continually improve their skills. Coaching is done at two levels—the team and the individual team members. The figure below shows how the emphasis switches from coaching the team as a whole at the iteration boundaries to coaching individual members during the iterations:[11]

Figure 4.12: Coaching at the Team and Individual Levels

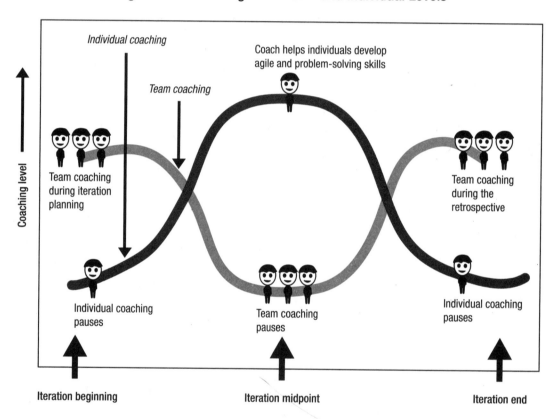

Team coaching happens more at the iteration boundaries because that's when the team assembles for events like iteration planning, iteration reviews, and retrospectives. These are great times to make sure the team's agile practices are working well and help the team embody their agile roles. Any process changes that arise from coaching the team should be made between iterations rather than in the middle of an iteration, because the iteration is the team's dedicated period for completing the work in a stable environment.

While the iteration is in progress, it's a good time to provide coaching to individual team members. Typically, a team member will have a problem or complaint, and ask for coaching or help. The coaching sessions should be one-on-one, confidential meetings in a safe, private environment. During the conversation, it's important to be frank, yet remain positive and respectful. After the meeting, the coach might need to follow up to make sure the issue has been resolved, or is at least getting better.

Lyssa Adkins outlines the following guidelines for one-on-one coaching:[12]

» **Meet them a half-step ahead.** Don't try to push people directly to the end point. Instead, coach them so that they move toward the end goal and take the next step from where they are now. For example, let's say our aim is to have the team members select their own work tasks, but they aren't there yet. Instead of just telling them bluntly that agile teams self-select their work, try to get them halfway there by asking them questions like, "Who can do this one?"

» **Guarantee safety.** Tell the team members that all coaching conversations will be kept confidential, and then make sure they are. People will be more willing to contribute and share if they aren't afraid that their concerns will be repeated out of context.

» **Partner with managers.** Often the team members' functional managers are not on the project team. These managers might not be using, or even aware of, agile methods. Because of this, the way the team members are measured in their functional roles might not match agile values. We need to partner with functional managers to align everyone's goals and ensure that the team members' project contributions are reported appropriately to their functional managers.

» **Create positive regard.** We might not personally like every individual we coach, but we do have to help them equally. If we dislike someone, this sentiment can show through, often in subtle ways. Therefore, it's important to cultivate a genuine compassion for others and a desire to help everyone improve in their roles.

Creating Collaborative Team Spaces

Now that we've explored various aspects of building agile teams, let's move on to examine how the physical location of the team members affects team performance. The location of the delivery team and the characteristics of their work environment have a significant impact on their capacity for collaboration and communication, and therefore their ability to use agile practices effectively.

Although agile methods promote using co-located teams and creating a collaborative team space, this approach also comes with challenges. Here we'll look at some important aspects of team location you should understand for the exam, including some of the challenges involved, and ways to address them.

Exam tip

All the ideas covered in this section on collaborative team spaces (co-location, team space, osmotic communication, distributed teams, etc.) are important agile concepts. You are likely to encounter multiple exam questions that test your understanding of these concepts, either directly or indirectly.

K&S Co-Located Teams

Since agile methods recommend face-to-face interaction as the preferred means of communication, it is no surprise that they also recommend that teams be *co-located*—in other words, that all the team members work together in the same location. It isn't enough to simply locate all the team members in the same city, or even on the same floor of the same building. An agile team is only considered to be co-located if all its members are working within 33 feet—10 meters—of each other, without any physical barriers such as walls or doorways between them.

As we'll see, co-locating the team has many advantages—however, it isn't always possible. Many agile projects use distributed teams in which the members are geographically dispersed, whether in different buildings, cities, or continents. In that case, the team will try to use digital tools to duplicate the effects of physical co-location as much as possible. This is known as virtual co-location. In this section, we'll talk about tools and best practices for both co-located and distributed teams—in other words, for physically and virtually co-located teams.

> ### ✓ Exam tip
>
> In this discussion, we'll use the term "co-location" to refer to teams located in the same physical space, unless otherwise specified. However, for the exam, you might see references to virtual co-location as well as physical co-location. Also, although the exam usually doesn't test specific numbers, you might need to know the definition of physical co-location given above: *all team members working in the same space, within 33 feet (10 meters) of each other, with no physical barriers between them.*

T&T Team Space

We've said that co-located team members need to actually be seated in the same *space*. This is their "team space" (or "war room")—the common area for collaboration and information sharing where they conduct their everyday work.

Agile teams often commandeer a large open area, like a conference room, to serve as their team space. In this area there should be plenty of wall space for whiteboards to be used during collaborative discussions, as well as room to post information radiators of the project metrics. To take full advantage of the benefits of co-location, the team space should also be supplied with the following tools and equipment:

» Whiteboards and task boards
» Sticky notes, sticky paper, flip charts
» Round table with screen/laptop
» Video conferencing capability
» No barriers to face-to-face communication
» Food, snacks, and toys!

The rationale for most of these items should be clear by now. For example, we've already discussed the use of whiteboards, sticky notes, information radiators, and face-to-face communication in talking about agile tooling and communication, and we'll be talking about the use of digital tools below. However, there are two other elements of agile team spaces that may be less familiar—"caves and common" and "tacit knowledge."

Caves and Common

One of the challenges of working in an open area without barriers is the lack of privacy and quiet time. To address this, agile team spaces follow a "caves and common" model. Most of the work is done in the large "common" area, where the team members work together as a group. However, they also have access to private spaces, called "caves," where they can go to make private calls, have one-on-one conversations, or work on their own for short periods of time. So if a team member is having difficulty concentrating in a noisy environment, he or she can retreat to a "cave" to work in quiet isolation for a short period.

Tacit Knowledge

"Tacit knowledge" is the unwritten information that is collectively known by the group, such as how to restart the printer when it jams. Although those instructions might not be written down anywhere, if everyone is working in the same space and has seen and heard people do it before, this knowledge will become shared and supported by the team. The benefit of tacit knowledge is that it leads to time savings since everyone on the team has a wide base of shared information.

Tacit knowledge works best for small, co-located teams—in fact, the emphasis on facilitating and maintaining tacit knowledge is one of the reasons why agile methods recommend limiting teams to 12 or fewer people. As teams grow in size, it becomes more difficult to maintain face-to-face communication with everyone, and tacit knowledge begins to break down. A larger team needs to commit more things to writing to keep everyone informed. Since it takes more time to create written documentation, some of the time savings of small team collaboration are lost.

T&T Osmotic Communication

Osmotic communication refers to the useful information that flows between team members who are working in close proximity to each other as they overhear each other's conversations. For example, let's say that Bob asks Jim how to restart the build server. Mary is just about to use that machine and overhears their conversation—as a result, she can intervene on the restart and save a potential conflict. This ability to pick up on things that would otherwise be missed is one of the major benefits of co-locating the team, saving them time and improving their teamwork.

Alistair Cockburn likens osmotic communication to energy fields that radiate from people. If you're too far away, you won't receive its benefits—you can only get the full benefit if you are in close proximity with no barriers, as shown below. In other words, to improve their osmotic communication we want to get people sitting and working closely together with few barriers between them.[13]

Figure 4.13: The Effect of Proximity on the Flow of Osmotic Communication

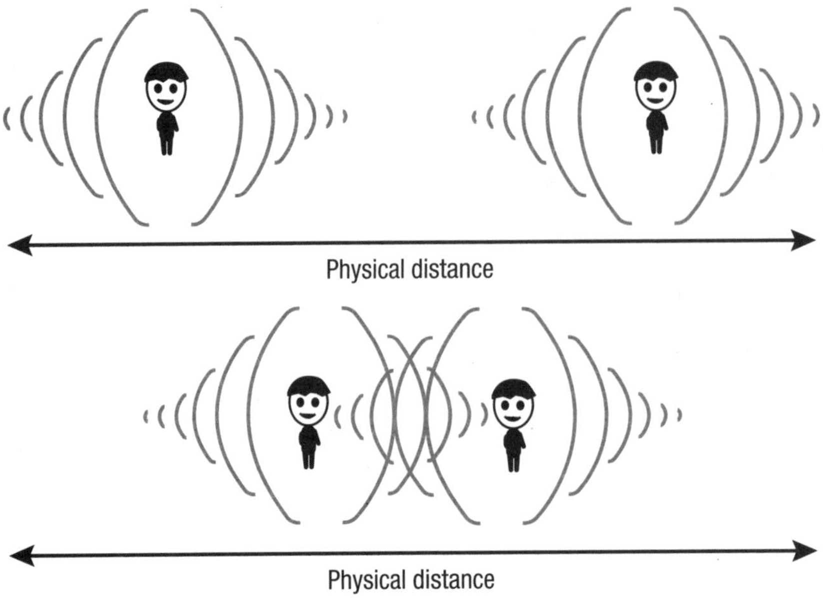

Physical distance

Physical distance

K&S Global, Cultural, and Team Diversity

As our communication options expand and become less expensive, and as our partner and customer bases widen, globalization, culture, and team diversity are becoming increasingly important factors for project teams. It's not uncommon to have team members from three or four continents working on the same project. Although different cultures bring challenges, they can also bring more efficiency to a project because there is a broader pool of resources to choose from.

Since knowledge work is invisible, agile teams rely on coordination and communication to share information. However, globally distributed teams face special challenges in this area that need to be addressed:

» **Different time zones**: Team members may be working at different times of day, which can impede their ability to get to know each other and achieve unity. If this is the case, we may need to spend more time coordinating their communication.

» **Different cultures**: Many societies are more hierarchical than Western culture, and they may consider it rude to make suggestions, give candid feedback, bring up problems, or disagree openly. In this case, we may need to explicitly encourage this behavior and explain that this kind of transparency is required for agile to work smoothly.

» **Different communication styles**: Different cultures may have very different communication norms. To address this, we may need to train the entire team about each other's communication styles. We can also watch out for misunderstandings, inappropriate jokes, culturally sensitive terminology, and clashing assumptions about gender roles.

» **Different native languages**: Different languages may lead to miscommunication about the work being done. To address this, we may need to take steps to ensure a shared understanding of the acceptance criteria and confirm that everyone on the team has the same understanding of what needs to be done. We may also need to provide more detailed instructions and documentation.

Even when people from different countries share the same native language, there are still differences to take into account. As an example, when I first made the transition from the United Kingdom to Canada, I discovered surprising differences in culture, expressions, and terminology, despite the supposedly common language. For instance, I remember someone saying to me, "We'll need biweekly status reports." Thinking that "biweekly" meant twice a week, I replied, "Okay, can we make that every Wednesday and Friday?"

Team member:	"Oh no, every two weeks."
Me:	"Wait, you mean every fortnight?"
Team member:	"What's a 'fortnight'?"

When conducting retrospectives, we should keep the team's cultural differences in mind. For example, people from some cultures might not feel comfortable raising problems with the current process in the presence of their managers. In such instances, it can be helpful to rephrase "What did not go well?" to "What areas could use improvement?" It can also be helpful to leave retrospective questions open at the end of the meeting so that people can respond to them in a private meeting room, via e-mail, or by filling out an anonymous feedback form.

Of course, diversity issues may need to be addressed on co-located teams, too. In this case, the XP practice of pair programming can be used. (On a non-IT project, developers can be paired off to complete specific tasks together, even if they aren't working together simultaneously, as in pair programming.) Working together on a task gives team members a great opportunity to get to know each other better, work as a unit, and become comfortable handling conflict, raising problems, communicating progress, and brainstorming.

Distributed Teams

Distributed teams are teams that have at least one team member working off-site. We've said that agile recommends that teams be co-located, and so far we've been talking about co-located agile teams as if they were the norm. Although this might have been true at one time, today the majority of agile teams are distributed. For example, in 2015 the Scrum Alliance surveyed 5,000 Scrum practitioners from 108 countries and across more than 14 industries. The resulting report ("The 2015 State of Scrum Report") found that only 26 percent of the respondents were working on a co-located team.[14]

So rather than being the exception, distributed agile teams are the norm. One reason distributed teams are becoming more common is that the Internet offers communication tools and reduced communication costs that make geographically distributed teams not only possible, but also cost-effective. As projects get bigger and require more specialized resources, it is inevitable that project teams will increasingly be distributed. And Highsmith asserts that agile methods work better for distributed teams than non-agile approaches because of the following factors:[15]

» The short iterations used in agile development force continuous close collaboration and coordination.
» The project will be easier to control because a releasable product is built each iteration.

In other words, if you have to manage a distributed team, it is best to use an agile approach, because the frequent feedback is more helpful than just piling on documentation to keep the team on track.

Exam tip

For the exam, you should understand that a distributed team is not the same as outsourcing. As Jim Highsmith points out, there is an important difference between distributed and outsourced projects. He explains, "Distributed projects basically have multiple development sites that can span buildings, cities, or countries. Outsourced projects involve multiple legal entities, therefore contracting, contract administration, and dealing with different development infrastructures are added to the team's workload."[16]

One aspect of distributed teams to keep in mind is that the team formation phases of Storming and Norming are more difficult when team members are not co-located. Some people are more likely to just disengage from an e-mail or VoIP debate, rather than stand up for their viewpoint the way they would if they were in the same room as the other person. If that person is out of sight, they find it easier to just dismiss them and their crazy ideas. On the other hand, certain people are in their element in a virtual debate, perhaps because they feel protected by distance and separation. (When you actually meet these people, you may be surprised at how very agreeable and meek they are in person.)

Storming and Norming are critical for helping teams build commitment to their decisions and results. Therefore, leaders of distributed agile teams need to ensure that there are enough debate and collective decision making early in the project for the team to fully work through these stages. This may mean introducing controversial or difficult pieces of work earlier in the project, just to get the team talking and working through the issues.

One best practice that will help the members of a distributed team communicate better is to start by holding a face-to-face kickoff meeting. It is so much easier to e-mail, call, or instant message someone whom you have already met in person. It's even better if we can extend this face-to-face kickoff event so that team members can work together for the first one or two iterations before they return to their respective countries, as shown below.

Figure 4.14: Bring Geographically Dispersed Teams Together Early to Improve Communications Later

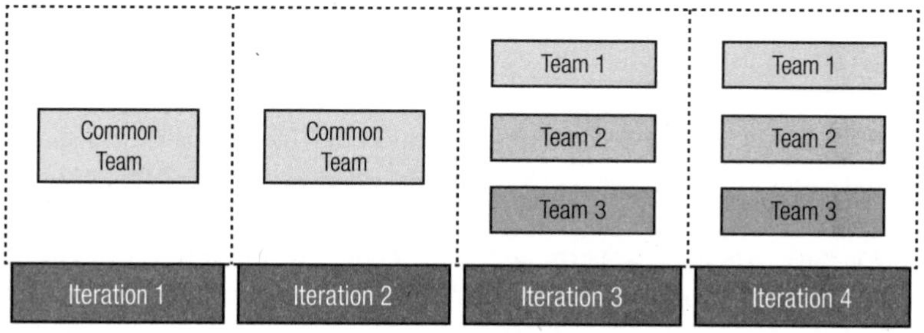

		Team 1	Team 1
Common Team	Common Team	Team 2	Team 2
		Team 3	Team 3
Iteration 1	Iteration 2	Iteration 3	Iteration 4

This model can be further extended by arranging for face-to-face release and planning meetings as the project progresses. Other models that work well include a rotating secondment (temporary assignment) of team members between locations to give the team members in each region an opportunity to work in person with people from the other teams and experience their cultures.

Background information

Tips for Managing Distributed Teams

In *Collaboration Explained*, Jean Tabaka offers the following tips for managing distributed teams:[17]

» **Maintain a metaphor**. Metaphors can help the team stay focused on the project mission or vision. For example, a metaphor for a project to build a security system may be "We are building the Great Wall of China."

» **Apply frequent communications**. When team members aren't in close proximity to each other, adding in more scheduled communications may help. For example, a distributed team may have two stand-up meetings a day plus scheduled one-on-one calls to compensate for the lack of spontaneous communications.

» **Intensify facilitation**. This may mean asking more questions, repeating responses more frequently when on conference calls, and working to keep everyone engaged.

» **Follow best practices for conference calls**. It is also important to keep conference calls effective and productive so people are willing to attend and contribute. Here are some basic guidelines to follow when facilitating conference calls:

  - *Keep on track*: There should be no fuzzy agendas.
  - *Keep on time*: As a general rule, keep calls to a one-hour limit.
  - *Keep track of who is on the call*: This can be done by maintaining a seating chart.
  - *Keep the decisions flowing*: Sending out agendas in advance can help achieve this goal.
  - *Keep the answers coming*: To do so, we need to engage participants with questions.
  - *Keep it fair*: This means we need to maintain fair telephone control.
  - *Keep it facilitated*: In other words, don't take control of decisions.
  - *Keep it documented*: Not only should we document the conference call, but we should also send feedback promptly.

Digital Tools for Distributed Teams

Distributed stakeholders present a challenge to both face-to-face communication and agile tools, such as information radiators, that rely on co-location to work best. However, this is a challenge that most projects have to deal with, since the majority of agile teams have at least one team member in a different location than the rest of the team.

The big challenge for distributed agile teams is finding ways to create "virtual co-location"—in other words, trying to replicate the benefits of face-to-face collaboration, osmotic communication, tacit knowledge, and improved relationships that come from working in close proximity to each other. Luckily the same factors that are making distributed teams more common are also providing tools that can help simulate the benefits of face-to-face collaboration. There are a host of electronic tools and gizmos that distributed teams can use to mimic the low-tech, high-touch tools favored by agile teams and establish osmotic communication. Let's look at some examples of such tools.

» **Videoconferencing, live chat, Skype**: These tools can be used to simulate a shared team environment and allow distributed stakeholders to chat and interact as if their colleagues were within earshot. Videoconferencing can be especially useful for stand-up meetings and retrospectives to provide a visual presence for distributed team members.

» **Interactive whiteboards**: These tools allow team members to share content with multiple locations and collaborate in a visual whiteboard-type environment that is much richer than a telephone conversation.

» **Instant messaging (IM) and VoIP (Voice over Internet Protocol) headsets**: These tools can make it seem like people halfway around the world are much closer. With VoIP headsets, for instance, team members can chat away with each other as if they were in the same (dark) room.

» **Presence-based applications**: These applications build on and extend IM capabilities by managing the "currently online" status of participants to create a virtual office environment for sharing information. These tools usually offer document and file management services, as well as a rudimentary project plan integration capability, to help team members collaborate.

» **Electronic task boards and story boards**: These tools can be used by people in multiple remote locations to mimic co-located information radiators.

» **Web-based meeting facilitators**: These can help keep track of team members and provide a central hub for connection and information.

» **Survey applications**: Online surveys can be an effective way of polling team members in real time or off-line to get answers to questions and opinions.

» **Agile project management software**: These tools can help with backlog creation and prioritization. They also produce velocity-tracking statistics and help the team own the process of tracking velocity.

» **Virtual card walls**: These virtual walls can mimic physical card walls, but they are accessible anywhere to team members with an Internet connection.

» **Smart boards**: These tools are great for capturing the output of design sessions without having to spend extra time creating formal documents.

» **Digital cameras**: Taking a digital picture is another way to quickly capture whiteboard sessions without the need to spend time polishing designs.

» **Wiki sites, document management tools, and collaboration websites**: Such tools allow stakeholders to create, view, amend, and discuss project elements.

» **Automated testing tools, automated build tools, and traffic-light-type signals**: These tools are typically used on software projects to show the status of the build. For example, if the build is stable and all tests are running, the tools show a green light. If some code has been checked in that breaks the build, they show a red light.

» **CASE tools**: These tools are now being used to reverse-engineer documentation from source code and data models. So rather than spending valuable time keeping systems documentation up to date, teams can now use CASE tools every week to perform tasks like reverse engineering new database models, depth-of-inheritance metrics, and code coverage statistics from the evolving system. This frees up developer time to focus on adding new features, rather than maintaining documentation.

Tracking Team Performance

In this final section of the chapter, we'll examine how agile teams monitor their progress and performance. This discussion is closely related to the concepts we'll cover in the next chapter, where we'll be talking about agile planning in detail. That's because the key tools (burn charts) and the metric (velocity) commonly used for tracking team performance are also used for planning and estimating. For the exam, you'll need to have a good grasp of these important agile tools.

Burn Charts

We've seen that agile teams rely on low-tech, high-touch tools displayed on highly visible information radiators—and burn charts are one of the most common tools displayed in this way. Burn charts are important because they make the team's progress visible at a glance and make it easy to forecast when the project (or a release within the project) will be complete.

There are two kinds of burn charts—burndown and burnup. Burndown charts show the estimated effort remaining on the project, and burnup charts show the features that have been delivered already. Both kinds of burn charts are easy to create in Microsoft Excel®.[18]

T&T Burndown Charts

A burndown chart tracks the work that remains to be done on a project. As work is completed, the progress line on the chart will move downward, reflecting the smaller amount of work that still needs to be done. The most common use of burndown charts is for measuring the team's progress in completing the project work.

Here's an example of a typical burndown chart:

Figure 4.15: Burndown Chart Using Time to Measure Progress

XYZ Project - Estimated Effort Remaining

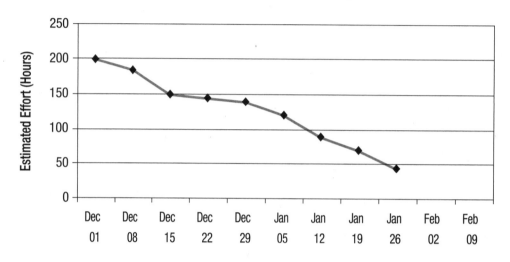

Looking at the above chart, it's easy to see at a glance that this project will be done in early February, since that's when the average slope of the progress line will reach 0 hours of estimated effort remaining. Typically, burndown charts show either the estimated time remaining, as shown here, or estimated story points remaining. (A story point is basically a unit of work that has been defined by the team, as we'll explain in chapter 5.)

Below is an example of a burndown chart that uses story points.

Figure 4.16: Burndown Chart (Story Points) with Ideal Progress

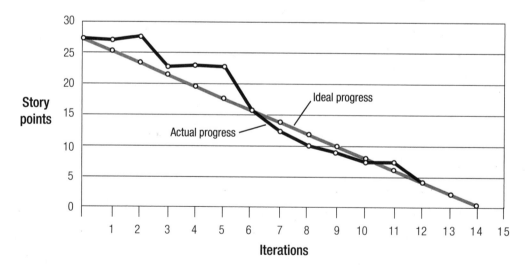

On this chart, the team has also forecast (in green) the effort they plan to do to complete the project on time—and we can easily see that they are on track to reach the target release date. That's because their average progress to date, as shown by the black line, is getting closer and closer to their ideal progress, as shown by the green line.

Although burndown charts allow us to quickly project when the work will be done, they do have one drawback—they make it hard to separate the impact of scope creep from the team's progress. If we look again at the first example above (figure 4.15), we can see that progress was slow (the line was flatter) between December 15 and December 29. This slowdown could have had several causes: people taking vacations over the holidays, the customer adding more scope, or the team discovering more tasks that needed to be done. We cannot discern the reason for the slow progress by looking at this chart.

T&T Burnup Charts

Burnup charts track the work that has been completed. Therefore, over the course of the project, the progress line on a burnup chart will move upward, showing the increasing amount of work that has been completed. The big advantage of using a burnup chart is that it can show changes in scope, making the impact of those changes visible.

Here's an example of a typical burnup chart:

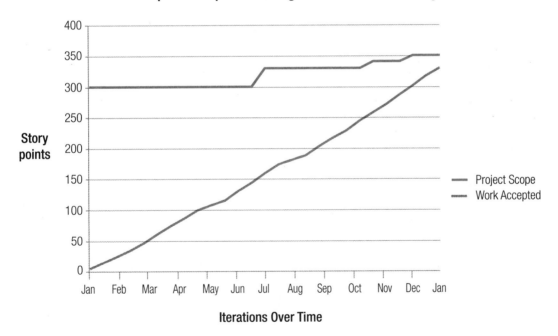

4.17: Sample Burnup Chart Using Points to Measure Progress

On this chart, we can see that the total scope started at 300 points in January; however, it increased to 330 points in July, and eventually reached 350 points at the end of the year. If the data in this chart was converted to a burndown chart, we would see areas where the rate of effort line would flatten—but we wouldn't be able to tell whether that flattening was due to an increase in scope or slower progress by the team. By separating progress and scope, burnup charts offer additional insight into the status of the project.

The burnup chart above shows work completed (or accepted). But what about work in progress? We can add work in progress to our burnup chart to track both work started *and* work completed. When we do that, we create a cumulative flow diagram (CFD), as discussed in chapter 2. Here is an example of a CFD that shows work completed, in progress, and not started:

Figure 4.18: Cumulative Flow Diagram

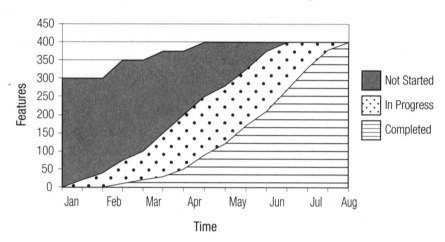

On this chart, we can see how the total scope (the top line) increased from 300 features to 400 during the project. We can also see the work in progress (the dotted section) and the completed work (the striped section). The team's true rate of progress is the gradient of the line at the top of the "Completed" section, which tracks the completion of features.

T&T Velocity

Velocity is defined as the "measure of a team's capacity for work per iteration." This powerful metric allows the team to gauge how much work they will be able to do in future iterations, based on the amount of work they completed in past iterations. This provides a way to track and communicate what they have accomplished, anticipate what they will be able to accomplish in the future, and forecast when the project (or release) is likely to be done.

In this discussion we'll talk about velocity in terms of story points. However, velocity is measured in the same units the team uses to plan the work, which could be anything—story points, user stories, hours, days, or even jelly beans! (Yes, jelly beans are actually used on many agile projects.)

Here's an example of tracking a team's velocity across iterations:

Figure 4.19: Velocity Tracking Chart

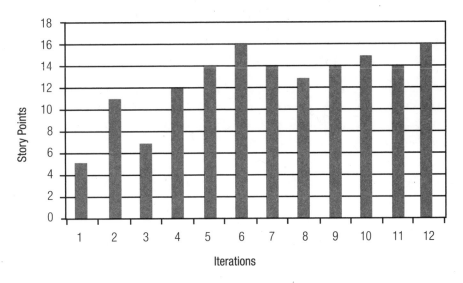

Velocity usually varies the most in the first few iterations and then begins to stabilize. This is because the team has to get used to working together, familiarize themselves with the project tools, and get comfortable interacting with the project stakeholders.

While it may seem logical to predict ever-increasing velocity as the team gains experience, velocity does typically plateau. One reason for this is that, as the product gets bigger, there is more to maintain, refactor, and possibly support if early versions of the product have been deployed. In general, knowledge work projects tend to increase in complexity as the work is being done.

Figure 4.20: Velocity Tends to Stabilize as the Project Progresses

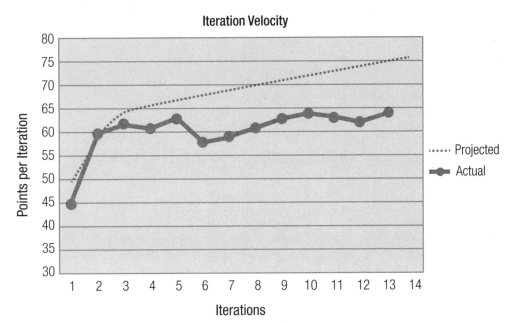

The fact that velocity tends to stabilize over time makes it a very powerful tool for planning and estimating. Once a team has tracked their velocity over multiple iterations, they can use their average velocity to estimate when the next release or project will be done. For example, let's say there are 50 story points of undeveloped work remaining in the backlog for the first release. The team has already completed ten iterations for that release, and they have been averaging about 20 story points per iteration. We can divide the remaining work by the team's average velocity per iteration (50 / 20 = 2.5) to estimate that it will probably take the team three more iterations to complete the release.

This kind of long-term track record allows an agile team to make pretty reliable estimates about how long it will take them to complete a given chunk of work. That's why velocity is the preferred metric for estimating and planning upcoming releases and iterations. (We'll delve more deeply into this topic in chapter 5.)

EXERCISE

Your team's velocity has been 40, 48, 52, 59, and 51 points per iteration since starting the project. Your backlog of remaining work has 600 points of functionality in it. Your sponsor wants to know when you expect to be done. How many more iterations of work will be required to complete the backlog?

40 + 48 + 52 + 59 + 51 = 250 / 5 = 50 sp velocity

$$\frac{600}{50} = 12 \text{ iterations}$$

ANSWER

You first need to calculate the average number of points per iteration:

(40 + 48 + 52 + 59 + 51) / 5 = 50

You then divide the remaining work by the average points per iteration:

600 / 50 = 12

Therefore, the backlog will likely require 12 more iterations of work to complete.

Chapter Review

1. As the product owner, what would you focus on?

    A. Facilitating the retrospectives and planning meetings
    B. Acting as servant leader to the team
    C. Organizing the development work
    D. Maximizing the value of the product

2. What would be a step forward in your team's evolution?

    A. From proficient to competent
    B. From Forming to Storming
    C. From Ha to Shu
    D. From self-organized to empowered

3. If it isn't possible to locate all your team members in the same room, they are likely to experience:

    A. Higher levels of conflict
    B. More privacy
    C. Less difficulty reaching convergence
    D. More communication challenges

4. As an agile team leader, you want to avoid:

    A. Prioritizing team goals over individual goals
    B. Providing rewards for expected behavior
    C. Finding out what motivates the team members individually
    D. Rewarding individual goals at the expense of project goals

5. What would a member of an agile delivery team focus on the most?

    A. Selecting which user stories to include in the product
    B. Building the product increment
    C. Communicating the project vision
    D. Testing the product increment to determine if it is done

6. If the team leader is effective, we would expect to see an agile team:

    A. Collaborate smoothly and harmoniously without disagreements
    B. Disagree with each other frequently when they are first learning to work together
    C. Learn to suppress their disagreements for the good of the project
    D. Develop stronger opinions and disagree more as the project progresses

7. What question would the sponsor of an agile project focus on?

    A. Will the project deliver the expected value on time and on budget?
    B. Does the team understand the project vision?
    C. Is the ScrumMaster prioritizing the features accurately?
    D. Does the product owner understand the end user's requirements?

8. What is the ScrumMaster responsible for?

    A. Directing and organizing the team
    B. Guiding the team's agile processes
    C. Prioritizing the user stories
    D. Managing the project

9. Your team needs to keep the product owner and other stakeholders informed about how the project is progressing. What tools will you use for this?

    A. Velocity chart and risk burndown graph
    B. Project roadmap and story maps
    C. The team's task board with WIP limits
    D. Prototypes, personas, and wireframes

10. At what team formation and development phase is conflict likely to be highest?

    A. Forming
    B. Fuming
    C. Storming
    D. Debating

11. Which of the following statements is true for using team velocity as a progress metric?

    A. Velocity is not accurate if there are meetings that cut into development time.
    B. Velocity measurements are disrupted if some project resources are part-time.
    C. To track velocity accurately we can't have any scope changes during the project.
    D. Velocity measurements account for work done and disruptions on the project.

12. Which of the following needs isn't relevant to designing agile team spaces?

    A. Allow team members to retreat to a cave for some quiet work.
    B. Help everyone understand how to operate and reset the printer.
    C. Let team members pair up at the same desk to work together.
    D. Let the sponsor easily follow the team's progress.

13. Self-organizing teams are best characterized by their ability to:

    A. Do their own thing
    B. Sit where they like
    C. Make local decisions
    D. Make project-based decisions

14. You have been assigned to lead a distributed agile team. To help them communicate, the best option would be to:

    A. Ask the team members to share photos of themselves.
    B. Require a common language for all project communications.
    C. Set up some initial face-to-face meetings for everyone to meet each other.
    D. Define common working hours so everyone can better communicate.

15. Agile team members are more effective at solving problems when they:

    A. Are motivated by rewards and punishments
    B. Are in the Storming stage of development
    C. Feel they have permission to make mistakes
    D. Are allowed to work on their own without anyone telling them what to do

16. Cockburn's Shu-Ha-Ri model tells us why:

    A. Agile teams that are empowered to make their own local decisions are more effective.
    B. Team members tend to argue a lot before they learn to work well together.
    C. Leaders need to understand how teams develop to lead them effectively.
    D. Teams new to agile should follow a method that has already been tested by others.

17. High-performing agile teams feature which of the following sets of characteristics?

    A. Consensus-driven, empowered, low trust
    B. Self-organizing, plan-driven, empowered
    C. Consensus-driven, empowered, plan-driven
    D. Constructive disagreement, empowered, self-organizing

18. Another agile team leader in your organization comes to you for advice. She's having trouble getting her team to take ownership of the project and get comfortable selecting the work to be done. She keeps finding herself making the decisions and directing their work. What do you advise her to do?

    A. At the stand-up meetings, assign a different person each day as the decision maker to get them all comfortable in the role.
    B. Implement an incentive plan and officially report any lack of participation to team members' functional managers.
    C. Meet them halfway and work with their functional managers to align each team member's goals with the project goals.
    D. Explain to them that agile teams self-select their work and tell them to get on with it.

19. The primary reason constructive disagreement is valued on high-performing agile teams is to:

    A. Weed out the weak.
    B. Test requirements for robustness.
    C. Generate buy-in for decisions.
    D. Build negotiation skills.

20. Osmotic communication works best when:

    A. The team members understand each other's cultures.
    B. The team is in the Performing stage.
    C. The team members' goals are aligned with the project goals.
    D. The team members work near each other.

Answers

1. **Answer**: D

   Explanation: The product owner's primary responsibility is maximizing the value of the product. It is the team coach or ScrumMaster who acts as a servant leader to the team and is most likely to facilitate the team's retrospectives and planning meetings. Agile team members organize their own work.

2. **Answer**: B

   Explanation: Although the transition from Forming to Storming might not always feel like a step in the right direction, this is the only one of the options that would take a team one step forward in their evolution. The steps "Shu" and "competent" precede the steps "Ha" and "proficient," rather than following them. Being self-organized is an aspect of being empowered, not an evolutionary step forward.

3. **Answer**: D

   Explanation: Team members who aren't co-located aren't likely to have higher levels of conflict since distance can make it easier to ignore disagreements (making it more difficult to complete the Storming stage). And since shared agreement (convergence) requires the team to first fully debate the various options (divergence), true convergence is also more difficult to achieve when the team isn't co-located, rather than less. If team members don't work in the same room, it doesn't necessarily mean they will have more privacy; it just means they won't all be together. The most likely outcome of this scenario is more communication challenges.

4. **Answer**: D

   Explanation: As an agile team leader, it might be helpful to focus on any of the activities listed here except for rewarding individual goals at the expense of project goals. An effective team leader understands the team members' individual goals and leverages them for the good of the project, rather than the other way around.

5. **Answer**: B

   Explanation: The best option here is "building the product increment." Selecting which features to include in the product is done by the product owner, not the team. (Although the team does decide which user stories can be completed in a given iteration, the question doesn't mention iteration planning.) Also, although the team tests the product increment, they can't decide if it's "done," since that's up to the customer. Finally, the team members are responsible for implementing the project vision, not for communicating it.

6. **Answer**: B

   Explanation: When agile team members are first learning to work together, we expect to see them move through the Storming stage, where they will disagree with each other frequently. Notice that the reference to the team leader's effectiveness is something of a distractor. Only one of these options would be expected to occur on an agile project—and that is a natural process of team formation, not based on the team leader's effectiveness.

7. **Answer**: A

   Explanation: The sponsor is focused on the big picture of value delivery; this person is responsible for ensuring that the project will deliver the expected value on time and on budget. Typically, it is the product owner who communicates the product vision to the team. The product owner is also responsible for ensuring that the end user's requirements are met, so if they don't know those requirements, they aren't doing their job; this isn't something the sponsor should be worrying about. Finally, the team's coach or ScrumMaster doesn't prioritize features; that is done by the product owner, customer, or value management team.

8. **Answer**: B

   Explanation: Agile teams direct and organize themselves. It is the product owner who prioritizes the user stories, and managing the project isn't one of the responsibilities of the ScrumMaster. That rules out all the options except for guiding the team's agile processes, which is one of the ScrumMaster's key responsibilities.

9. **Answer**: A

   Explanation: Although all the tools listed can be used to communicate project information to stakeholders, only velocity charts and risk burndown graphs track progress over time. Velocity charts show how much work the team has completed in each iteration, and risk burndown graphs show how well the team is managing the project risks.

10. **Answer**: C

    Explanation: The stage where conflict is highest is the Storming phase. When the team is in the Forming stage, they are still getting to know each other. Fuming and debating are not team formation and development phases.

11. **Answer**: D

    Explanation: A team's velocity measures the actual work done (and therefore the team's capacity). Therefore, this metric includes all the interruptions and anything else that has occurred on the project, including the meetings, part-time resources, and scope changes that are common on projects.

12. **Answer**: B

    Explanation: Agile team spaces are designed to incorporate the concepts of caves and common, collaborative workspaces, and information radiators of the team's progress. Although the team's space may include a printer, the instructions for using it aren't relevant for designing the space.

13. **Answer**: C

    Explanation: This question can be confusing because teams may have the opportunity to make some project-based decisions, but remember to always look for the best choice in the options presented. Self-organizing teams primarily have control over local decisions related to the project execution. For example, they may decide what to do next and how to solve a technical problem. Sponsors typically make external decisions, such as increasing the budget or extending the schedule. Doing their own thing or sitting where they like may or may not occur, but these choices are not readily associated with characteristics of a self-organizing team. Therefore, choice C is the best option.

14. **Answer**: C

    Explanation: If possible, setting up some initial face-to-face meetings for everyone to meet is an effective way of improving remote communications later in the project. Once people have met face-to-face, it is generally much easier to follow up with e-mail, phone calls, etc. Defining common working hours or a common language might appear to be helpful, but these actions might also be viewed as disrespectful. Sending photos is also unlikely to assist much, and it is certainly not the best option.

15. **Answer**: C

    Explanation: When we feel threatened, it's harder to find smart answers because we have reduced cognitive resources for problem solving. As for the other options, rewards and punishments aren't the agile approach to motivating team members (nor are they likely to improve problem-solving skills). While the Storming stage of development certainly presents problems to solve, team members aren't necessarily skilled at doing so yet. And agile approaches are highly collaborative, rather than leaving people to work on their own.

16. **Answer**: D

    Explanation: Cockburn's Shu-Ha-Ri model states that when mastering a new process, we start by following the rules or guidelines established by others. We need to first master that process before we will have the expertise required to tailor our approach. While the other options are all correct statements, they aren't related to the Shu-Ha-Ri model.

17. **Answer**: D

    Explanation: Through a process of elimination, we can determine that the correct answer is choice D. High-performing agile teams work in high-trust, rather than low-trust, environments, and they are consensus-driven, not plan-driven.

18. **Answer**: C

    Explanation: In this situation, the team leader should assume a coaching role to help the team members get to the point where they are comfortable selecting their own work. This will include meeting team members a half-step ahead, guaranteeing safety, partnering with their managers, and building positive regard. Assigning someone as a decision maker at stand-up meetings is incorrect since agile teams are consensus-driven and decisions aren't made in stand-up meetings. Incentive plans can be useful, but what this team really needs is guidance, not rewards and punishments. Simply explaining that agile teams self-select their work isn't enough to get team members comfortable assuming more ownership of the project.

19. **Answer**: C

    Explanation: Constructive disagreement is a form of healthy conflict, where team members work through issues to find a solution that is right for the team and the project. As a result, constructive disagreement generates team buy-in for decisions. Agile methods do not seek to "weed out the weak," and while constructive disagreement could help clarify requirements and build negotiation skills, those are not the primary reasons it is valued on a project.

20. **Answer**: D

    Explanation: Osmotic communication is the useful information that flows between team members who are working in close proximity to each other as they overhear each other's conversations. Therefore, it works best when the team members are sitting near each other.

CHAPTER 5

Adaptive Planning

Domain V Summary

This chapter discusses domain V in the exam content outline, which is 12 percent of the exam, or about 14 exam questions. This domain deals with sizing, estimating, and planning, including adaptive planning, progressive elaboration, value-based analysis and decomposition, and release and iteration planning.

Key Topics

- » Affinity estimating
- » Agile discovery
- » Agile sizing and estimating techniques
- » Daily stand-ups
  - Ground rules
  - Three questions
- » Defining and testing acceptance criteria
- » Estimating initial velocity
- » Estimating tasks
- » Fast failure

- » Ideal time
- » Iteration planning process
- » Planning poker
- » Product roadmap
- » Progressive elaboration
- » Relative sizing
- » Release planning process
- » Rolling wave planning
- » Slicing stories
- » Spikes
  - Architectural spike
  - Risk-based spike

- » Story maps
- » Story points
- » Timeboxing
- » T-shirt sizing
- » User stories
- » User story backlog
  - Refining (grooming) the backlog
  - Requirements reviews
- » Value-based analysis and decomposition
- » Wideband Delphi

Tasks

1. Use progressive elaboration and rolling wave planning to plan at multiple levels.
2. Make planning transparent and involve key stakeholders.
3. Manage expectations by refining plans as the project progresses.
4. Adjust planning cadence based on project factors and results.
5. Inspect and adapt the plans to changing events.
6. Size items first, independent of team velocity.
7. Adjust capacity for maintenance and operations demands to update estimates.
8. Start planning with high-level scope, schedule, and cost range estimates.
9. Refine ranges as the project progresses.
10. Use actuals to refine the estimate to complete.

This chapter is about planning, but not just any kind of planning; agile projects call for *adaptive planning*. An adaptive approach acknowledges that planning is an ongoing process and provides multiple mechanisms to proactively update the plan. This is what distinguishes agile planning from a more predictive, static approach in which most of the plan is created up front and replanning is done primarily in response to exceptions to the plan and change requests.

This flexible approach to planning provides the "increased adaptability" component of the agile value proposition. We've already discussed three elements of this proposition in chapters 2 and 3—early delivery of business value, early reduction of risk, and increased visibility. In this chapter we'll explain how agile teams are able to offer increased adaptability over the course of the project life cycle, as well.

Figure 5.1: Agile Value Proposition

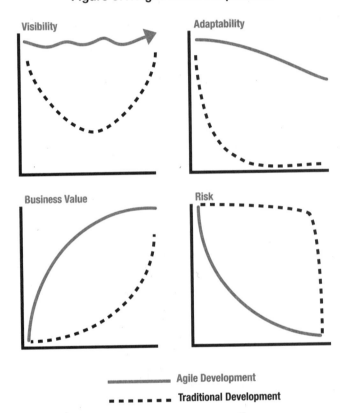

Visibility

Adaptability

Business Value

Risk

———— Agile Development

- - - - - Traditional Development

In agile, planning isn't a one-time stand-alone activity; iterative planning efforts continue throughout the development process, even down to the team's daily check-in meetings (daily stand-ups). As a result, the topics we'll discuss in this chapter are closely interwoven with the other domains—especially domain II (value-driven delivery) and domain III (stakeholder engagement). To help organize our discussion of these wide-ranging exam topics, we'll group them into three themes—agile planning concepts, tools for sizing and estimating, and the processes of release and iteration planning.

© 2015 RMC Publications, Inc • 952.846.4484 • info@rmcls.com • www.rmcls.com

Exam tip

The topics discussed in this chapter cover a lot of ground, and many of them—such as progressive elaboration, timeboxing, user stories, product backlog, and release and iteration planning—are fundamental agile concepts that are very likely to be covered in the exam questions, either directly or indirectly. However, as for the other chapters, don't focus on details and definitions that won't help you answer situational questions. Instead, try to understand how the concepts discussed in this chapter embody and support the agile mindset, values, and principles in practice.

Agile Planning Concepts

In this section, we'll explain some of the key ideas that provide the foundation for the agile planning process. You can think of these concepts as an extension of the agile mindset to the planning effort. If we start from the values and principles laid out in the Agile Manifesto, how does that shape how we plan our projects? What fundamental planning practices are common across agile methods, even if they are implemented in different ways? We'll begin this discussion by explaining the idea of adaptive planning.

Adaptive Planning

As we've discussed throughout this book, agile methods are *value-driven*—they aim to maximize the delivery of business value. This is why the backlog is prioritized with high-business-value items at the top and why releases aim to maximize the value of the functionality being delivered.

As part of this focus on value delivery, agile methods also try to minimize any nonvalue-adding work. Since planning activities don't directly add business functionality, they could be considered waste. So to minimize waste, we should take the most efficient approach to planning, right? From an efficiency standpoint, doing the necessary planning for a project only once and then not returning to the planning effort would seem like the best thing to do. Unfortunately, this is not an effective approach for knowledge work projects due to their high rates of change, nor is it safe or responsible to plan only once.

The only way we can be successful on such projects is to *plan to replan*. This approach is called adaptive planning, and it is based on the conscious acceptance that early plans are both necessary and likely to be flawed; therefore, replanning and adaptation activities should be scheduled into the project. Uncertainty drives the need to replan.

To illustrate this concept, think about a project as a journey. If we are going to a well-known destination and traveling over well-known terrain, we can use a map and a GPS unit to create a detailed, reliable up-front plan for the entire trip. Our rate of progress may vary from the original schedule, but that's why we have status reviews—to compare where we are to where we thought we would be by now, always referring back to the baseline plan. Taking the variances from the plan into account, we can recalculate our consumption and estimated time-of-arrival projections. Perhaps we'll encounter a roadblock or diversion, or maybe we'll have to go back to the steering committee and ask for permission to go a different way, but such events will be exceptions—infrequent deviations from the plan.

In contrast, undertaking a novel knowledge work project is more like a journey across a deserted island that few people have ever visited, using a rough treasure map as a guide. It is quite likely that no one on the team has ever been to the island before. There are no GPS capabilities or drive-time averages to help us plan our journey. Instead, as a team we discuss the goal, agree on a general plan, and set off in the direction we think is most likely correct. If we come to an obstacle, such as a cliff or a lake, that isn't on our map, we don't blindly attempt to follow our original plan. Instead, we accept that it was flawed and adapt our journey based on the new information we have learned.

And so it is with many agile projects—the unprecedented nature of the work means that we frequently discover issues and encounter a high rate of change. Therefore, we go into these projects with the expectation that we will be adapting the plan. To quote Alfred Korzybski, "The map is not the territory."[1] This expression speaks to the need to adapt our approach as we encounter unplanned obstacles.

Agile versus Non-Agile Planning

Before we turn to the concepts used in planning agile projects, let's take some time to compare the planning approaches of agile and non-agile projects. Agile planning differs from traditional planning in three key ways:

» Trial and demonstration uncover the true requirements, which then require replanning.
» Agile planning is less of an up-front effort, and is instead done throughout the project.
» Midcourse adjustments are the norm.

Let's explore each of these points in more detail.

Trial and Demonstration Uncover the True Requirements, Which Then Require Replanning

When undertaking uncertain endeavors like knowledge work projects, getting stakeholder agreement is critical to the project's success. But how do we get stakeholders to agree on the requirements? It is often best to have an initial discussion about the vision for the product and iteratively progress from there with a tangible prototype. We can then allow stakeholders to adjust the project based on their experience with the prototype, rather than expecting them to fully describe what the product should look like and what it should do when there is still uncertainty about the scope and the proposed solution.

So instead of creating very detailed specifications and plans, agile projects build a prototype to better understand the domain and use this prototype as the basis for further planning and elaboration.

Agile Planning Is Less of an Up-Front Effort, and Is Instead Done throughout the Project

The *PMBOK® Guide* has a strong emphasis on upfront planning; it recommends making detailed plans for the areas of project scope, schedule, budget, quality, human resources, communications, risk, procurement, stakeholders, change management, configuration management, and process improvement—all before we begin work on the project.[2] In contrast, agile methods recognize that the level of risk and uncertainty on knowledge work projects makes extensive up-front planning problematic, so they distribute the planning effort throughout the project's life cycle. Spreading it out like this allows the team to better adjust to emerging information.

As we've seen, knowledge work projects are often intangible, and the project requirements are difficult to articulate. Rarely do we build the exact same system twice, which makes it difficult to create analogies to existing functionality. These issues can lead to a "gulf of evaluation," where mismatches develop between the team's interpretation of the original requirements and the customer's true goals.

© 2015 RMC Publications, Inc • 952.846.4484 • info@rmcls.com • www.rmcls.com

Executing a knowledge work project is a complex, creative, and high-risk endeavor. Unlike many predictable manufacturing projects, the work is often a research-and-development-based process. Attempts to create detailed, task-oriented plans for knowledge work teams are likely to lead to fragile plans that are quickly abandoned. Or, if the plans are not abandoned, a great deal of time has to be spent updating them, rather than completing the project. Unfortunately, many projects undertake the bulk of their planning too early in the life cycle, when little concrete data is known about the problem domain, the business environment, and the team dynamics.

The planning emphasis for a traditional project management approach is illustrated below. As indicated by the green region, planning is predominantly done early in the project life cycle.

Figure 5.2: Planning Efforts Over the Project Life Cycle—Non-Agile Approach

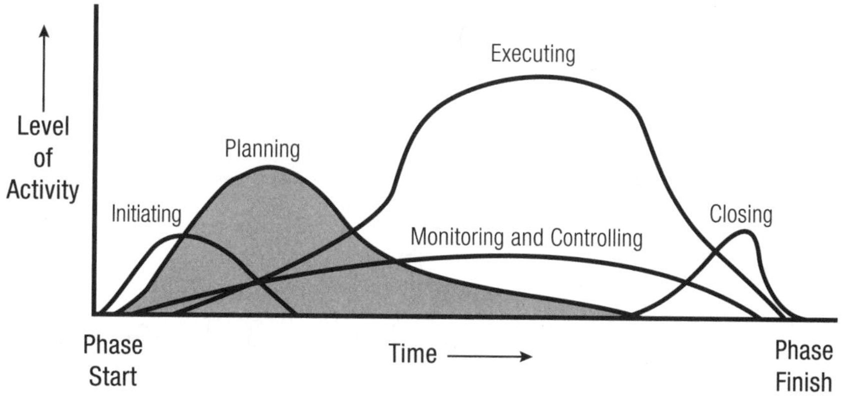

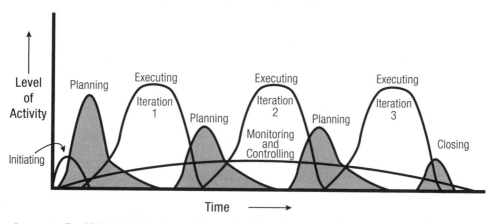

Image originally published in "Collaboration over Command and Control" by Mike Griffiths on gantthead.com on April 3, 2007, copyright © 2007 gantthead.com. Reproduced by permission of gantthead.com.

Agile methods accept the realities of knowledge work projects and deliberately make planning a more visible and iterative component of the project life cycle. In the figure below, the repeated green sections show how planning happens throughout the life cycle of an agile project.

Figure 5.3: Planning Efforts Over the Project Life Cycle—Agile Approach

Image originally published in "Collaboration over Command and Control" by Mike Griffiths on gantthead.com on April 3, 2007, copyright © 2007 gantthead.com. Reproduced by permission of gantthead.com.

If we could add up the green areas on these two diagrams, we would probably find that the total amount of planning on the agile project is more than the total on the traditional project. As a rule, on agile projects, we end up doing more planning, not less—but the planning activities are distributed differently over the project life cycle.

No matter what the endeavor, there is always a responsible amount of up-front planning that should be done. Barry Boehm illustrates the risks of not doing enough up-front planning like this:

Figure 5.4: "Planning Risks" Graph

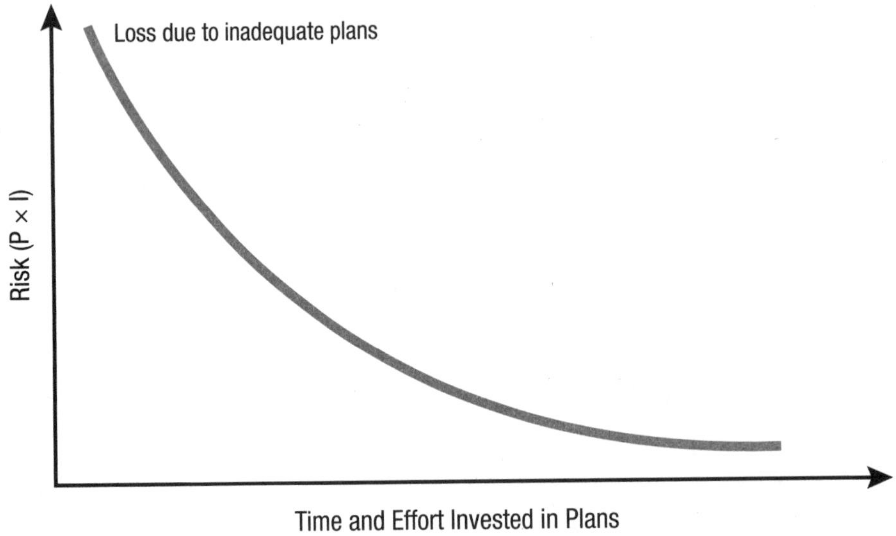

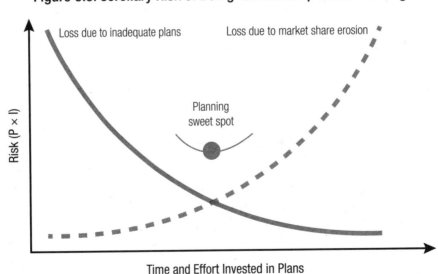

The above chart shows that when little time and effort is invested in planning, the risk of oversight and delays due to rework is high. As more time and effort is invested into planning, this risk drops away. However, what people sometimes miss is that there is a corollary risk of doing too much up-front planning. To reflect that risk, Boehm adds another line, like this:

Figure 5.5: Corollary Risk of Doing Too Much Up-Front Planning

The dotted line on this diagram represents the risks involved in doing too much up-front planning. This risk starts off low, but as more time and effort is invested in up-front planning, the risk of creating a very detailed, brittle plan increases—as does the risk of delaying the delivery of value and return on investment because of the amount of time spent planning.

Boehm's conclusion is that we should aim for the sweet spot where the sum of these two risks is lowest. We need to do enough up-front planning to minimize the risk of duplication and rework; at the same time, we need to avoid overplanning to minimize the risk of delayed value delivery and a brittle project plan.

The curves illustrated above hold true for any kind of project. However, due to the changeable nature of knowledge work, the risk that an agile plan will break escalates sooner than it would in a more stable, predictable project environment. Therefore, the amount of up-front planning that is optimal for agile projects is typically less than it is for other projects.

Midcourse Adjustments Are the Norm

When aiming at a static target, the best approach is to aim very carefully and then fire directly at the fixed target. However, when aiming at a moving target that isn't following a predictable path, we need more of a guided-missile approach, making a lot of mid-flight adjustments to ensure our efforts reach their goal.

A guided missile is pointed in the general direction of its moving target and then, once it is released, a sophisticated system of sensors, thrusters, and feedback systems takes over, guiding the missile to the target. This analogy is appropriate for agile projects, because they often have moving targets. The product the business wants may be transformed by late-breaking change requests that are triggered by changes in the business environment or offerings from competitors. To stay on target, agile methods use sophisticated adaptation systems to gather feedback and make adjustments to the backlog and plans as the project work is being done.

Now, I do realize that missiles have a lot of negative connotations—they are destructive, and they blow up and hurt people—which isn't a good way to describe a project! With this analogy, however, we're focusing on the missile's ability to steer itself dynamically toward a moving target. So let's imagine that instead of exploding once it reaches the target, the missile will deliver something pleasant, such as a bunch of flowers.

Figure 5.6: Agile Methods Take a Guided Missile Approach

Aiming directly at a static target:

Using a guided missile approach to reach a moving target:

There's another great way to explain this concept that I learned from my first mentor, who used to say, "You cannot chase a dog with a train!" It comes to roughly the same thing—a train has to stay on its tracks and cannot adjust to the movements of a free-running dog. If our project and goals are likely to change direction unpredictably, we will need a nimble, lightweight planning structure to keep up with them, not a large, monolithic structure that is slow to turn or change direction.

The point to take away from these analogies is that agile methods don't shy away from planning—they just take a different approach that is better suited to a quickly changing environment. Although to a casual observer it may look like the planning stage has been largely skipped and that the team just goes ahead and starts developing things, a continually evolving planning process is actually taking place.

Planning occurs on agile projects first with a high-level plan, and then at regular points throughout the project for subsequent releases and iterations. Agile teams also factor a lot of feedback into these ongoing planning processes. For example:

» Backlog reprioritization affects iteration and release plans.
» Feedback from iteration demonstrations generates product changes and new requirements.
» Retrospectives generate changes to the team's processes and techniques.

EXERCISE

Test your knowledge of the concepts we've just discussed by completing the following True or False quiz.

Statement	True or False
Agile projects typically do more up-front planning than traditional projects.	False
Agile projects typically do more overall planning than traditional projects.	True
If we create our plans at the last responsible moment, they will not change.	False
Midcourse adjustments on agile projects are not common.	False
Knowledge work projects tend to have high rates of change.	True
If the project diverges from the original plan, this could be a sign our initial plan was flawed.	True

© 2015 RMC Publications, Inc • 952.846.4484 • info@rmcls.com • www.rmcls.com

ANSWER

Statement	True or False
Agile projects typically do more up-front planning than traditional projects.	False Although the total amount of planning over the course of the project is likely to be more than on a traditional project, the up-front planning is likely to be less.
Agile projects typically do more overall planning than traditional projects.	True When you add up all the replanning that is done, agile projects plan more than traditional projects.
If we create our plans at the last responsible moment, they will not change.	False Creating plans at the last responsible moment certainly helps avoid some changes, but can't prevent all changes.
Midcourse adjustments on agile projects are not common.	False Midcourse adjustments are accepted as normal on agile projects.
Knowledge work projects tend to have high rates of change.	True Knowledge work projects have high rates of change because they often build new, novel, and unprecedented solutions, which prevents us from easily applying plans and lessons from similar projects in planning them.
If the project diverges from the original plan, this could be a sign our initial plan was flawed.	True Initial plans for knowledge work projects are always likely to be flawed.

Principles of Agile Planning

Next, let's look at some high-level principles for agile planning. The questions on the exam will assume that you understand and can apply the following planning concepts:

» Plan at multiple levels.
» Engage the team and the customer in planning.
» Manage expectations by frequently demonstrating progress and extrapolating velocity.
» Tailor processes to the project's characteristics.
» Update the plan based on the project's priorities.
» Ensure encompassing estimates that account for risks, distractions, and team availability.
» Use appropriate estimate ranges to reflect the level of uncertainty in the estimate.
» Base projections on completion rates.
» Factor in diversions and outside work.

Let's briefly review these nine principles to help reinforce your understanding of them.

» **Plan at multiple levels**. At a high level, we plan the overall scope and what "done" looks like for the project. We plan releases in a little more detail, painting the map of which features we would like delivered when. We then plan the iterations in even more detail, identifying which stories will be built in each iteration, and planning the tasks needed to complete them.

» **Engage the team and the customer in planning**. On a knowledge work project, it's unlikely that the leader or manager will have all the information required to satisfy the customer's needs. So we have to engage both the team and the customer in planning. This will take advantage of the team's knowledge and technical insights and also generate their buy-in and commitment for the plan that is developed, so that we don't have to "sell" it to them afterward.

» **Manage expectations by frequently demonstrating progress and extrapolating velocity**. Agile teams show what has been built as the process goes along. This keeps stakeholders up to speed on what is happening and manages their expectations about what can be built by the end of the project. We also use current rates of progress, rather than plans or hopes, to predict completion dates and costs.

» **Tailor processes to the project's characteristics**. Large projects will need more planning than small projects. If the requirements and technologies are well understood, it is safer to do more up-front planning than when these factors are unknown. If there are more uncertainties, the team will need to plan in spikes to explore options and confirm that the proposed technological approach will work.

» **Update the plan based on the project's priorities**. The business sets the project priorities, and these are reflected in the backlog priorities created by the product owner, in collaboration with the development team. Whenever we learn of a change, we need to re-examine our backlog and release plans to see if it means that anything else needs to be changed. For example, if we are told that completing the English version of the website is now the project's top priority, does that mean that all the internationalization and translation work has been demoted?

» **Ensure encompassing estimates that account for risks, distractions, and team availability**. Understandably, sponsors want to know when things will be done; therefore, estimates that don't take into account known variables are unrealistic and unhelpful. To produce better estimates, we start with base historical averages (such as velocity trends), and then factor in future team availability and the distractions, diversions, and other calls on the team's time that will inevitably occur.

» **Use appropriate estimate ranges to reflect the level of uncertainty in the estimate**. When we are pretty certain about a short-term, well-understood goal, we can safely give a narrow estimate range. For example, "Recording my time should take 15 to 20 minutes, and I will do it on Friday." But when there is a lot of uncertainty and a long timeline, there will be many opportunities for issues to arise. In that scenario, we need to manage expectations and provide a broader estimate range. For example, "I hope to complete the portrait painting of your family in 5 to 8 days, depending on everyone's availability and my schedule."

» **Base projections on completion rates**. Whenever possible, our plans and projections should be based on actual completion data for the project, because those numbers show our real, rather than ideal, rate of progress. They already have distractions, defect remediation rates, and nonproductive time "baked in," and so are more likely to be replicated in the future than plans based on theory rather than reality.

» **Factor in diversions and outside work**. People get called back to support old projects; they go on vacation; they have both planned and unplanned absences. So our plans should not assume year-round availability or 100 percent dedication to a project; we need to factor in some nonproductive time.

© 2015 RMC Publications, Inc • 952.846.4484 • info@rmcls.com • www.rmcls.com

Exam tip

Don't try to memorize the nine points listed above; instead, read through them and think about their implications for projects. Then, as you read the rest of this chapter, consider how each topic we discuss relates back to these principles. This will help you develop the mindset you'll need to answer questions on the exam. For example, let's say the answer choices for a question imply either that "Planning is done by the project manager" or "Planning is done by the team and the customer." When you see these choices, you should know that option B is correct—not because you have memorized the second principle described above, but because you have internalized the agile "mindset" for planning.

K&S Agile Discovery

The concept of "agile discovery" is an umbrella term that refers to the evolution and elaboration of agile project plans in contrast with an up-front, traditional approach to project planning. It covers topics such as:

» Emergent plans and designs versus predictive plans and designs
» Preplanning activities to gather consensus on the best approach to follow
» Backlog refinement (grooming) and how it is performed
» Estimating uncertain work versus certain work
» The characteristics of new product development versus well-understood and repeatable projects

So if you see agile discovery mentioned in an exam question, think of it as a mindset pointer. For example, "A team is engaged in agile discovery and has produced some estimates; how should they best be presented to stakeholders?" Here the use of the term "agile discovery" is a cue that the agile mindset of emergent plans and incomplete initial requirements is in effect. An answer that refers to using a broad estimate range to reflect uncertainty is more likely to be correct than an answer that implies more certainty in the estimates.

T&T Progressive Elaboration

Progressive elaboration refers to the process of adding more detail as new information emerges. Agile methods rely on progressive elaboration to first create, and then refine, many kinds of project assets to make them increasingly accurate as the team learns more about the project. These "progressively elaborated" assets might include:

» Plans
» Estimates
» Risk assessments
» Requirements definitions
» Architectural designs
» Acceptance criteria
» Test scenarios

At the beginning of a project, the team needs to size and estimate the work involved to determine how big the endeavor is likely to be and to create a reasonable strategy and execution approach. However, we also know that the beginning of a project is when we know the least about the endeavor. At this early point, there has not yet been any "learning by doing" on the project. So it would be foolish to limit our planning and estimation activities to the start of the project. Instead, we must continually refine our plans and estimates as the project progresses and new details emerge. This process of continual updates is the essence of progressive elaboration.

The concept of progressive elaboration is illustrated in the next two figures. In the first diagram below, the "Now" arrow indicates that we are at the start of the project and are creating the up-front plans. The diagonal lines in the early iterations show how detailed our plans for those iterations are. As you can see, at this point the first iteration has been planned in a lot of detail, but the subsequent iterations have been planned in increasingly less detail.

Figure 5.7: Level of Planning Early in a Project When Using Progressive Elaboration

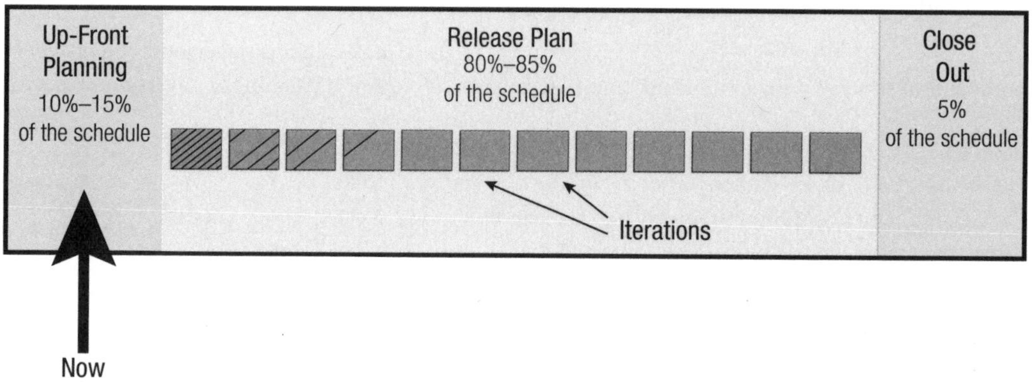

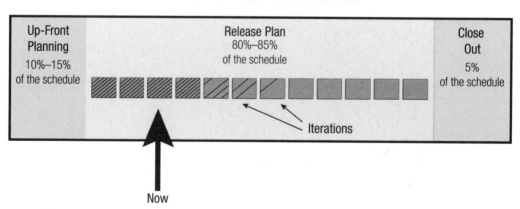

The next diagram shows how we progressively elaborate our plans and add details as the project continues and new information emerges. Here, the "Now" arrow indicates that we are in the third iteration. At this point, the fourth iteration has been planned in detail, while the plans for the fifth, sixth, and seventh iterations have decreasing levels of detail. This process of refining our plan as we get closer to working on an iteration continues throughout the project.

Figure 5.8: Level of Planning Later in the Project When Using Progressive Elaboration

Progressive Elaboration versus Rolling Wave Planning

Many people are confused about the difference between "rolling wave planning" and "progressive elaboration." Since either or both of these terms could appear on the exam, let's spend a moment to explain the difference between them. Basically, one of these terms refers to a strategy and the other to its implementation.

» *Rolling wave planning* is PMI's term for the strategy of planning at multiple points in time as more information about the project becomes available.[3] Rolling wave planning is the game plan that says "We won't try and do all our planning up front. We recognize that it will be better to plan a bit, and then revisit and update our plans multiple times throughout the project."

» *Progressive elaboration* is what we do to incorporate new information into our plans as we progress with the project. It is how we implement the rolling wave planning approach.[4]

K&S Value-Based Analysis

Agile planning is based on value-based analysis. This is the process of assessing and prioritizing the business value of work items, and then planning accordingly. This process continues over the full life cycle of a project, impacting how we scope, plan, schedule, develop, test, and release work. At every stage of the project, we are asking, "What is the business value of this item or practice?" and "What items in this set have the highest business value?" We then prioritize the work to deliver the highest-value items first.

Figure 5.9: Analyzing the Business Value of Work Items

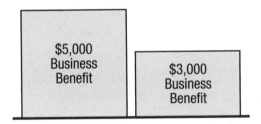

To fully understand an item's value, we also need to understand its development and delivery cost. For example, a feature that is worth $5,000 to the business but costs $4,000 to develop is not as valuable as something that delivers $3,000 in value to the business and only costs $1,000 to develop.

Figure 5.10: Considering Both Business Value and Development Cost

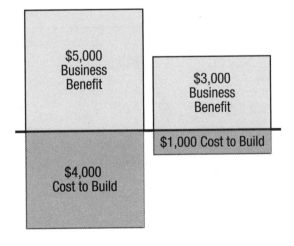

Therefore, when evaluating the value of a work item, we need to factor in its likely development cost. In order to do that, agile teams will create high-level estimates of the backlog items early in the project. This allows them to make cost-benefit value comparisons that the customer can use in prioritizing items for development. For example, to maximize the value delivered in the next release, the team might plan (with the customer's buy-in) to complete several medium-value features that they can develop quickly rather than one higher-value item that would consume all the remaining development time.

Two other factors that need to be considered are payback frequency and dependencies:

» Will the work item generate business value every week or month (such as a time savings for staff) or just once (such as a compliance check)? When evaluating the return over a one- to five-year period, we should take an extended view to assess the true business value.

» Sometimes a high-business-value item is dependent on a low-business-value item. In that case, the team will need to undertake the low-business-value item first, to allow them to deliver the high-business-value item as soon as possible.

With these considerations in mind, the team and the customer will analyze the real business value of features, prioritize the features appropriately, and plan the work.

K&S Value-Based Decomposition

Value-based decomposition is a continuation of the process of value-based analysis—here, the team elicits requirements from stakeholders; groups, breaks down, and ranks those requirements; and then pulls the prioritized requirements into the development process. The figure below shows what this process might look like on a typical agile project.

Figure 5.11: Example of Value-Based Analysis and Decomposition Process

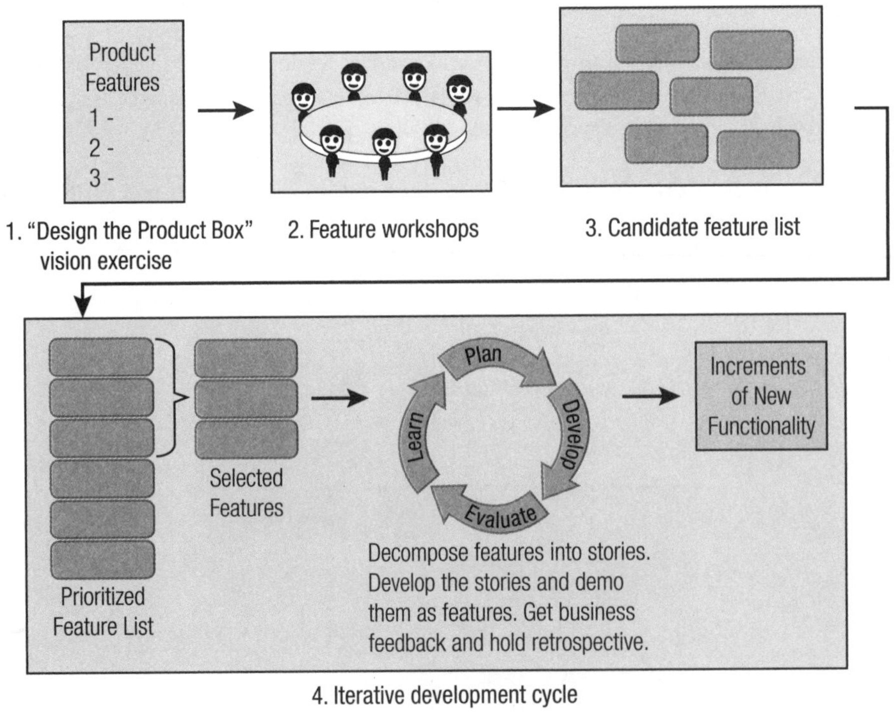
Product Features
1 -
2 -
3 -
1. "Design the Product Box" vision exercise
2. Feature workshops
3. Candidate feature list
Prioritized Feature List
Selected Features
Plan
Develop
Evaluate
Learn
Increments of New Functionality
Decompose features into stories. Develop the stories and demo them as features. Get business feedback and hold retrospective.
4. Iterative development cycle

In this macro-level example of the process, we can see several examples of value-based decomposition:

» The top three features or objectives that the product or solution must deliver are captured in the visioning exercise, as explained below.
» The major functional elements of the vision are drawn out in the feature workshops.
» The features are then prioritized and consumed by the development cycle.

The first step in this process will be some kind of effort to define the vision. The way a project is kicked off may vary from project to project—in the above example, the stakeholders use a visioning exercise called "Design the Product Box." In this activity, they design an imaginary "product box" that is a metaphor for the system they are building, on which they identify the product's top three features. This exercise helps the stakeholders capture the high-level vision for the project and reach consensus around a common mission, goals, and success criteria.

Figure 5.12: Design the Product Box Example

On the front of the box:
• Name of product or solution
• Logo or graphic that represents the idea
• Top 3 key features or objectives

On the back of the box:
• Product description
• List of remaining features
• Operating requirements

Continuing with our example of the decomposition process (figure 5.11), in step 2 the stakeholders hold a series of feature workshops in which the project vision is broken down into the potential features of the system. These workshops result in a candidate feature list (step 3). In the final step, that list is prioritized based on business value and risk to get the prioritized backlog that the team will use to complete the iterative development process (step 4).

Progressive Elaboration in Action

As the team gets into development and tries to implement the features, they will discover supporting features and other elements that previously had not been considered. So throughout the project, the process of identifying new requirements, grouping them or breaking them down into functional elements, and prioritizing those elements is repeated at an increasingly refined and more detailed level, like a fractal leaf pattern.

Figure 5.13: Fractal Leaf Pattern

This iterative process is another example of how agile methods use progressive elaboration. In the value-based decomposition process, we pull forward elements of the project—in the level of detail that is appropriate for the project phase we are in and how much we currently know about the project—into a format that we can work with for the next process. We then continue to refine and elaborate those elements, adding increasing levels of detail and transforming them into a format that can be further refined in subsequent processes. The end product of this process is a highly detailed deliverable that is still true to the original design objectives.

"Coarse-Grained" Requirements

Agile methods don't attempt to specify fully detailed requirements up front. Instead, we initially keep the requirements "coarse grained," and then progressively refine them as the planning process continues. This approach has a number of advantages:

» It helps keep the overall design balanced so the product doesn't become lopsided by overdevelopment in any particular area.

» It delays decisions on implementation details until the "last responsible moment." This means we are not rushing to develop things that may later need to be changed as a result of new information or late-breaking change requests.

Delaying decisions like this is one of the many paradoxes or balancing acts involved in agile projects. We aim to make decisions as late as we can, to incorporate changes, while also building increments of the system as early as we can, to gain feedback—which can then result in system changes. Although this approach may sound illogical, it is really an effective way to mitigate risks and find out about problems and changes within the friendlier project environment, where changes cost less, than in the more hostile production environment, where changes are much more expensive to make.

T&T **Timeboxing**

Agile planning and development are based on the concept of *timeboxing*. A timebox is a short, fixed-duration period of time in which a defined set of activities or work is undertaken. If the work planned for the timebox isn't complete when the time runs out, we stop what we're doing and simply move the uncompleted work into another timebox. Timeboxes are what allow agile teams to adjust their scope to achieve the highest-priority, best-quality product within a fixed cost and timeframe, as depicted in the agile triangle of constraints.

Figure 5.14: Inverted Triangle Model

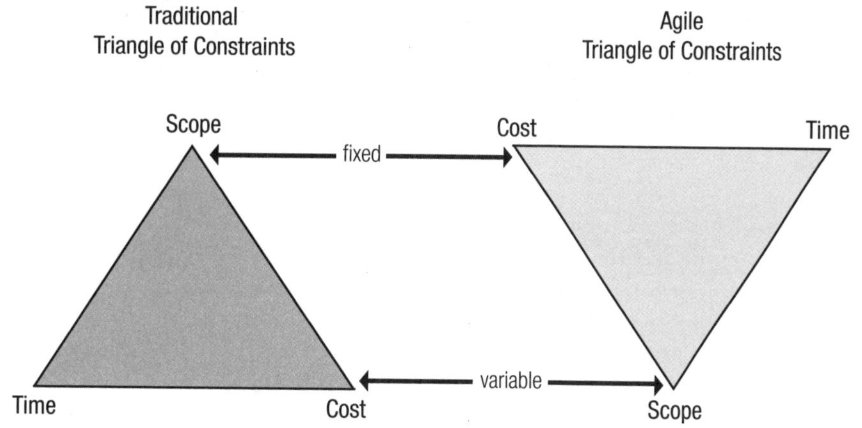

Agile timeboxes can last anywhere from a few minutes to a few weeks. For example:

» Daily stand-up meetings are timeboxed to 15 minutes.
» Retrospectives are often timeboxed to 2 hours.
» Iterations and sprints are typically timeboxed to one to four weeks.

Let's explore the concept of timeboxed iterations a little further. Imagine that a team has planned to complete ten work items within a timeboxed iteration, but at the end of the timebox, they have only finished eight of the items. Even though they haven't completed all the planned work items, they don't extend the timebox to get that work done. Instead, they report that eight items were completed, and then return the remaining two items to the backlog to be considered in planning the next timeboxed iteration.

We've seen that agile projects typically have a lot of uncertainty, especially if the domain of the project is novel to the organization or if new technologies are being used on the project. Timeboxes help bring some level of order and consistency to an otherwise highly variable work environment. They offer regular opportunities to assess results, gather feedback, and control the costs and risks associated with an endeavor. This is why timeboxes have been referred to as "the control in the chaos." They provide frequent checkpoints where the team can gauge their progress and replan their ongoing approach.

Another way to think about timeboxed iterations is to imagine an actual box that represents the team's capacity to accomplish work, as illustrated below.

Figure 5.15: Assigning Work to a Timeboxed Iteration

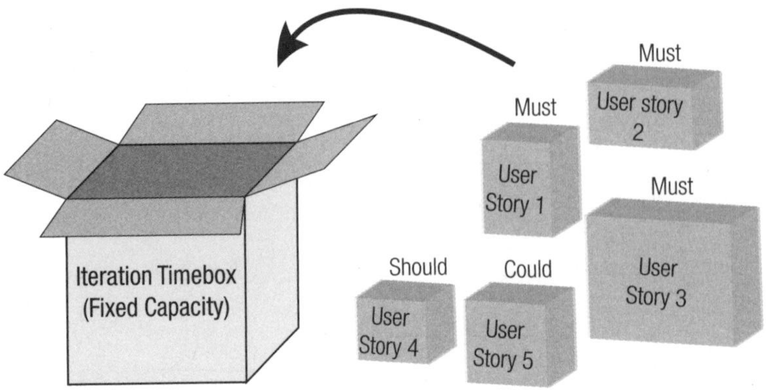

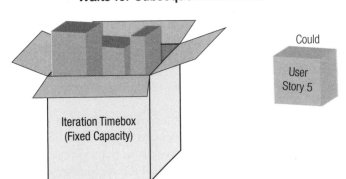

The team loads the work items (user stories) into the timebox in order of priority. If all the items don't fit into the box—in other words, if the team doesn't have the capacity to complete all the work within the iteration—then the lowest-priority item will have to wait. It will be returned to the backlog and considered in planning the subsequent iterations.

Figure 5.16: Work That Doesn't Fit into the Timebox Waits for Subsequent Iterations

Timeboxes can also serve as powerful tools for completing focused work. For example, some people use Pomodoro timers (25-minute timers, often shaped like a tomato) to help keep themselves focused and on task for a short period of time. The 25-minute setting allows for two work sessions per hour, with a few minutes outside of the focused session to check e-mail, get a drink, stretch your legs, etc. The reason this is effective is that people tend to get distracted and multitask inefficiently. Instead of answering every e-mail or message as it comes in, it may be better to work through without any distractions or interruptions for 25 minutes, and then take a short break.

Exam tip

You can also use the timeboxing approach in preparing for the PMI-ACP exam. Get a one-hour sand timer, and commit to setting aside one solid hour each evening for study, with no distractions. During that time, don't check your e-mail or phone or let yourself get sidetracked from studying. The advantage of a sand timer over a digital timer is that you can check the volume of sand in each half to get a quick visual reading of the time remaining. Also, the falling sand generates a sense of rapidly depleting time that is lost with a digital counter. A sand timer constantly reminds you that "time is passing by, so you'd better get on with it."

Background information

Parkinson's Law and Student Syndrome

Agile practices such as timeboxes are designed to minimize the effects of Parkinson's Law and Student Syndrome.

» Parkinson's Law states that work tends to expand to fill the time available. In other words, if we have three months designated for gathering requirements, we will definitely spend all three months gathering requirements. Because agile projects focus on short timeboxes, such as iterations of one or two weeks, there is less time to fill.

» The idea behind Student Syndrome is that when people are given a deadline, they tend to wait until they have nearly reached the deadline before they start working. The short timeboxes on agile projects mean that the next deadline is never far away, and this helps keep people focused.

Estimate Ranges

Estimating agile projects has some important similarities to and differences from estimating traditional, noniterative projects. For example, near the start of an agile project when we know least about the project, our estimation techniques are quite similar to traditional estimation methods. We apply heuristic (expert-knowledge-based) approaches—such as looking at data from projects that created similar products or that were about the same size and complexity—to see how long those projects took. We can also perform parametric (calculation-based) estimates, such as bottom-up estimates that are based on the number and complexity of user stories.

However, agile projects are typically more difficult to estimate than other types of projects. These efforts are typically complex and have a high degree of uncertainty—the organization might not have undertaken a similar project before, the approach or technology being used might be new, or there might be other significant unknowns. This combination of complexity and uncertainty makes it more problematic to provide estimates for knowledge work projects.

To help manage this uncertainty, agile teams avoid using single-point estimates; instead, we present estimates in ranges to indicate our level of confidence in the estimate and manage our stakeholders' expectations. For example, saying "This project will cost $784,375.32" gives the impression that we can reliably predict the outcome down to the penny. If we don't have that degree of certainty, it's better to say, "We believe this project will cost between $775,000 and $825,000." When giving estimates in ranges, the width of the range should reflect our confidence in the accuracy of the estimate to manage our stakeholders' expectations.

Background information

Another Way to Manage Expectations

I once did some consulting work with an engineering team, and was surprised to see that they presented their two project estimates as "Project A: $8,000,000.88" and "Project B: $3,000,000.99." I teased them about the 88 cents and the 99 cents, asking if they'd had to add on postage or something. They politely explained that this was their standard notation for first- and second-order estimates. For Project A, the ".88" meant the estimate was likely plus or minus 20 percent, and for Project B, the ".99" meant it was plus or minus 10 percent. This was a clever way to reflect the team's confidence in their estimate ranges and manage expectations.

When working with estimates, we need to understand the quality of our input variables—it is all too easy to apply math to a bunch of highly speculative variables and then begin to believe there is more accuracy to our estimates than is realistic. Our estimate ranges should be narrower when we are more certain about the estimates, and wider when we are less certain.

The diagram below shows Barry Boehm's Estimate Convergence Graph, which shows how the estimates for software projects typically move from a very broad range early in the life cycle to a more manageable range once the scope and specifications are understood and agreed upon—and then continue to narrow as the team learns more about the project. Beneath the graph, I have added in green text how the stages of an agile project typically map to our degree of confidence, as reflected in the estimate ranges.

Figure 5.17: Estimate Convergence Graph

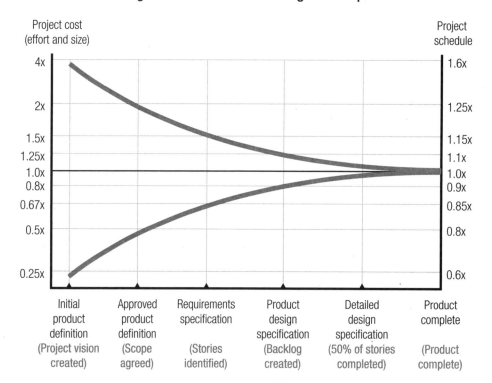

© 2015 RMC Publications, Inc • 952.846.4484 • info@rmcls.com • www.rmcls.com

Background information

Key Points for Agile Estimation

Here is a summary of some key points about agile estimating, as background information for the exam. We'll be going into more detail about some of these ideas in the next section.

» **Why do we estimate?** Estimates are necessary for scheduling the project and determining which pieces of work can be done within a release or iteration. (In the initial planning effort, high-level estimates are also needed to calculate the expected financial return and get the project approved.)

» **When do we estimate?** Agile teams continuously refine their estimates until the last responsible moment before the work is done. Up-front estimates are certainly necessary, but they are also the least accurate since that is when we know the least about the project. So agile teams progressively elaborate their estimates throughout the project, factoring in actual costs and velocity to date to create increasingly more accurate estimates going forward.

» **Who estimates?** In agile methods, the team members who will be doing the work are responsible for estimating their own work. They know the most about the technology, and their involvement gains their buy-in and commitment to the estimates.

» **How are estimates created?** Estimates are created by progressing through the stages of sizing, estimating, and planning (as explained below). To create a holistic estimate, the development, rollout, and maintenance costs also need to be factored in.

» **How should estimates be stated?** There is always some degree of uncertainty in our estimates. Therefore, estimates should be stated in ranges (e.g., "$4,000 to $4,500," or "16 to 18 months") that show their degree of certainty, to manage stakeholder expectations.

T&T Ideal Time

When asking the team to estimate their work, the question of how best to factor in interruptions, diversions, and nonproject work usually crops up. Such nonproject activities may include attending staff meetings, checking e-mails, or the occasional medical appointment. For example, for salaried team members who have a 40-hour workweek, should we estimate 35 hours per week for them to work on the project and allow 5 hours for them to do these other activities? And what about contract staff? They have fewer staff meetings and seem to come in when sick anyway, so should we count on 38 hours for them?

Putting stereotypes and our own expectations aside, we can simplify this issue by estimating in "ideal time." This means estimating as if there were no interruptions. In an eight-hour day, we assume that all eight hours will be available for work. Although this obviously won't be a realistic estimate of duration, the purpose of estimating in ideal time is to simplify the estimation process and take the variable of availability out of the equation. In doing so, we can get a more accurate sense of the effort involved in the work.

An ideal time estimate tells us how long a task will take if all other peripheral work and distractions are removed. It assumes that the work item being estimated is the only thing that is being worked on, that there will be no interruptions, and that we have everything we need to complete that work item—in other words, we aren't waiting for someone else to deliver other work or provide information.

EXERCISE

For this exercise, assume that after you have slept and performed essential activities like washing, preparing food, and eating, you have 12 available hours left in your day. Now let's say that studying for the PMI-ACP exam will take you 60 hours. (Note: This is a made-up number; the amount of time you will need to prepare for the exam will vary based on your experience.) Using this information, calculate the ideal time estimate and the likely time estimate for studying for the exam.

Ideal Time:	*5 days*
Likely Time:	*8 days*

ANSWER

Ideal Time:	If preparing for the exam theoretically takes 60 hours and we have 12 hours a day available, then we do the following calculation:
	$$60 / 12 = 5$$
	So in ideal time, it should only take us 5 days to complete all studying required to pass the PMI-ACP exam.
Likely Time:	To come up with a likely time estimate, we take other factors into account, like work and family and the fact that there are nights when we arrive home and are too tired to think about anything. Perhaps it is more realistic to assume that, on average, we will likely get 4 hours per week to spend on preparing for the exam. Using this information, we can do the following calculation:
	$$60 / 4 = 15$$
	So at this rate, it will take 15 weeks to prepare for the exam.

This exercise brings forward another point to consider—the effectiveness of the timeline. Even if we did have 12 hours a day for 5 straight days to study for the exam, we would not be able to retain or absorb all the information in that time. So studying over a more extended period of time may not only be more realistic, but in this example, it is also probably more effective. In the same way, we need to be careful the timeline isn't too long, to avoid the problem of "knowledge leak"—the slow decline of knowledge that we don't use every day. While too short of a timeline can be a problem, so can too slow of a process.

© 2015 RMC Publications, Inc • 952.846.4484 • info@rmcls.com • www.rmcls.com

K&S **Tools for Sizing and Estimating**

We've said that agile planning is based on three assumptions—the details will emerge as the project progresses, we'll need to adapt our plan based on feedback, and frequent reprioritization will be the norm. This acceptance that things will change, and the resulting realization that it is better to embrace uncertainty than to resist it, is the underlying driver of all agile planning practices.

In this section, we'll examine the specific tools and techniques that agile teams use for adaptive planning. We'll start by explaining how sizing and estimating fit into the planning process, and then discuss the tools used for these efforts roughly in the order they are used, from decomposing requirements to estimating the work items. However, for now just focus on understanding the tools. The overall process will become clearer in the next section, when we discuss release and iteration planning.

Sizing, Estimating, and Planning

So far we've been talking about "planning" at a high level. But now that we're getting into specific planning tools, we'll need to break down agile's planning process into its three steps—sizing, estimating, and planning. In agile, the process of converting requirements into estimates and plans goes through a specific sequence, as shown below.

Figure 5.18: Agile's Planning Sequence

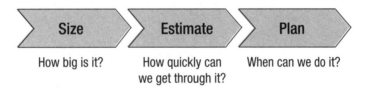

As shown here, we size the work first, then we estimate it, and then we plan it. If you think about it, this is common sense. We need to determine how large the task is first, then estimate how quickly we'll be able to complete it, then finally determine when it can be done—considering how many people we have to do the work and what other work or delays may occur.

So to begin with, we need a way to break down large chunks of work into smaller units that we can size, estimate, and plan. With that in mind, let's turn to the agile approach to decomposing requirements.

Decomposing Requirements

A large project in its entirety is a complex, overwhelming amount of work that cannot be accurately sized, estimated, or planned. So we need to decompose, or break down, the requirements into smaller, increasingly refined chunks of work, until we reach a level that is small enough to actually estimate and plan. As we've seen, on agile projects this breakdown is called value-based analysis and decomposition, and it is done through iterative planning. Here we'll look at the typical structure of this decomposition.

A basic requirements hierarchy for an agile project has three levels, as shown below:

Figure 5.19: Basic Structure for an Agile Requirements Hierarchy

Although there isn't one "right" way to structure a requirements hierarchy, this diagram depicts the simplest approach. Here, the product features are first broken down into smaller chunks of work called user stories. Then, as the team further refines their plan, they break down the user stories into smaller chunks, called tasks. In other words, features are larger than user stories, which are larger than tasks. So far, so good.

However, some teams extend this basic model by adding *epics*, which are large user stories that span one or more iterations—and there is no single, universally agreed-upon way to do this. Some teams put epics above the user story level. In this case, the epics might be above or below the features—or "epic" might just be used in place of "feature." The diagrams below illustrate three different ways that epics might be positioned in a requirements hierarchy.

Figure 5.20: Three Ways to Structure a Requirements Hierarchy with Epics

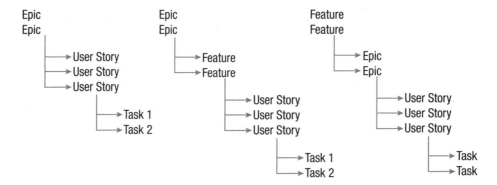

Exam tip

Don't worry about these inconsistencies in how epics are used. For the exam, you should know that the product features are broken down into user stories, which are broken down into tasks. However, you won't be asked where epics fall in the requirements hierarchy. It's still useful to understand how they may be used, however. If you are involved in a project that uses both epics and features, you'll want to find out what structure, or order of precedence, is being used.

Requirements Are Decomposed "Just in Time"

As we've discussed, the project requirements are broken down "just in time" or at the "last responsible moment" as the team gets closer to doing the work. As shown below, the highest-level requirements (such as epics) are progressively refined into features, stories, and then the most detailed pieces of work—such as tasks, business rules, acceptance tests, and so on.

Figure 5.21: Requirements Decomposition Process

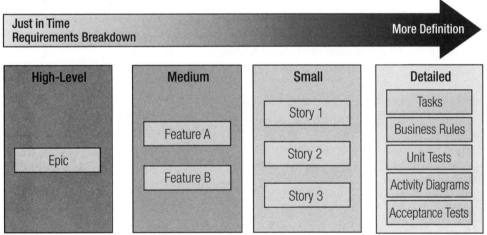

T&T User Stories

Next, we turn to one of the most basic agile planning tools—user stories. As an experienced agile practitioner, you probably already have an understanding of this term (and we've already used it many times). However, it's useful to review the basics to make sure your understanding of this concept is aligned with the way it will be used on the exam. So let's take a closer look at user stories.

A user story is defined as a "small chunk of business functionality within a feature that involves roughly 1 to 3 days' worth of work." (This time frame can be defined more specifically as "4 to 40 hours' worth of work.") Agile teams typically break the product features down into user stories and write them on index cards or enter them into a requirements management tool. They also use a prioritized backlog of these stories to align the team's priorities with the customer's needs.

Creating the User Stories

Let's see how user stories are developed. The team starts by coming up with a list of the potential stories ("candidate stories") that they think are involved in building each feature. For example, here are the three candidate stories that a team has identified for the "Sell movies" feature of an online movie service:

Figure 5.22: Candidates for User Stories

The team has estimated that each of these candidate stories will take about three days to build. Although that's on the large side for a user story, it's fine for initial planning purposes. (As we'll see when we get to release planning, large stories are typically broken down into smaller stories later so they can be estimated in more detail as the team gets closer to working on them.)

However, these "candidate stories" aren't actually user stories yet—at this point they are just requirements that have been decomposed into smaller pieces. To turn them into user stories, the next step is to document them from the perspective of the user or customer and write them on an index card. That's important because a "user story" is, literally, a "story" written from the perspective of the "user" of that functionality.

Although there is no one "right" template for user stories, they are often written in the following format:

"As a *<Role>*, I want *<Functionality>*, so that *<Business benefit>*."

Example: "As a MoviesOnline customer, I want to search movies by actor, so that I can more easily find movies I would like to rent."

The advantage of this template is that it forces the team to identify the user ("Who is asking for this?") and the benefit to the business or user ("Why are we doing this?") for every piece of functionality. That's helpful because it's all too common for projects to become bloated with requirements that have no clear owner or universally understood benefit.

Background information

Choosing the Actor for User Stories

In writing a user story, often there are two (or more) roles that we could use for the actor in the story. For example, our "complete order" candidate story could be written from the perspective of the end user ("...I want to complete an order, so I can download and watch the movies") or from the standpoint of the business ("...I want to complete orders, so I can process customer payments and collect revenue").

So how do we decide which actor to use? One approach I use to determine the best actor is to ask which story would have the greatest value. In other words, "What's the strongest case for doing that story that would move it up the backlog so it gets worked on first?"

If we think about the above example from this perspective, the business's need for payment would be the most compelling format for the story, since that is their reason for operating (and completing the project). However, if we look at the "search for movies" candidate story, the customer's perspective would probably be the strongest case. So in this example, I would write the "complete order" story from the perspective of the business, and the search story from the perspective of the customer.

"Given, When, Then" is another user story format that can be used to document nonfunctional or system-based requirements and then also used for acceptance tests.

> **Given** the account is valid and the account has a MoviesOnline balance of greater than $0,
> **When** the user redeems credit for a movie,
> **Then** issue the movie and reduce the user's MoviesOnline balance.

Regardless of the format, user stories need to be written in a way that can be tested, to give the team a straightforward way of knowing when the story is finished. Testing involves a binary choice; agile user stories are either done and accepted by the customer, or not—there is no "partly done" or "done except for" category on agile projects.

EXERCISE

Rewrite the following requests so they are in the user story format: "As a *<Role>*, I want *<Functionality>*, so that *<Business Benefit>*." You will need to make some assumptions to complete this exercise. For example, I have not identified who is requesting the requirement in each item below, so you will need to assign a logical role.

1. Code the system to support banner advertisements from external advertisers to drive website revenue.

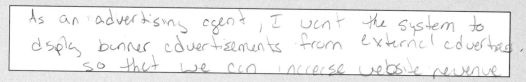

As an advertising agent, I want the system to dsply banner advertisements from external advertsed. so that we can increase website revenue

2. System needs to allow reports to be run on membership trends, including age and location demographics.

> As an executive, I want to run membership trend reports including age and location demographics, so that I can employ targeted marketing strategies

3. All financial transactions should be handled via the SecureServe system.

> As finance exec, I want all transactions to be handled via the SecureServe system so that we ensure customer security

ANSWER

The following are examples of user stories for each item. Depending on the assumptions you made, your answers may be different.

1. As a business manager, I want the system to support banner advertisements so that we can generate additional website revenue.
2. As a marketing manager, I want membership reports including age, location, and trends so that we can run better-targeted incentive programs to drive up membership, referrals, and revenue.
3. As CFO, I want all financial transactions to be handled via SecureServe so that credit card information is protected and fraud is reduced.

The Three C's

At this point you might be thinking, "Okay, we have a bunch of index cards now. But where do we write down all the details needed to complete each story and specify what needs to be built?" The answer is that we don't. In fact, the text on a user story card is intentionally kept as brief as possible. It is simply meant to serve as a placeholder for a richer form of communication—a conversation.

That's because a user story isn't just a written statement—each user story consists of three elements, which are known as the "3 Cs"—the card, the conversation, and the confirmation. Let's look at each of these components in more detail.

» **Card**: The user story card includes just enough text to identify the story because the card is simply a token that represents the requirement for planning purposes. It doesn't provide all-inclusive requirements—we could describe it as a "contract for a conversation" with the product owner. The card is the team's reminder to have a discussion with the businessperson who knows about that requirement. As planning proceeds, the team might add notes to the card about the story's priority and cost. The card is often given to the developers when the story is scheduled to be done, and returned to the customer when the story is complete.

© 2015 RMC Publications, Inc • 952.846.4484 • info@rmcls.com • www.rmcls.com

» **Conversation**: The details of the story are communicated via a conversation—a verbal exchange of ideas, opinions, and information between the customer and the development team. This discussion takes place both when the story is being sized and estimated (during release planning) and when the story is prepared for development (during iteration planning). This conversation might be supplemented with documents, which will ideally take the form of actionable examples.

» **Confirmation**: This refers to the customer's confirmation that the story has been implemented correctly. "Confirmation" means that the product increment passes the customer's acceptance tests and meets the agreed-upon definition of "done." These criteria are defined at the start of the iteration, when the customer and the team develop the examples that will provide the basis for acceptance-testing the outcome.

If these three elements seem like a strange way to document our project requirements, recall that Agile Manifesto principle 6 states that "The most efficient and effective method of conveying information to and within a development team is face-to-face conversation."

INVEST: Characteristics of Effective User Stories

Now that we've looked at the three components of user stories, let's examine how we can write our stories to make them most effective. There are several characteristics of good user stories, and fortunately there is an easy way to remember them—the INVEST mnemonic shown below.[5]

Figure 5.23: Characteristics of Effective User Stories

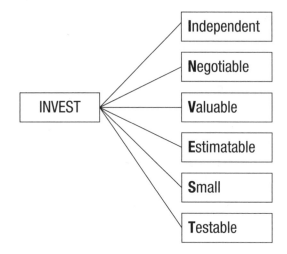

Let's look at these characteristics in more detail:

» **I – Independent**: Ideally, we want to be able to reprioritize and develop our user stories in any order. This is hard to achieve, but it is a goal. We try to create independent user stories that can be selected on merit, rather than dragged into the release because other user stories are dependent on them.

» **N – Negotiable**: The team should be able to discuss user stories with the product owner and make trade-offs based on cost and function. For example, does "Print" mean "Hit the Print Screen button and paste the image into a word-processing program," or does the business need a fully formatted Crystal Reports document with headers and footers? Negotiating user stories leads to an improved understanding of the true requirements, costs, and acceptable compromises.

» **V – Valuable**: If we cannot determine the value of a requirement, then we should question why it is part of the project. Even if a user story is phrased in some way other than the role, function, and business benefit format, it should in some way state the value, or benefit, of the requirement. User stories without clearly understood business benefits will be difficult to prioritize, since backlogs are usually ranked on business value.

» **E – Estimatable**: Although "estimatable" is not really a word, it conveys the idea that we have to be able to estimate the effort of a user story. If it's not possible to say whether the story will take a day or will require two weeks' worth of work, we won't be able to prioritize it based on its cost-benefit trade-off.

» **S – Small**: Small user stories are easier to estimate and test than large user stories. As a unit of work gets larger, our ability to reliably estimate it decreases. Also, by the time a large user story is reported as late, it may take an equally long time to correct or redo the work. Therefore, user stories should be kept small—typically from 4 to 40 hours' worth of work. Another advantage of smaller user stories is that they can usually be completed within one iteration. However, this doesn't mean, "the smaller, the better." Extremely small user stories, such as those that will take 2 hours or less to complete, should also be avoided, since there is an overhead cost involved in creating and tracking stories.

» **T – Testable**: If we cannot test something, how will we know when it is done? Having testable user stories is important for tracking progress because agile teams often measure their progress based upon how many user stories have been successfully accepted.

To summarize, we are looking for small, independent, valuable chunks of work that we can readily estimate and test—the functions of which can be negotiated with the business to find the right level of cost versus performance.

Background information

Software Stories Should Include All Relevant Architectural Layers
When developing software systems, user stories should typically be vertical slices of functionality that cut through all the relevant architectural layers of the system, as shown below.

Figure 5.24: Software User Stories Should Include All Architectural Layers of the System

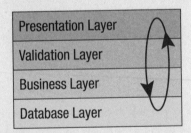

For example, a user story that involves saving and retrieving a customer balance is preferable to one that focuses on mocking up the customer account screen. The first example uses all the components of the system that the final solution will use (the presentation, validation, business, and database layers), whereas the second example just refers to the user interface (presentation) layer. Because it only focuses on one layer, creating a mock-up of the customer account screen won't reveal any issues in the other layers. In addition, the user story won't act as a reliable indicator of true development speed, since other work still remains to be done to achieve the desired functionality.

T&T User Story Backlog (Product Backlog)

After the user stories are written, they are listed and sorted to create the user story backlog (also known as the "backlog" or "product backlog") for the project. This is a single, visible master list of all the functional and nonfunctional work that has been identified for the project, sorted by priority.

Figure 5.25: User Story Backlog Example

User Story	Story #	Priority
As a customer, I want to search for products so that I can buy them.	4	1
As a customer, I want to add products to a shopping cart so that I can pay for them.	2	2
As CFO, I want to complete an order so that I can receive payment.	3	3

Since the backlog is the list of work that needs to be done, as stories are completed, they are removed from the list. (The team might want to electronically flag the completed stories so they will show up in a separate list of completed work—but they will no longer be in the backlog.)

The backlog is organized by priority from the top down so that the highest-value stories are always at the top of the list. This tells the development team exactly where to focus their attention next.

Figure 5.26: Prioritization of the User Story Backlog

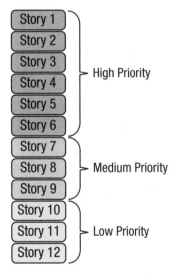

© 2015 RMC Publications, Inc • 952.846.4484 • info@rmcls.com • www.rmcls.com

273

In addition to guiding the team's priorities, the backlog also serves as a planning tool for managing releases and iterations and helps the team balance change requests and risk remediation efforts with feature development (since the backlog includes all of that work). This tool helps the stakeholders coordinate the project and keeps everyone working toward the agreed-upon mission, while still allowing for revisions and updates as new and better information arises.

Managing all the project requirements in a single document like this is one of the most fundamental agile practices. On a fast-moving project, having a single source of information about all the work being done is essential for effective communication and provides highly visible documentation of the project's scope and its status.

Exam tip

On an agile project, there is only one master list of the project work—the user story or product backlog. However, agile teams sometimes use the term "backlog" more loosely to refer to any list or category of work that needs to be done. For example, Scrum refers to the stories selected for the current iteration as the "sprint backlog." However, that isn't actually a separate list; it just refers to the stories the team has selected from the product backlog to complete in a sprint. So although the term "backlog" may be used loosely, agile projects have only one backlog, or visible master list of work.

The more casual uses of "backlog" are unlikely to come up on the exam. So if you see a reference to the "backlog" in an exam question, you can assume that it is referring to the team's master list of work—unless the "sprint backlog" is indicated by the wording of the question.

T&T Refining (Grooming) the Backlog

As expressed in Agile Manifesto principle 2, staying flexible to meet the customer's changing business needs is fundamental to agile, and the backlog is one of the key ways that agile teams ensure this adaptability. The backlog is always evolving, and it needs to be kept continually updated with the latest requirements and new information. Even after a project is well under way, the customer will continue to reorganize and reprioritize the backlog. This process of keeping the backlog updated and accurately prioritized is called refining (or grooming) the backlog.

Exam tip

We briefly mentioned the Scrum activity "backlog refinement" in chapter 1; this activity used to be called "grooming the backlog," and the exam content outline still uses the older term. This is understandable, since the term "grooming the backlog" has been accepted by generic agile and is still used widely, not just in Scrum. For the exam, these two terms are interchangeable, and either of them might appear in an exam question.

Refining the backlog involves progressively adding more detail and adjusting the estimates and priorities of the backlog items, as well as adding new work and removing items that are no longer important. A well-groomed backlog is the foundation of an agile project and essential for the team to successfully deliver value to the customer. The diagram below shows how new work is added the backlog, based on priority.

Figure 5.27: Grooming the Backlog

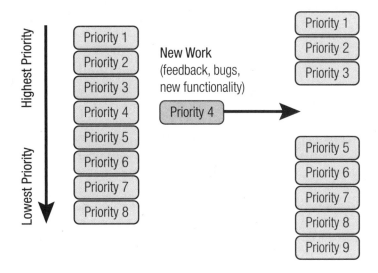

There are three types of changes that may need to be made to the backlog:

» New stories may be added
» Existing stories may be reprioritized or removed
» Stories may be sliced into smaller chunks or resized

As a rule, the first two kinds of changes are done by the customer or value management team. As new work is discovered—whether based on feedback, testing, requests for new functionality, or changing business needs—the customer or value management team will insert a new item into the backlog at the appropriate priority level.

The third kind of change, slicing the stories, is typically done by the development team. (We'll discuss slicing stories in more detail under iteration planning later in this chapter.) However, the development team doesn't prioritize any of the work; to ensure that value is being delivered to the customer, prioritization is done only by the customer or the customer's representatives.

Changes to the backlog can arise from different sources, or reflect collective decisions made in planning meetings. In general, any changes to the backlog should be discussed in the next planning meeting to ensure that they are understood by everyone.

T&T Requirements Reviews

The tools and techniques listed in the exam content outline also include the term "requirements reviews." If you encounter this term on the exam, be aware that it is essentially another name for the process of refining or grooming the backlog.

> ### Exam tip
>
> For the exam, keep in mind that the customer is responsible for setting the priorities and making sure the backlog is up to date. The delivery team is responsible for estimating the work so that the customer can prioritize the work based on a cost-benefit assessment—but the team doesn't do any of the prioritization.

T&T Relative Sizing and Story Points

Studies show that in general, people are not very good at making absolute estimates—we just aren't very good at predicting exactly how much effort it will take to complete a task. Also, often what should be trivial work takes much longer to complete than anticipated, and sometimes new and unforeseen tasks appear. The whole process usually just takes longer than we expected. We could try to include a buffer for these issues in our estimates, but then we would be criticized for padding them. This apparent no-win situation is why estimating has become so unpopular. To compound the problem, the actual time required to complete a knowledge work item often depends on the details of the work and who will be doing it—and that information often isn't known until shortly before the work is done.

So how does agile approach this challenge? It turns out that while people are not very accurate at making absolute estimates, they are better at (and more comfortable with) making comparative (relative) estimates. Therefore, to estimate more efficiently and accurately, agile teams rely on *relative sizing*. They do most of their estimating not in hours or days, but in a relative unit called "story points." Estimating in terms of relative size rather than absolute size allows us to make useful estimates where trying to predict the exact effort required for each story would be next to impossible.

The difference between absolute and relative estimates can be illustrated by giving directions. Using an absolute approach, I would say "You can get to the grocery store by traveling 1.3 miles southeast." To use a relative approach instead, I would say "Go straight out the door for about eight or nine blocks until you reach the park; the grocery store is another five blocks straight ahead." This second set of directions is easier to both explain and understand; it uses recognizable landmarks and relative measures (blocks) and is not stated in absolute measurements (miles).

Like estimating distance via blocks rather than miles, if we have known chunks of work already done, we can estimate new pieces of work more quickly and accurately by referencing the known entities. Using a software development example, imagine we have already developed a simple input screen and have given that task a relative size of 2 story points. We can then estimate the remaining tasks by comparing them to the input screen. For example, we might assign 1 story point to a simple fix or change because we think it will take about half as much work as developing the screen. We might estimate other simple screens as 2 story points, and bigger pieces of work as 3 points or 5 points. In this example, "2 story points" has become the equivalent of a city block, in that it now serves as the reference that we are using to estimate other things against.

Another reason for using relative sizing is that people work at different speeds. A very skilled and experienced person may work two to three times faster than a junior person who needs to spend more time checking their approach and validating the interim steps. When estimating project work we want to understand the work size independent of people's working speeds. Story points help us do this. A 3-point story might be completed by an experienced developer in one morning, or by a slower worker over the course of a day.

Estimating in points rather than hours removes discussions like "this would only take me two hours" and allows the team to focus on the work. It's better to simply say that a task is 3 units of work, rather than having to remember that it would be one hour for Fast Frances but two hours for Medium Mike.

For example, let's say someone says to you, "Hey, let's go for a half-hour run." You have to consider, can you run the same distance they can in 30 minutes, or will that run take you 45 minutes? It's better to know the length of the run—let's say six miles—rather than someone else's estimate based on their own velocity. It's better to say that a task is 1 story point rather than saying it will take Fast Frances 30 minutes to complete. That second estimate isn't helpful if you don't know whether Frances will be assigned to do the work.

Now, of course, taking a comparative approach to estimating does not stop weird things from happening or keep tasks from taking longer than anticipated; however, switching our estimation unit from hours to story points does make that problem easier to accept. Rather than thinking that we are just bad at estimating, we recognize that a couple of the stories took longer than we thought they would. We can see that because those estimates weren't made in a vacuum—they were based on how well we understood how the new activities compared to the work we had already completed.

Estimating in hours can also create issues if some people have other demands on their time and never really stand a chance of delivering 40 hours' worth of functionality in a week. Using a relative measure like story points rather than hours focuses attention on the real issue—the work itself—rather than how much time the work "should" take or how much work we "should" get done.

Another advantage of estimating in story points is that we remove the artificial ceiling of "hours per week." If we have completed 40 hours' worth of user stories by noon on Friday, what's our incentive to take the next story off the stack and carry on working that afternoon? Hour and day estimates are so closely aligned with our workweeks and sense of duty and accomplishment that this connection can actually cause productivity issues. So instead we say, "Last week the team delivered 42 story points, and this week they delivered 45." Can you see how this feels different than saying, "Last week the team did 120 hours of work, and this week they did 130 hours"?

Although story points are the relative unit most commonly use in agile methods, the term for the relative unit of measure doesn't matter—it could be "story points," "points," or "gummy bears." The main idea is to get away from estimating in hours.

The Fibonacci Sequence

Story point estimating is typically based on the Fibonacci sequence of numbers, or some variation of this sequence. This is a common and naturally occurring progression that describes how things grow. The growth of rabbit populations, sea shells, and tree branches all follows the Fibonacci sequence—as do problem sizes and the efforts required to solve them.

Figure 5.28: Fibonacci Sequence

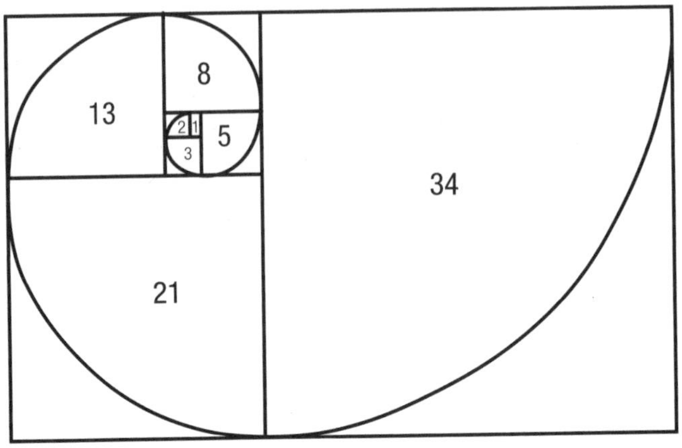

The Fibonacci sequence is derived by adding the previous two numbers in the sequence together to get the next number, starting with 0 and 1. So it starts out 0 + 1 = 1, 1 + 1 = 2, 1 + 2 = 3, 2 + 3 = 5, etc. which creates the sequence "1, 2, 3, 5, 8, 13, 21 . . ." Using this sequence, the size of a user story could be estimated as 1, 2, 3, 5, 8, 13, or 21 story points, but not any number in between. Why would we want to limit our estimates like that? Because it makes the estimating process much faster and more efficient. There is enough variation in the numbers of this sequence to show natural size groupings while eliminating most of the squabbling over small differences in estimates.

Guidelines for Using Story Points

For the exam, you should be familiar with the following guidelines for using story points to estimate user stories:

» **The team should own the definition of their story points.** The story point sizing for a given project should be created and owned by the team. For example, the team can decide that 1 story point equals the effort involved in creating a simple screen, or that 1 story point is equal to the average amount of work they can get done in an ideal day (with no interruptions for meetings, e-mail, etc.). The unit the team chooses is what should be used. By accepting the team's decision about the estimating unit, we reinforce their ownership of the estimates. It doesn't matter if another team's story point unit is sized differently—or if that other team is using an ideal week, rather than an ideal day. The story point unit doesn't have to be consistent across the organization; the team just needs to use it consistently for all the work in their current project. (They could change it for the next project.) This means that we can't compare velocities between teams, because the unique composition and definition of each team's story point make such comparison meaningless.

» **Story point estimates should be all-inclusive.** We shouldn't need to add extra time to the project for testing or refactoring. Instead, our story point estimates should include all known activities required to complete the stories. Otherwise, we end up trying to shoehorn in extra tasks or resort to multiplying the estimates by a "fudge factor" (such as 1.5) to account for additional work. Inclusive story point estimates are preferred over a "magic multiplier" approach because inclusive estimates are more accurate and transparent. In contrast, multipliers can mistakenly be applied several times or forgotten about, and they are harder to defend.

» **The point sizes should be relative.** The size of one point should be *relative* (i.e., consistent) across all the estimates for a given project. So a 2-point user story should involve about twice as much effort as a 1-point story. A 3-point story should take roughly three times as much effort as a 1-point story, and be equal to the effort for a 1-point story and a 2-point story combined. Completing four 5-point user stories should be equivalent to completing twenty 1-point user stories. Although this may sound obvious, it's still worth emphasizing. With a relative scale, the base unit always needs to be consistent so that we can use it to calculate velocity, compare iterations, and validate our release plan.

» **When disaggregating estimates, the totals don't need to match.** When decomposing the work, such as breaking a user story into its component tasks, the sum of the smaller units (the tasks) doesn't need to add up to the estimate of the larger unit (the user story). That's because one of the reasons for breaking down the larger unit is to check our estimate of how long it will take to complete that work. As we break down the user story and learn more about the work involved, we should expect the estimates for the tasks to reflect that new information. If the sum of our estimates for the tasks doesn't equal our estimate for the user story, that is important to know. (And if there is a dramatic difference, that might be a red flag we need to investigate.)

» **Complexity, work effort, and risk should all be included in the estimates.** The total time required to complete a story is a function of its complexity, its level of risk, and the amount of effort involved. (For example: "Does the work require analysis, or is it likely to result in surprises? Do we have to add a single data field, or thirty fields? If we can't complete the work using plan A, will we have to rethink this piece of the project?") When estimating user stories, the team should be sure to assess all three of these attributes.

Background information

Best Practices for Estimating User Stories
In his book *User Stories Applied: For Agile Software Development*, Mike Cohn asserts that the best approach for estimating user stories is one that:[6]

» Allows us to change our mind whenever we have new information about a story
» Works for both epics and smaller user stories
» Doesn't take a lot of time
» Provides useful information about our progress and the work remaining
» Is tolerant of imprecision in the estimates
» Can be used to plan releases

EXERCISE

Test your knowledge of agile estimating concepts by answering the true or false questions in the table below:

Question	True or False
Agile estimates are all encompassing; they should include time for documentation and testing.	*True*
Agile estimates are timeboxed; once the estimates are set, they cannot be altered.	*False*
Agile estimates are created by the product owner.	*False*
Story points are preferable to ideal days, because story points better match estimate characteristics; they are called "stories" because not all stories are true.	*False*
Agile teams create their own estimates.	*True*
Risk should not be factored into user story estimates.	*False*
Teams new to agile should rely on an experienced project manager to create the estimates for them.	*False*

ANSWER

Question	True or False
Agile estimates are all encompassing; they should include time for documentation and testing.	True
Agile estimates are timeboxed; once the estimates are set, they cannot be altered.	False
Agile estimates are created by the product owner.	False
Story points are preferable to ideal days, because story points better match estimate characteristics; they are called "stories" because not all stories are true.	False
Agile teams create their own estimates.	True
Risk should not be factored into user story estimates.	False
Teams new to agile should rely on an experienced project manager to create the estimates for them.	False

T&T Affinity Estimating

Affinity estimating is a technique that involves grouping items into similar categories or collections—i.e., "affinities." In agile, we can use this technique for many purposes, but one of the most important is to make sure our story point unit remains consistent for all our estimates over the duration of the project. Affinity estimating is a form of triangulation—it offers a comparative view of the estimates and provides a reality check. By placing our user stories into size categories, it is easier to see whether stories with similar estimates are in fact comparable in size. This helps us make sure we have not gradually altered the measurement of our story point during the estimating process.

Let's look at how to do this. After each round of estimating, we create columns on a whiteboard that represent different story point sizes of user stories—column 1 represents the 1-story-point category, column 2 represents the 2-story-point category, and so on. If we have already done multiple rounds of estimating, we start by placing stickies for the stories we've estimated in previous rounds into the appropriate column.

Figure 5.29: Affinity Estimating with Story Points

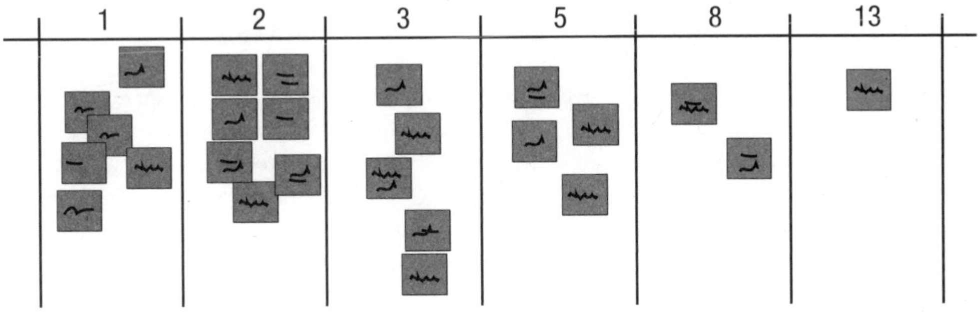

Next, we start placing the user stories we have just estimated into the appropriate columns. As we place each card, we compare it to the cards that are already there (from this round or previous rounds). If the new user story looks like it is about the same size as the other stories already posted in that column, then great—our estimation scheme hasn't warped. If that story looks different, however, we will need to do a checkpoint to discuss what exactly a story point means, and then recalibrate the estimates.

If this is our first round of estimating, then after checking that all the stories within each column are comparable, we can also look across the columns to see if the stories appear to be proportionately sized across the columns. For example, are the stories in column 2 (2 story points) about twice the size of the stories in column 1 (1 story point)? Are the stories in column 8 about four times the size of the stories in column 2? And so on. If we find any discrepancies, we will again need to revisit our definition of a story point and recalibrate our estimates. The outcome of this process should be consistently sized story points across all the estimates that have been made to date.

T&T T-Shirt Sizing

T-shirt sizing is a high-level estimating tool that is used to do the initial "coarse-grained" estimates of the product features and user stories during the initiation stage of a project. At that point, we aren't trying to generate detailed estimates; we're aiming to do "just enough" estimating to map out the overall effort that we expect will be involved in the project and get the work started. This is referred to as coarse-grained estimating, since these early estimates will be progressively refined as the project continues.

To see how this works, let's look at an example project. For an online movie service, we have identified six product features:

- » Rate movies
- » Browse movies
- » Rent movies
- » Sell movies
- » Review movies
- » Sort movies by year

Preventing Story Point Inflation

Affinity estimating is a great way to discover and control the effects of "story point inflation." This refers to the common tendency for a team's story point estimates to increase over the course of a project. This upward trend is usually an unconscious side effect of stakeholder scrutiny of how much work is getting done in each iteration and how much work remains. Since stakeholders are happy when velocity targets are met, agile teams have a tendency to gradually inflate their point estimates.

Here's an example of how this happens. The team is in an estimating session and can't agree on whether a story is 3 points or 5 points. Then someone says, "Look, let's just call it a 5 and be done with it." Although this might not be anyone's conscious intention, when the team's velocity is calculated at the end of the week, they will get credit for 5 points rather than 3—"Look how much work we're getting done, yay!" True, that makes everybody happy, but the trouble is that we are now using an elastic tape to measure our progress. Since we can't effectively compare our current estimates with our past velocity, how can we objectively assess how far through the project we are, and when we will be done?

Affinity estimating keeps us honest. After the estimating session, we put our new 5-point story up on the board and compare it to our old 5-point stories, asking, "Is this new story really as much work as the old stories?" If it isn't, then we are having some point inflation, and we need to discuss and reset the size of each of our story point categories.

Our next step is to roughly size the six features for our movie service, so that we can assess the relative effort involved in developing each one. Since we haven't done any work on the project yet, we'll have to make these estimates based on our experience with similar work items in the past. To reflect the uncertainty involved in those estimates, our estimating unit will be T-shirt sizes, ranging from Extra Small (ES) to Extra-Extra Large (XXL). We won't try to estimate the absolute size of each category, or even how much bigger or smaller each size is compared to the other sizes. All we will know is that Extra Small is smaller than Small, which is smaller than Medium, and so on.

Figure 5.30: Features by T-Shirt Size

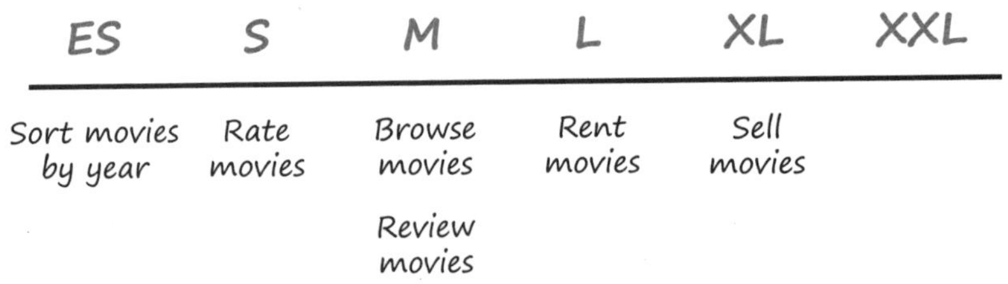

ES	S	M	L	XL	XXL
Sort movies by year	Rate movies	Browse movies	Rent movies	Sell movies	
		Review movies			

The results of our sizing effort are shown in the above figure. As you can see, we've decided that:

» The functionality that allows users to sort movies by year will require the least effort to build, so this feature is Extra Small.

» The online shopping cart we need to sell movies will require the most effort to build, so this feature is Extra Large.

» There are two features that we think will take Medium effort to build—"Browse movies" and "Review movies."

» None of these features will require an Extra-Extra Large effort.

Based on our collective experience and expertise, we are pretty sure that these relative sizes are correct. But so far we don't have a way to check or confirm these guesstimates. To do that, we need to decompose these six product features into the next level of detail—user stories. Once we do that, we end up estimating that it will take 45 user stories to build the six features, as shown below.

Figure 5.31: Features and User Stories

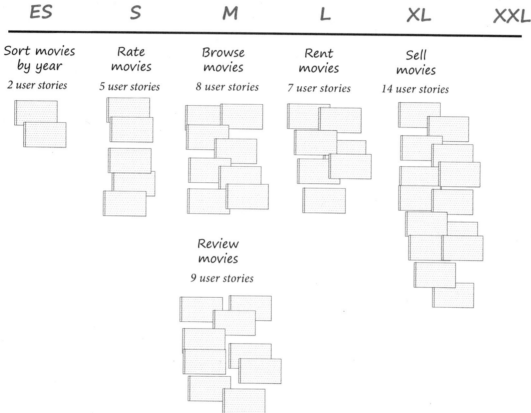

In other words, these 45 user story cards represent all the work that we estimate will need to be done to build the product. Of course, these user stories may change as we proceed—but this is our initial working plan. Just based on the number of user stories in each feature, this breakdown appears to confirm that "Sell movies" will take the most effort to build, and "Sort movies by year" will take the least effort, as we originally guessed. However, it also appears that "Rent movies," which we had sized Large, might actually be smaller than "Browse movies" and "Review movies," which we sized Medium.

Or maybe not—because we haven't determined the relative size of the 45 stories yet. Some of them might be very small, and others might be very large. At this point, the largest user stories might actually contain several smaller stories that we'll have to break down as we get closer to building them. So our next step is to estimate all the user stories in T-shirt sizes—just like we did for the features. (After we do that, we can also use affinity estimating to double-check that the stories in each category are comparable in size.)

Our organization of the 45 user stories based on T-shirt sizes might look something like this:

Figure 5.32: User Stories by T-Shirt Size

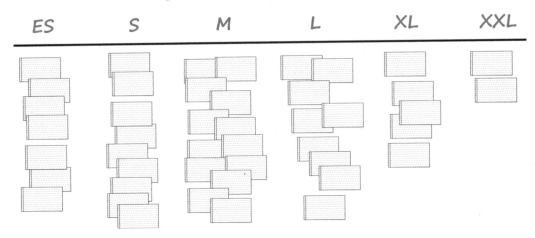

Now that we have sized all the stories, we can use the relative sizes of the stories in each feature to refine our T-shirt estimates of the features. Let's say that in this example we find out that, on average, the user stories in "Rent movies" are materially larger than the stories in "Browse movies" or "Review movies." As a result, we can see that "Rent movies" really will require more effort than the two Medium features, just as we had originally guesstimated.

Remember that agile estimating is done in stages, using progressive elaboration. The purpose of using T-shirt-size estimates is to come up with the initial plan that will be continually refined during release and iteration planning throughout the project.

T&T   Story Maps

A story map is a high-level planning tool that agile stakeholders can use to map out the project priorities early in the planning process, based on the information available at that point. A story map is essentially a prioritized matrix of the features and user stories for the product that is being built. Once created, it can be easily adapted to serve as a "product roadmap" that shows when the features will be delivered and what will be included in each release. The team will then refer back to this roadmap as they progressively elaborate and refine their plans during release and iteration planning. (We'll discuss product roadmaps next.)

To create a story map, we start by listing groups of features (or sometimes a usage sequence) for the product horizontally across the top of a matrix, from left to right. Down the columns, we arrange the user story cards in each feature in descending order of priority. (Story maps are another example of low-tech, high-touch agile tools—anyone can move the cards around on the map.)

Figure 5.33: Story Map

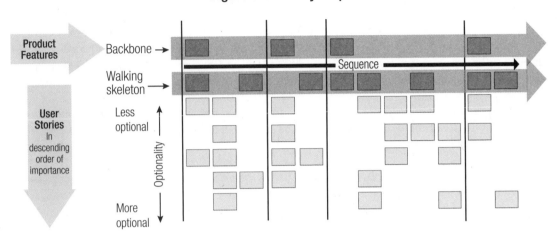

Although some of the feature columns in this diagram are wider than others, that's just because they include more user stories. At this point, we don't know yet how many iterations will be required to build each feature, or how the releases will be timed. (That will be determined in release planning.)

The first two rows of user stories at the top of the diagram are especially important, and they have special names. On the top row of the story map, we place the stories that describe the essential functions needed for the system to work. This line is called the *backbone*. For example, let's say that the customer is asking for "a smartphone with a high-end camera." We will put on this first line all the essential elements that are required for such a product to function—including a screen, a battery, an outer casing, a microphone, speakers, a charger port, and a camera. Notice that we can't meaningfully prioritize these elements. It's pointless to ask, "Is the screen or the battery a higher priority?" We need both. So the items in the backbone are just given, based on the product that is being built.

On the next row, we place the stories that describe the smallest version of the system that will meet the customer's most basic needs. This line is called the *walking skeleton*. Unlike the items in the backbone, which are simply required, the items in the walking skeleton will depend on what the customer considers to be a minimal viable product. In our example, we need to ask, what are the most basic features that will provide value to the customer? Does the minimal version need to have touch-to-shrink-and-expand capability? To know which stories to put on this line, we have to talk to the customer and find out which features are essential in the most minimal version of the product, and which features can be added in later iterations.

Finally, we place all the other user stories below the walking skeleton, in descending order of priority to the customer. So now we have a high-level picture of the product, by feature, in order of business value.

Story maps show what the customer considers to be absolutely essential for the system (the backbone) and necessary for the minimal solution (the walking skeleton), along with the other features that will comprise the later releases. This helps the stakeholders align their expectations and discover any misunderstandings or mismatches. As such, these tools are often used in stakeholder communications or posted on a wall as an information radiator of the project plan. They have the benefit of engaging people in a different way than the same information depicted on a Gantt chart, since the story map format is more visual and inviting.

T&T **Product Roadmap**

A product roadmap is a visual depiction of the product releases and the main components that will be included in each release. This is a communication tool that provides project stakeholders with a quick view of the primary release points and intended functionality that will be delivered. Although there are various ways of depicting a product roadmap, story maps are a commonly used approach, as popularized by Jeff Patton.

With this approach, after placing the features on the story map according to their importance and sequence (as described above), the team balances the customer's priorities with their projected capacity, and outlines what they plan to deliver in each release.

For example, we can convert the story map we created above to a product roadmap by outlining the user stories we plan to deliver in the first, second, and third release. We can show this to the product owner to communicate our plan for these first three releases.

Figure 5.34: Product Roadmap (Story Map Showing the Project Plan)

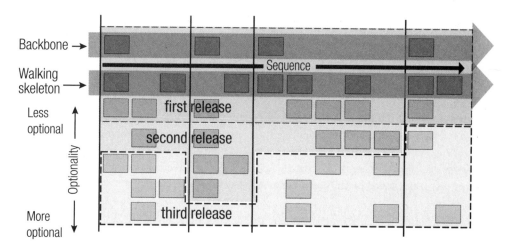

Using this approach, the product roadmap for a project will typically consist of one or more story maps showing what will be delivered in each release.

Although the product roadmap shows what we plan to deliver in each release, remember that this is a high-level planning tool—and we know there will be changes. So in planning each release, we will go back to the roadmap and confirm that the basic plan will still work, or make any necessary adjustments. We add up the story points we planned for the release, and check whether the plan is still viable, given our velocity trends and the amount of time designated for the release. If the plan no longer looks viable, or if it looks too risky, then we might request a backlog review with the product owner to see what can be changed.

Figure 5.35: Confirm the Plan during Release Planning

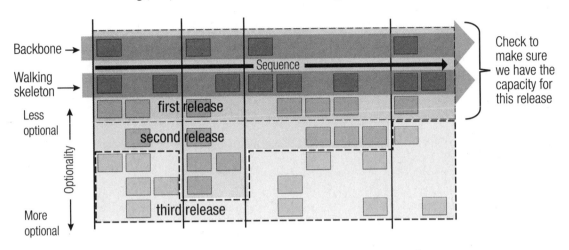

The topics we've discussed so far—such as affinity estimating, T-shirt sizing, story maps, and roadmaps—are sizing tools used for high-level planning. We'll conclude this section by looking at two tools for estimating user stories—wideband Delphi and planning poker.

T&T Wideband Delphi

Wideband Delphi is a group-based estimation technique in which a panel of experts submits estimates anonymously so that none of the participants know who has made which estimate. This anonymity produces improved estimates because it minimizes the cognitive and psychological biases that can result in flawed estimates, including:

» **The Bandwagon effect**: People tend to converge around the viewpoint that is gaining the most adherents, even if it doesn't reflect their own opinion.

» **HIPPO decision making** (HIPPO = HIghest-Paid Person's Opinion): People gravitate to the ideas of experts or superiors, rather than judging ideas on their own merit. Nobody wants to disagree with the boss.

» **Groupthink**: People make decisions to maintain group harmony rather than expressing their honest opinions.

Exam tip

Although wideband Delphi is a complex process, for the exam, you won't need more than a broad familiarity with this approach, at most. This isn't an important tool for agile teams, and it might not even be mentioned on the exam. From an agile perspective, the most important point about wideband Delphi is how it uses anonymous rounds of estimating to combat the biases listed above and reach convergence. These elements have influenced the much more popular agile estimating tool called planning poker that we'll discuss next.

The wideband Delphi estimation process starts with a planning effort to define the "problem" that is being analyzed—i.e., the work that is being estimated. Instead of estimating the whole project at once, the expert group breaks it down into more manageable chunks. They create a problem specification, identify the assumptions and constraints, and outline the process for the estimation rounds. This plan specifies details such as the measurement units that will be used (e.g., person weeks, hours, dollars, etc.), whether the estimates should include documentation time, and the exit criteria that they are aiming for (e.g., we want to get to +/- 20 percent tolerance on the estimate range).

Before the experts begin creating estimates, they read the problem specification and discuss any qualification questions. They each list the tasks involved in the part of the project that is being estimated and enter their estimates for each task. After the first round of estimation, the facilitator gathers their estimate sheets and plots them on a chart, being careful not to identify which estimate is from which estimator.

Figure 5.36: Wideband Delphi—Round 1 Estimates

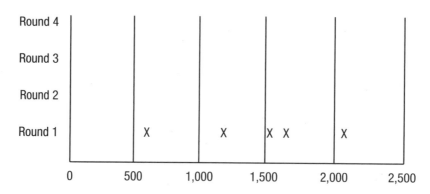

Here we can see that after the first round of the estimation process, the group's estimates range from approximately 600 person hours to 2,100 person hours. The group doesn't know if the 600-hour estimate came from the sponsor or if the 2,100-hour estimate came from the most experienced developer. All of this information is kept anonymous.

The participants then discuss the tasks they included and any assumptions or other significant factors that influenced their estimates. For example, Bill might report, "I added two weeks for regression testing of downstream applications, because we are amending the accounting table; the last time we did that, it broke the billing system, so we need to allow some time for additional testing and possible remediation."

Once they have discussed all the tasks, assumptions, and significant factors, the group does another round of anonymous estimating. After several rounds, a consensus usually starts to emerge in the estimates, as shown below.

Figure 5.37: Wideband Delphi—Getting Closer to Consensus

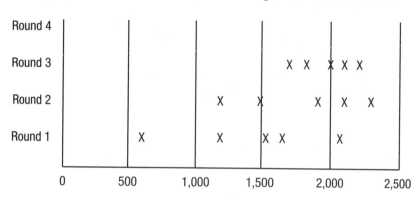

Here, we can see that with each round, there has been a "northeast drift" toward a tighter set of estimates. This drift reflects the emerging consensus that the part of the project being estimated will require a higher range of person hours than most of the group originally estimated. Once the estimates are close enough together that they reach the exit criteria (in our example, the highest and lowest estimates must be within a range of +/- 20 percent of the median estimate), the estimating process is complete.

At this point, a single master task list is created from everyone's task lists. If any tasks were excluded from the estimation process (such as documentation), then those tasks are added to the master list. The team reviews the results to make sure everyone agrees on the final task list and estimate range.

Wideband Delphi estimation reflects agile values, because it is:

» **Iterative**: The process is repeated several times, until a consensus is reached.
» **Adaptive**: Team members have a chance to update and improve their next round of estimates based on discussion with other participants.
» **Collaborative**: A team-based collaborative process improves participants' buy-in to the results.

Despite these agile characteristics, wideband Delphi isn't widely used by agile teams. However, it is important to understand how and why this method works, since it the basis for another technique that *is* widely used in agile—planning poker.

T&T Planning Poker

When an agile team needs to come up with the estimates for their user stories, the most common way of doing this is a collaborative game called planning poker. Planning poker is designed to provide a faster, more efficient process that has all the advantages of wideband Delphi—since planning poker is iterative, adaptive, collaborative, and anonymous enough to minimize most bias.

Planning poker uses playing cards that show numbers based on the Fibonacci sequence. Each participant receives a set of these cards, as shown below. The numbers on these cards represent the relative units that will be used for the estimates, such as story points, developer days, or jelly beans—whatever the team is using to size its estimates.

Figure 5.38: Planning Poker Cards

Once everyone has a set of cards, the facilitator—who could be the product owner, the ScrumMaster, or any team member—will read a user story. For example, let's say the first story is, "As a customer, I want to be able to change my password so that I can ensure my account is secure." The group will discuss the story briefly—and then each estimator will select a card to represent his or her estimate of the size of that story, without showing it yet.

When everyone is ready, the facilitator will count to three, and then everyone will lay down their cards at the same time, showing their estimates of the effort required to develop that user story. They all turn over their cards simultaneously to ensure that the initial estimates are not influenced by other members of the group.

Let's say there are four estimators, and three of them lay down cards with the number 5 and one person lays down a card with the number 3. In this case, the estimate would be recorded as a "5"—when the range is small and there is a rough consensus on the estimate, the largest estimate is selected. After that, the team will move on to the next story to keep the game moving quickly.

However, what if there are three cards with the number 5 and one card with the number 13? In this case, the team will discuss the outlier (13)—and the conversation might go something like this:

Facilitator:	Okay, Bob, can you tell us why you think this is a 13?
Bob:	Well, we are using LDAP authentication to keep the passwords synchronized with other applications. Password changes will need to be pushed back to the LDAP server, and we don't have security permissions to do that. So we need to get the security group engaged, and that takes time.
Facilitator:	Hmm, I didn't know that. Okay, let's estimate this again.

At that point, everyone picks up their card and estimates again, taking the new information into account. This time, we might see a new consensus emerge around the number 13, similar to the "northeast drift" we saw in wideband Delphi.

Planning poker is another example of the participatory decision models discussed in chapter 3. It is faithful to the wideband Delphi approach in that it is iterative, adaptive, and collaborative; minimizes bias; and supports the emergence of convergence and consensus.

Studies at Motorola and Microsoft show that many project teams find planning poker to be not only as accurate as their previous approach to estimation, but also quicker and more enjoyable.[7] That said, the goal of this technique is not to create precise estimates. Instead, it aims to help the team quickly and efficiently reach consensus on reasonable estimates so the project can keep moving forward.

Release and Iteration Planning

Agile projects are divided into releases and iterations. As we've discussed earlier in this book, an iteration is a short, timeboxed development period, typically one to four weeks in duration. A release is a group of iterations that results in the completion of a valuable deliverable on the project. An agile project will have one or more releases, each of which will contain one or more iterations, as illustrated below.

Figure 5.39: Project Broken into Releases and Iterations

Iterations (each with an iteration plan)

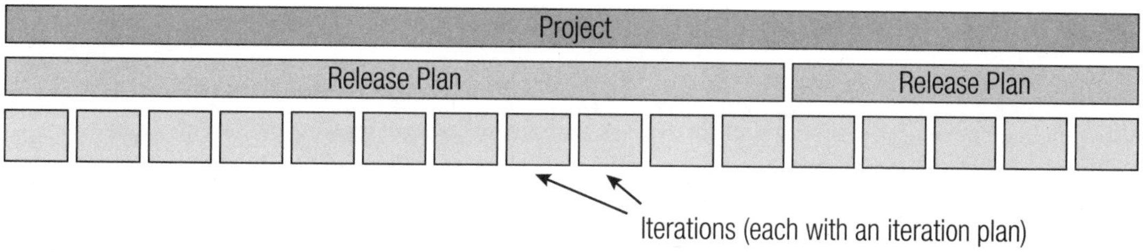

This diagram shows a single project with two releases to production. The first release contains eleven iterations, and the second contains five iterations. We start planning releases and iterations early in the project life cycle and progressively refine the planning effort multiple times as the project progresses.

On agile projects, we select from the top-priority backlog items to come up with our next iteration goal. We then decompose the iteration goal into user stories to get the iteration plan. We continue planning by decomposing those user stories into tasks. While the work is being done, we discuss the details of the work in the daily stand-up meetings. These are all practical examples of how agile teams decompose, prioritize, and plan the project components at the last responsible moment.

Types of Iterations

The activities done in the iterations are the heart of the agile development process. However, not all iterations are equal—some play a specialized role in the project. The most common types of iterations are the development timeboxes during which the team builds increments of the product, as explained above. However, for certain projects, the team might need to schedule other types of iterations as well. These iterations are planned and implemented just like the development iterations; the difference is that they don't focus on building a product increment.

To explain the different types of iterations, we'll use this diagram, which shows when each type of iteration occurs along the project timeline:

Figure 5.40: Types of Iterations

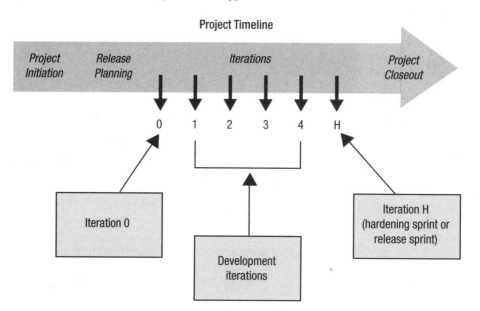

This timeline shows a very simple project that has a single release and four development iterations. Before the first development iteration, the team has scheduled an "Iteration 0." This is an optional iteration that we can use to set the stage for our development efforts. By definition, Iteration 0 typically doesn't involve building any deliverables for the customer. Other than that, its nature, length, and scope depend completely on the needs of the project. But in any case, the work undertaken in Iteration 0 should be limited to "just enough" for the first development iterations to be successful.

Once this preparatory work has been completed, the development iterations begin with Iteration 1. In this example, there are four development iterations. At the end of Iteration 4, we have finished building the product increment, and the customer has accepted it. However, let's say that for this project we need another effort after the deliverable is accepted to prepare it for release. This specialized kind of iteration is variously called "Iteration H," "hardening sprint," or "release sprint." The work done in this wrap-up iteration might include stabilizing the code, documenting the product, compiling the final assemblies, or completing additional testing. If a project has multiple releases, we might need to schedule a hardening iteration before each release.

Bear in mind that Iteration 0 and Iteration H are optional elements of agile. Many agile teams prefer to incorporate any preparatory or hardening work that is needed into their development sprints, rather than scheduling them as separate efforts. Many agile projects get along just fine without them. However, if these specialized iterations are needed, we have to decide this in advance and factor them into the timeline for our release plan.

Next, we'll look at "spikes," which are similar to iterations in that they are another kind of short, timeboxed effort devoted to a specific purpose on an agile project.

Spikes

Spikes are a key tool that agile teams use to head off problems and resolve them as early as possible. A spike is a short effort (usually timeboxed) that is devoted to exploring an approach, investigating an issue, or reducing a project risk. Although spikes can be done at any time during a project, they often take the form of brief exploratory iterations or proof-of-concept efforts that are done at the start of a project, before the development effort begins.

When used in this way, a spike might look very similar to Iteration 0. But it isn't really the same thing. That's because a spike is more flexible than Iteration 0; Iteration 0 simply focuses on setting the stage for the initial development efforts, but a spike can be used to investigate a wide range of issues at any time. In fact, it wouldn't be unusual for a project team to schedule both a spike and Iteration 0 at the start of a project, if circumstances warrant it.

Below we'll define two terms for specialized kinds of spikes that you might encounter on the exam: architectural spikes and risk-based spikes. However, bear in mind that in the real world these two terms are somewhat flexible—they aren't necessarily used in the same way by different teams, and the generic term "spike" can be used instead.

T&T Architectural Spike

An architectural spike is a short, timeboxed effort dedicated to "proof of concept"—in other words, checking whether the approach the team hopes to use will work for the project. For example, we might say, "We'll spend one week testing the performance of the native database drivers before making a decision on connectivity." Or, "We need to test the viability of using straw as a renewable home insulation product." The idea is to explore the viability of an approach or candidate solution in a short time frame. We are seeking to answer questions like "Can we do it?" or address stakeholder concerns like "Prove it will work!" to see if we can use the approach we have in mind.

T&T Risk-Based Spike

A risk-based spike is a short, timeboxed effort that the team sets aside to investigate—and hopefully reduce or eliminate—an issue or threat to the project. These short experiments to investigate risky portions of the project are a key tool for risk management.

As an example, imagine we are building a container-based algae production unit to capture carbon dioxide (CO_2) emissions. This production unit calls for underwater lighting using cheaply available parts. In this scenario, we might undertake a spike to test waterproofing options for the fluorescent tubing. By testing various options in short, small-scale experiments, we can reduce the risk associated with a key component of the project (developing cost-effective underwater lighting) early and at a low cost. If this exercise is successful, then we can eliminate that risk, and the project's overall risk profile will be reduced, as shown below. Here we can see that our risk-based spike in January to test the underwater lighting approach was successful, so we were able to eliminate that risk. (This is a "risk burndown" graph, which we will discuss in the next chapter.)

Figure 5.41: Algae Project Risk Profile

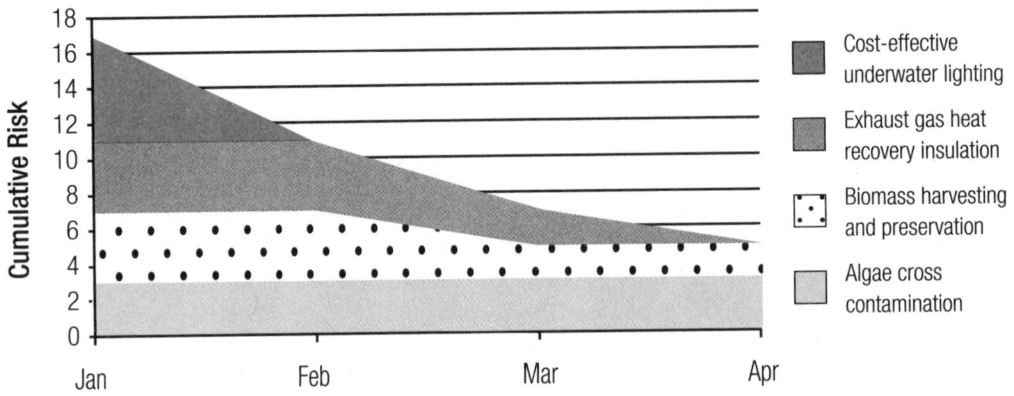

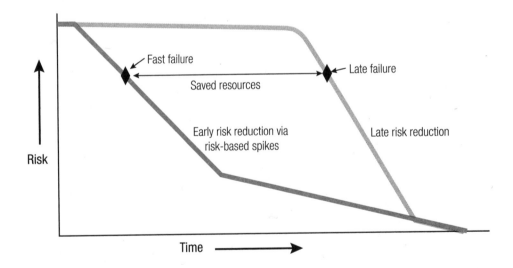

Risk-based spikes are often used on software projects to test unfamiliar or new technologies early in the project before we proceed too far with development. For example, imagine that our analysis of a medium-priority feature raises the question, "Can we interface with the old legacy system from a mobile device?" We might use a risk-based spike to investigate this question before we continue too far down the path of analyzing, designing, and developing this feature.

Fast Failure

If a proof-of-concept effort isn't successful, we can try a different approach. But if none of the approaches we try are successful, we reach a condition known as "fast failure." Although it sounds bad, fast failure can be a good outcome; rather than continuing on a project that would have eventually failed, the remaining funds and resources from the project can now be directed to other projects that are awaiting resources.

Figure 5.42: Fast Failure

Research and development companies that have a business model based on new product development recognize that a large percentage of their projects will not be successful. They generally operate with the philosophy that if something were easy, other companies would have done it by now—so they look for a breakthrough new approach instead. In this kind of environment, fast failure can greatly reduce the sunk costs invested in nonviable projects, which allows the company to try more approaches, thus increasing the odds of finding a revolutionary new product.

Fast failures can occur on a smaller scale on successful projects, too. Maybe our architectural spike on using straw for insulation found that straw would allow too many bugs and rodents to live in the walls. As long as we learn this quickly and haven't based the whole project upon this approach, we can move on and try some other material for renewable insulation.

EXERCISE

Think about how each of the following issues might be investigated through a risk-based spike, and describe what short exercise you would undertake to prove the concept or explore the risk.

1. We are not certain whether the remote data center can validate new credit card applications in the two-subsecond response time needed to ensure a reasonable customer experience.

2. We do not know if Project Alpha has solid management support, and we have concerns about getting approval for buying the development and production machines when it comes time to do so.

3. Project Beta will require working with Ted in the database group again, and the last time we had to do that, there were major arguments about approach and approvals.

ANSWER

1. This is a classic technical risk that can be investigated by a small-scale test. We need to do some performance benchmark tests for various times of the day and under various activity loads and then measure the response times to see if the two-subsecond response time is achievable.

2. People often associate risk-based spikes with technical work, but there is nothing stopping us from also using this approach for business and human resource risks. So for this scenario, if we are not sure whether there is management support for Project Alpha, let's try getting approval to order those development and production machines in the first iteration of the project and surface the issue sooner rather than later.

3. This is another nontechnical risk that we can again use a risk-based spike to address. How about scheduling some work with Ted as early as possible in the project to see if we can find a way to improve the relationship and get things to go more smoothly this time? If the approach fails, then at least we know about the issue and can consider it in planning the work, so that this problem doesn't jeopardize the success of the project. Exploring the issue early gives us time to help resolve it within the project or escalate it to others outside of the project who can address it.

High-Level Planning (Visioning)

Before we can start planning the first release, we need to complete the high-level visioning process. This involves identifying and roughly sizing the product features and user stories. Here, we are trying to do "just enough" estimating to map out the overall effort required for the project and get the work started. To do that, we create our initial coarse-grained estimates—using tools such as affinity estimating, T-shirt sizing, story maps, and the product roadmap—that we will progressively refine as the project continues.

The participants in the high-level estimating process are likely to include the product owner and sponsor, as well as key members of the delivery team, and possibly other major stakeholders. This differs from the later, more-detailed stages of agile estimating, which are primarily done by the development team. That's because this early planning effort is essentially an extension of the project visioning process; it doesn't try to define the specific work items or delve into the nitty-gritty details of the work. It might provide just enough detail for the team to commit that the project is achievable.

Outputs of High-Level Planning

Before we can move on from high-level planning and start the release-planning process, the following elements need to be in place:

» **An updated, prioritized backlog of user stories and risk response actions:** To kick off the planning process, the customer needs to update and rank all the features and risk actions related to the upcoming release by their business value, so that we can pull the most important stories from the top of the backlog.

» **High-level (coarse-grained) relative estimates for each user story:** Since release planning builds on high-level planning, that process needs to be completed first. We need to have already sized the user stories and risk response actions related to the release goal in coarse-grained terms, and grouped them into relative T-shirt sizes.

© 2015 RMC Publications, Inc • 952.846.4484 • info@rmcls.com • www.rmcls.com

» **A release goal (i.e., deliverable) focused on customer value:** With the above elements in place, we can define our goal for the upcoming release in terms of the business value that will be delivered. For example, "The initial release will deliver the features of renting and buying movies to get some revenue coming in." Although we already mapped out the goals for each release during high-level planning (on the product roadmap), we will need to carefully re-examine those goals, taking into account what we have learned since that plan was made. We will ask, "What can we realistically deliver in the first release, and how will we go about doing that?"

» **A target date for the release:** As we discuss the deliverable for the upcoming release, we also come up with a target date for completing it. This date might simply be based on the customer's requirements, or it might take into account the team's projected velocity and coarse-grained estimates.

T&T Release Planning

Release planning is done in a meeting in which all the stakeholders are represented; such a meeting is held before we start work on each new release. The goal of these meetings is to determine which stories will be done in which iterations for the upcoming release—and, in less detail, for the subsequent releases. (As we have seen, agile teams plan the near-term iterations and releases in the most detail, and outline the later portions of the project more roughly.) During the release-planning meeting, we will:

» Assess the prioritized backlog and review the sizing of stories, resizing them as needed.
» Sort the stories by release, selecting those that will be in the upcoming release, the next release, and future releases.
» Refine our initial outline or roadmap for the upcoming release, changing it as needed
» Slice the stories that will be done in the upcoming release into pieces of work that are small enough to be completed within one iteration.

As we slice the stories, we will map out a rough outline of how the release will be accomplished and how many iterations will be needed. At the end of this meeting, we should have:

» A shared understanding of the release goal
» A list of the stories to be done in the upcoming release, sliced into manageable chunks
» A roadmap for completing those stories (i.e., a plan for what we will accomplish in each iteration)
» A rough outline of what will be done in the future releases

By the way, other than prioritizing the stories and risks based on business value, the product owner should not influence the delivery team's planning and estimates of what they can accomplish. It is the team members—not the customer, ScrumMaster, or value management team—who should determine what is feasible in developing the product.

Next, we'll look more closely at some of the key activities involved in release planning: selecting the user stories, calculating how much the team can get done, estimating the team's initial velocity, and slicing the stories.

Selecting the User Stories for the Release

Releases are planned around delivering useful and valuable functionality to the customer. A release may be *date driven* ("We need something to demo at the trade show") or *functionality driven* ("Once we can capture and process customer orders, we want to go live; management reporting and account renewals can come later"). Whatever drives the release, we need to determine the functionality that can be developed and turned over or delivered for the planned release.

To put it another way, when planning a release we ask, "What proportion of the user story backlog can be delivered in this release?"

Figure 5.43: Planning a Release

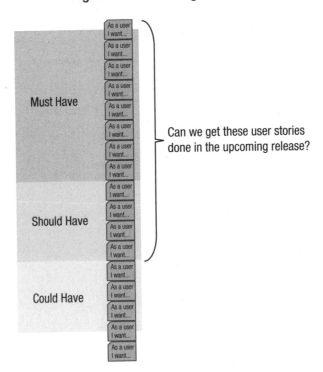

In the above figure, the functionality that represents all the user stories ranked as "must haves" and "should haves" in the backlog has been selected for the upcoming release. So the next question is, "How likely is it that we will be able to complete this work by the release date?"

How Much Can We Get Done?

To plan how much work can be completed in a given iteration, agile teams generally use the velocity trend that has emerged over their previous iterations. To see how this works, we'll use this velocity tracking chart for the first eight iterations of a project:

Figure 5.44: Using Velocity to Plan a Release

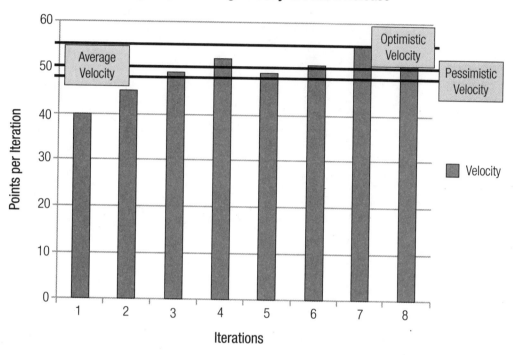

In this example, our velocity started with 40 points completed in Iteration 1. In Iteration 4, we broke past the 50 points' mark, and in Iteration 7, our velocity reached 55 points—but this was a one-time occurrence. If we want to calculate the best-case delivery date, we could use our maximum velocity of 55 points to make an optimistic prediction of how many story points we'll deliver in every iteration going forward, but it doesn't look very likely that we'll be able to achieve that consistently, given our velocity for the other iterations. So it would be safer to assume a lower velocity for the release. Based on this data, 48 story points per iteration would be our most pessimistic estimate. However, for the most likely estimate, we want to use our average velocity, which in this example is 50 story points per iteration.

Now let's say the product owner is asking for a release in two iterations' time. Can we get that done? To find out, we need to determine how much of the backlog we'll be able to complete in that timeframe. We start by calculating 2 iterations x 50 points per iteration = 100 points' worth of functionality. We then add up the story points for the backlog items, working from the top down, until we reach 100 points:

Figure 5.45: Applying Average Velocity to the Backlog

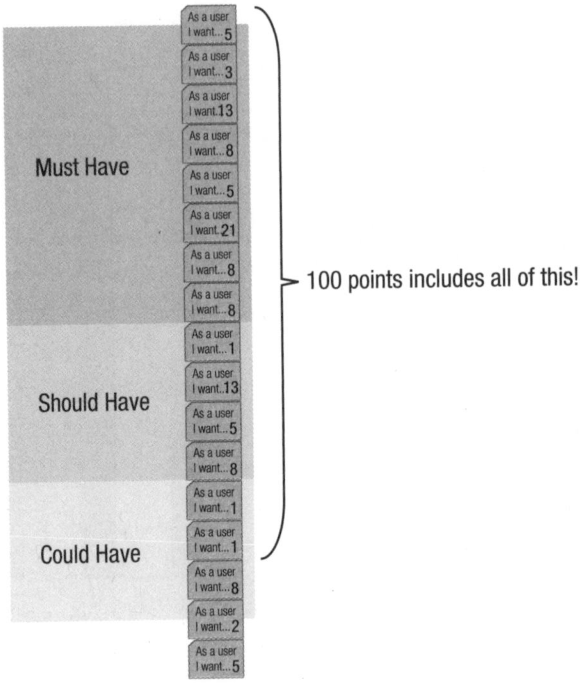

In this example, 100 points takes us all the way through the "must-have" and "should-have" functionality, as well as a couple of small "could-have" stories. So barring any unforeseen circumstances, it looks possible to deliver all the must-have and should-have stories for the release. These calculations help ensure we do not try to commit to more functionality than we can deliver within a release.

Estimating Velocity for the First Iteration

As we've seen, plans are usually based on the team's average velocity. But when estimating the first few iterations of the first release, we don't have any velocity data yet. So this is a special case. To estimate our velocity for the initial plans, we need to gather the team, look at the backlog, and ask:

> *Based on each person's capacity, which of the top-priority stories in the backlog for this release can we realistically commit to finishing in the first iteration?*

The total points for these stories become our initial velocity estimate. For example:

> *We believe we can complete the top four stories in the backlog during the first iteration, and these stories add up to 15 story points. So for initial planning purposes, we will set our velocity at 15 story points.*

Here, we are basically thinking of the first iteration as a timebox, and pulling as many stories from the backlog into the timebox as we think will fit. Once we are satisfied with our estimate, we use the total story points in that timebox as our initial velocity for planning purposes. Although this initial velocity estimate is needed to

© 2015 RMC Publications, Inc • 952.846.4484 • info@rmcls.com • www.rmcls.com

set the deadline for the first release, we know that it is just a starting point—it isn't a valid metric for making long-term predictions. So it will only be used for the immediate planning needed to get the project going.

In general, teams tend to be overly optimistic in setting their initial velocity. And even if that estimate is accurate for the first iteration, as the project unfolds it may not turn out to be realistic over the long term. Because of the uncertainty of estimating initial velocity, some teams add a buffer to their velocity estimates for the first few iterations. This gives them an extra cushion of safety during the period when their velocity is likely to be most volatile. However, this is only a temporary measure; as each new iteration is completed, the team's actual velocity will become an increasingly accurate planning metric.

EXERCISE

Using the following information, calculate how many iterations will be needed to complete the release.

The team's velocity remains fairly stable, averaging 20 points per iteration. They have 200 points' worth of functionality left in the backlog for this release. However, in each iteration, they have been discovering there is about 10 percent more work than anticipated, consisting of change requests and new functionality. The sponsors would like to know how many more iterations it will take before the release will be done.

11 iterations

ANSWER

If we expect to complete 20 points per iteration and we have 200 points left in our backlog, we can do the following calculation: 200 points / 20 points per iteration = 10 iterations. So at first glance, it seems like we should estimate another 10 iterations' worth of work. However, we also need to consider the 10 percent of additional work that we are likely to have: $(200 \times 0.1) + 200 = 220$. We then calculate 220 points / 20 points per iteration = 11 iterations. Therefore, we expect the release to require 11 more iterations.

Slicing the Stories

One of the key activities done in the release planning meeting is slicing stories. This simply means breaking down any stories that are too large to be completed within one iteration. There are two ways to slice a story into smaller chunks, depending on whether the original user story is *compound* or *complex*.

Slicing Compound Stories

A compound story includes other independent stories within it—in other words, it includes multiple goals. In the example shown below, the story, "As a customer I want to purchase a movie so that I can watch it," can be sliced into three separate sub-stories, each of which is small enough to be done in one iteration. Although these smaller stories are all part of building the original story, they are relatively independent of each other.

Figure 5.46: Slicing a Compound Story

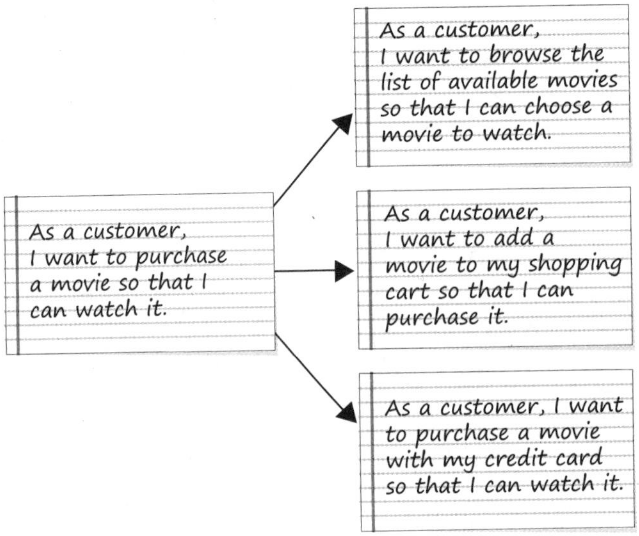

As a customer,
I want to purchase
a movie so that I
can watch it.

→ As a customer,
I want to browse the
list of available movies
so that I can choose a
movie to watch.

→ As a customer,
I want to add a
movie to my shopping
cart so that I can
purchase it.

→ As a customer, I want
to purchase a movie
with my credit card
so that I can watch it.

Slicing Complex Stories

A complex story, on the other hand, is just one really big or complicated story; it doesn't include separate goals or sub-stories—but it's so large or complex that it can't be completed in one iteration. Therefore, the team needs to break it into smaller pieces to get it done. For example, the goal of creating the movie database might be so large that it has to be divided into smaller chunks of work that are done over multiple iterations. Those smaller pieces won't be separate stories, they'll just be the original goal, divided into segments. Although we will plan how much we want to accomplish in each iteration, the functionality won't be usable until it is completely done.

Figure 5.47: Slicing a Complex Story

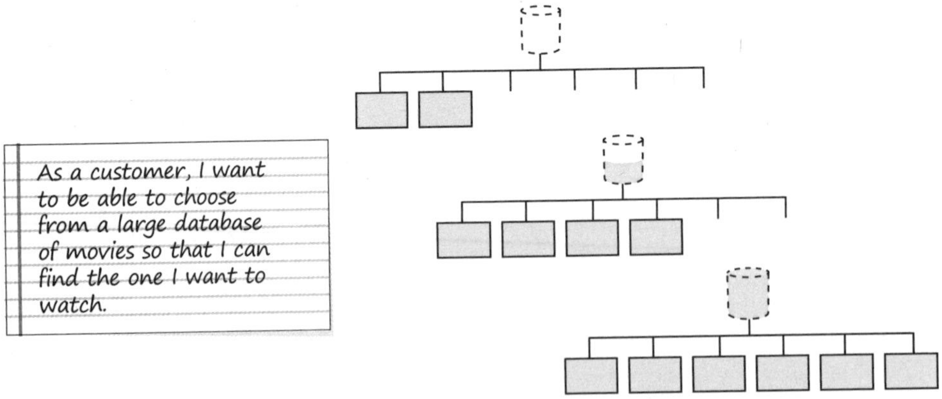

As a customer, I want
to be able to choose
from a large database
of movies so that I can
find the one I want to
watch.

The team is done slicing when all the stories in the backlog slated for the upcoming release are small enough to be completed in one iteration. Notice that this process doesn't involve dividing the user stories into the tasks needed to build them; that final level of detail will be added in the next step, iteration planning.

T&T Iteration Planning

Iteration planning begins with a meeting that includes the delivery team, the product owner, and possibly other stakeholders or subject matter experts as needed. However, the planning done in this meeting is highly detailed, and as a rule this meeting is primarily run for, and by, the delivery team. The customer's role in the discussion is usually limited to answering questions and making sure the team's understanding of the requirements is correct. For example, the customer might speak up if the team seems to be over-engineering the product above and beyond what is actually needed for the solution.

Like release planning, iteration planning requires a backlog that has been freshly prioritized by the product owner. The other input required for this meeting is a goal for the iteration. Usually this goal will have been set forth in the release plan, although it might need to be updated if circumstances have changed.

Exam tip

There isn't one "right" or standard way to do iteration planning—in this area, there is wide variation across the agile universe. Our discussion of this topic draws upon the practices defined by Scrum, since that's the most common process for planning iterations (or "sprints," in Scrum lingo). To make this discussion easier to follow, we are presenting it as a specific series of steps. However, you won't be tested on the exact process described here, and it isn't helpful to think of this as the one "correct" process or sequence of steps. Instead, focus on understanding the underlying concepts so that you'll be prepared to logically think through any exam questions on iteration planning.

In the first half of the meeting, the product owner describes the backlog items they'd like to see developed in the sprint, and based on that, the team members select a set of items that they think are achievable. The different planning responsibilities of the customer and the team are important to know for the exam:

> *The customer has the final say on the priorities for the iteration.*

> *The development team has the final say on the amount of work that can be accomplished in the iteration.*

In the second half of the meeting, the team breaks down the selected backlog items into the smallest unit of work—tasks—to come up with a list of the action items for the iteration. They then discuss how the work will be done, and make a commitment to undertake the work within the sprint timebox.

Note: Although the process we're presenting is loosely based on Scrum practice, it has been adopted by generic agile and isn't limited to Scrum teams. So in this discussion we will continue our usual practice of using terminology from various methodologies and generic agile interchangeably, to prepare you for the different terms you will encounter in the exam questions.

The Iteration Planning Process

Here's a closer look at what we need to accomplish during the iteration planning process—both during and after the iteration planning meeting:

» **Discuss the user stories in the backlog.** The planning meeting begins with a discussion of the high-priority user stories in the backlog to confirm our understanding of those stories with the customer. As a group, we also discuss any factors, such as dependencies, that might prevent us from just grabbing all the stories from the top of the backlog.

» **Select the user stories for the iteration.** With that in mind, we select the top-priority stories that we will commit to deliver in the upcoming iteration. This is often a balancing act—we need to include enough story points to stay on track to meet the release goal in the planned number of iterations. But at the same time, we also need to be realistic about what we can accomplish, based on our average velocity to date. So this decision may involve some negotiation and discussion. (Remember that the customer or ScrumMaster shouldn't try to influence this decision. It is up to the team members to decide how much they can accomplish.)

» **Define the acceptance criteria and write the acceptance tests for the stories.** Next, we work with the customer to develop the acceptance criteria for the stories we have chosen. How will we know when a story is "done"? How can we test our work to make sure it meets the customer's needs? Basically, we are looking for as many objective acceptance criteria as possible, so that we can write acceptance tests for our work.

» **Break down the user stories into tasks.** In the next step, we decompose each user story into its component tasks—the smallest chunks of work required to build the story. You might be wondering, "Hey, didn't we already break down and estimate the stories in release planning?" The answer is yes, we did—but remember the concept of progressive elaboration. In the release-planning meeting we decompose our stories as needed for that level of planning. However, those breakdowns are still fairly coarse-grained, and meant to be preliminary. The ultimate details of the stories—the tasks—are broken down at the "last responsible moment" during iteration planning.

» **Estimate the tasks.** After the planning meeting, we can estimate the effort required to complete each of the tasks in real time, since our estimates so far have been in story points. Estimating the tasks in real time is an optional step, but it does help confirm that our plan for the iteration is really workable. Remember that our total estimate for the tasks that make up a given user story don't need to add up to our estimate for the user story. So once we have that total, we might need to reconsider the number of stories we can complete in the upcoming iteration—or even in the release as a whole. Some teams also add a buffer to their iteration plan based on realistic expectations (not arbitrary padding) for completing user stories that turn out to be larger than anticipated or resolving problems.

Iteration Planning Summary

To help you envision this process, here's a graphical depiction of the iteration planning steps:

Figure 5.48: Summary of Iteration Planning

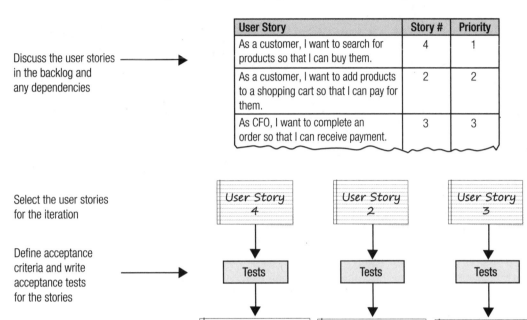

Discuss the user stories in the backlog and any dependencies →

User Story	Story #	Priority
As a customer, I want to search for products so that I can buy them.	4	1
As a customer, I want to add products to a shopping cart so that I can pay for them.	2	2
As CFO, I want to complete an order so that I can receive payment.	3	3

Select the user stories for the iteration

Define acceptance criteria and write acceptance tests for the stories →

Break the stories into tasks →

Estimate the tasks in ideal time, then assign and schedule them in real time →

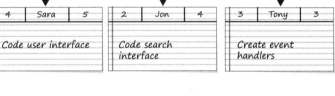

✓ Exam tip

You should know that the "conversation" and "confirmation" parts of a user story (two of the three Cs) take place during the "discuss the user stories" and "define acceptance criteria" steps in this process—although the customer's final confirmation will come later, in the iteration review.

» **Conversation**: Align the team's understanding of the story with the customer.
» **Confirmation**: Identify the customer's acceptance criteria for the story.

To complete our discussion of iteration planning, we'll look more closely at some of the activities involved in this process: selecting the user stories, determining the acceptance criteria, estimating the tasks, and using actual results to refine estimates.

Selecting the User Stories

To select the user stories for an iteration, we follow the same process as for release planning. In iteration planning, we are selecting user stories that the customer has indicated are high priority and that we believe can be developed, tested, and delivered within the iteration. Let's return to the example we used for release planning, where the team was averaging 50 story points per two-week iteration—so we will use 50 story points as our iteration-planning goal.

It would be easy for the product owner or team leader to just select the next 50 points' worth of functionality off the top of the backlog and then tell the team to get on with the work. However, this approach would undermine the agile principles of team self-organization and empowerment that we have worked so hard to nurture. So it's important for the team to discuss the goals for the next iteration with the customer and then select the user stories they will commit to deliver as part of the next iteration.

Remember that the team members are the closest to the technical details of the project, and may have access to information that the product owner or ScrumMaster doesn't know. For example, they may have some refactoring (fixing or improving previous work) scheduled for this iteration that might reduce their capacity for new work. Or conversely, they may be confident that they can complete additional work in the upcoming iteration based on an improvement initiative they have recently undertaken. So while the team leader should question any iteration projections that vary from the team's average velocity, it is important to let the team plan their own iterations. (Besides, that's what retrospectives are for, right? To ridicule the team for their poor estimating? No, not really! We'll discuss retrospectives in chapter 7.)

Defining the Acceptance Criteria and Writing the Acceptance Tests

Each user story needs to have clear acceptance criteria, and establishing a shared understanding of these criteria is an important part of the iteration-planning meeting. Acceptance criteria are the "confirmation" part of the three Cs—discrete business-rule tests that show whether a completed feature is working as intended. In other words, acceptance criteria are explicit statements of how the customer will define "done" for each story. Also, as stories are split into smaller stories or tasks, their acceptance criteria will make clear exactly what has been split off.

Below is an example of five "discrete business-rule tests" for the credit card payment functionality of an online shopping cart—the five criteria being used to determine if this story is working correctly. At this point, the tests have been run once, and the "Trial 1" column shows the results.

User Story: "As a customer, I want to pay for shopping cart items with a credit card so that I can buy movies."

Criteria	**Required**	**Trial 1**
Test with Visa, MasterCard, and American Express.	*Pass*	*Pass*
Test with Diners Club.	*Pass*	*Fail*
Test with a bad card number that is missing 3 digits.	*Fail*	*Fail*
Test with an expired card.	*Fail*	*Fail*
Test with a purchase amount over the card limit.	*Fail*	*Fail*

© 2015 RMC Publications, Inc • 952.846.4484 • info@rmcls.com • www.rmcls.com

The important point to notice here is that a successful outcome (as shown in the "Required" column) may be either "Pass" or "Fail," depending on the rule. For example, we *want* to fail the last three tests—we wouldn't want the system to accept a bad number or an expired card, or to exceed the card limit. So although only one "Pass" is shown in the results for this first trial, this user story actually passed all the acceptance tests except the one for using a Diners Club credit card.

Estimating the Tasks

After discussing the stories and breaking them down into their component parts, the team will estimate the tasks, and this is typically done in real time (hours or days). Although the stories have already been estimated in story points, it can be helpful to create more detailed bottom-up estimates in real time to schedule the work and confirm our iteration plan. Let's walk through how this is done.

Exam tip

As an experienced agilist, you may be aware that some agile teams just estimate their stories in points and leave it there. In the Kanban community, going further to estimate tasks in hours is considered to be waste, and the practice is becoming less popular. So estimating tasks in real time isn't a mandatory step. However, you should understand how this is done, in case it appears on the exam.

First, we prepare a "task card" for each task in the iteration, using index cards. On the card for each task, we record the story number, a brief description of the work, and our "ideal time" estimate for how long it will take to complete that work, in hours or days. Remember, ideal time estimates are made in terms of effort, not duration. They reflect the time required to complete the task in an ideal scenario without any interruptions, distractions, delays, or availability issues.

Figure 5.49: Task Card Example

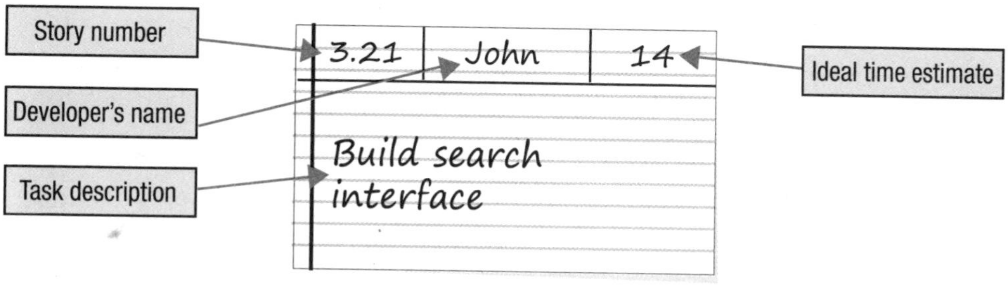

Story number	3.21	John	14	Ideal time estimate
Developer's name				
Task description	Build search interface			

On this card, notice that the task isn't written in the user story format (role/functionality/business benefit). That's because the tasks don't need to be understandable by the customer. So the team will record them in the technical terminology used by those who will be doing the work, rather than the user story format.

Once we have filled out a card for every task, we add up the time estimates for all the tasks and compare that total to the time available for the iteration. At that point, we might also tentatively assign each task to a specific team member and add the name of the person who will do the work to each card. However, such assignments are highly likely to change—so they will only be finalized at the "last responsible moment" before the work is done.

Although these initial task assignments could change later, we can use them as a starting point for translating our ideal time estimates into actual calendar time. Assigning the tasks to specific people allows us to determine that person's availability during the upcoming iteration, taking into account their nonproject commitments, vacations, holidays, and so on. It's helpful to do this because we have committed to deliver the selected backlog stories within the sprint. To be confident we can deliver this work, we need to calculate our capacity and adjust it for any planned absences.

EXERCISE

Project Beta is being estimated in ideal days, assuming 8-hour days and 5-day workweeks. Each of the five people on the team averages 30 hours of availability for the project each week. How many ideal days' worth of work can they commit to deliver in their next 10-day iteration?

> $5 \times 30 = 150$ hours / week
> $150 \times 2 = 300$ hours / iteration
> $300 / 8 = 37.5$ ideal days

ANSWER

To get the answer, we do the following calculations:

  5 people × 30 hours per week = 150 hours per week

  150 hours per week × 2 weeks (for a 10-day iteration) = 300 hours per iteration

  300 hours per iteration / 8 hours per day = 37.5 ideal developer days' worth of work

Therefore, the team should not commit to completing anything over 37.5 ideal days' worth of work.

Use Actual Results to Refine Estimates

As an agile project progresses, the iterations provide hard evidence of real progress. We can now start factoring in the velocity of completed iterations to better judge our true progress and estimate the remainder of the project.

The reason why the velocity of completed iterations can be used to estimate project progress fairly accurately is because iterations involve all disciplines of development. Remember that in software development, user stories should be vertical slices of functionality that cut through all the relevant architectural layers of the system. This means we will get exposure to the different aspects of the development effort in each iteration—and therefore, we will quickly be able to rely less on our up-front estimates and instead increasingly use our emerging velocity. We can also compare our velocity to the remaining backlog to estimate the timeline for project completion.

Figure 5.50: Agile versus Traditional Estimation Approaches

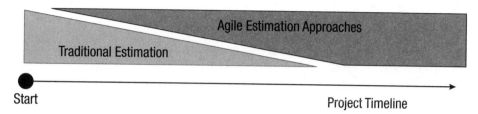

Image originally published in "Estimation for Agile Projects," by Mike Griffiths on gantthead.com on January 1, 2008, copyright © 2008 gantthead.com. Reproduced by permission of gantthead.com.

For example, if after the first three or four iterations, our velocity has stabilized and averages 50 points per two-week iteration, and the backlog of remaining work contains 500 points' worth of functionality, it is reasonable to assume that it will take 10 more iterations to complete the remaining work (500 / 50 = 10).

Background information

Calculating Burn Rate

With the data given for this example, we can also calculate the salary burn rate per iteration. Let's say we have a three-person team of Bob, Juan, and Tina:

» For Bob, we multiply 10 days by 8 hours by his rate ($50 per hour), for a total of $4,000.
» For Juan, we multiply 10 days by 8 hours by his rate ($80 per hour), to get $6,400.
» For Tina, we multiply 10 days by 8 hours by her rate ($95 per hour), to get $7,600.

Now if we add these numbers together ($4,000 + $6,400 + $7,600), we get an iteration burn rate of $18,000. Since we assume it will take 10 more iterations to complete the remaining work, we can calculate how much it will cost to complete the remaining backlog features: $18,000 x 10 iterations = $180,000.

This frequent feeding back of actual results into the estimates is a valuable reality check for the project. Underachievement is uncovered early, since agile projects measure progress by the number of accepted user stories, rather than an estimate of "percent complete" against analysis or design deliverables. It is very easy to be overly optimistic in interpreting "percent complete" on deliverables, whereas measuring progress based on acceptance is a much more solid indication of work accomplished. Uncovering underachievement is never pleasant, but it is best to uncover it early in the project, when we still have most of the project left to take corrective action.

T&T Daily Stand-Ups

Once the work is underway during an iteration, the team members continue the planning process on a micro level by holding daily stand-up meetings (or "daily scrums"). These short, focused team meetings are a core agile planning tool. They keep everyone focused on the agreed-upon scope and iteration goal, and provide an early warning system for issues, allowing them to be resolved before they can derail the team's progress.

These meetings are called "stand-ups" because the team members stand throughout the meeting to make sure they aren't tempted to talk any longer than necessary. Daily stand-ups are timeboxed to 15 minutes, although they often finish well before the timebox is up. During the meeting each team member, in turn, briefly answers the following three questions.

1. What have you worked on since the last meeting?
2. What do you plan to finish today?
3. Are there any problems or impediments to your progress?

This tells the other members of the delivery team what that person has been working on, what they are working on now, and whether anything is blocking their progress. The meeting is held at the team's task board, and while they talk, the team members may move the stickies on the task board to show their progress.

Once each member of the delivery team has answered the three questions to report their status, the stand-up ends. There is no further discussion in that meeting. Any issues or impediments that require further discussion are handled in follow-up conversations or meetings. This keeps the meeting focused and within the 15-minute timebox.

To stay productive, daily stand-ups follow the ground rules shown below, which should be posted prominently in the meeting area:

Figure 5.51: Ground Rules for Daily Stand-Ups

Ground Rules for Daily Stand-Ups

- If you have a task, you must attend.
- Only those who have tasks can talk.
- Speak to the team, not the coach or ScrumMaster.
- No side conversations.
- Create a new sticky note for each new task that is started.
- Discuss issues after the meeting.
- Solve problems off-line

Notice that these rules keep the meeting focused on the members of the delivery team and the tasks they are doing. The team's coach or ScrumMaster will attend the stand-up, but won't take part unless he or she is also working on development tasks. The stand-up is for the delivery team and their coach; other project stakeholders don't participate in it—although the product owner or other interested parties might attend it and observe. But fundamentally, the daily stand-up is "for the team" and "by the team."

The purpose of the daily stand-up is to provide a daily review of tasks and issues to keep the team up to date and focused on their iteration goal. Since this meeting ensures that each team member knows what everyone else is doing, it helps them coordinate their efforts so that the work flows smoothly. It also keeps their coach informed about issues and impediments on a daily basis, so that he or she can take action to resolve them. Another benefit is that daily stand-ups negate the need for writing weekly status reports or holding traditional "go-around-the-room" status meetings, which tend to take a lot longer than 15 minutes.

Exam tip

For the exam, remember the daily stand-up ground rules and the point that these meetings are "for the team" and "by the team." The team members speak to each other, not to the ScrumMaster, coach, or team leader. You should know that the stand-up discussion is strictly limited to these three questions and timeboxed to 15 minutes.

EXERCISE

Consider the following conversation snippets and indicate whether they are valid topics to discuss at a stand-up meeting by placing a check mark in the appropriate column.

Conversation Snippet	Valid Topic	Invalid Topic
"My PC still needs more RAM."	✓	
"I finished testing the launcher."	✓	
"I think we should add a turbo booster."		✓
"I just finished adding the supercharger."	✓	
"Bill from accounting did not approve my trip to see the users."	✓	
"Wendy from marketing won't go on a date with me."		✓
"I am still stuck trying to attach the nose cone."	✓	
"If you thread it backwards, the nose cone should go on easily."		✓

ANSWER

During daily stand-up meetings, we need to be strict about keeping people focused on reporting progress, work planned, and impediments. Anything else is supplemental and should be taken off-line.

Conversation Snippet	Valid Topic	Invalid Topic
"My PC still needs more RAM."	✓ This is an impediment and is therefore a valid topic.	
"I finished testing the launcher."	✓ This snippet is about project progress and is therefore a valid topic.	
"I think we should add a turbo booster."		✓ This conversation is about suggested new scope, so the discussion should be taken off-line.
"I just finished adding the supercharger."	✓ This comment is a report on project progress, which makes it a valid topic.	
"Bill from accounting did not approve my trip to see the users."	✓ This is an impediment, so it is an appropriate topic for a stand-up meeting.	
"Wendy from marketing won't go on a date with me."		✓ This comment is completely off-topic and doesn't belong in the daily stand-up meeting.
"I am still stuck trying to attach the nose cone."	✓ This is an appropriate topic, because it's an impediment.	
"If you thread it backwards, the nose cone should go on easily."		✓ This is related to solving the issue, rather than simply reporting it. The discussion should be taken off-line.

Chapter Review

1. Your team has just discovered that your story point has drifted. What should you do next?

    A. Re-estimate the product backlog.
    B. Perform Fishbone Analysis to find the root cause.
    C. Use affinity estimating to compare all the estimates made so far.
    D. Postpone the next sprint and instead perform a spike to resolve the issue.

2. The team is slicing their user stories. What are they doing?

    A. Sizing stories for the next iteration
    B. Gathering user requirements
    C. Estimating how much they can get done in the next iteration
    D. Dividing stories into pieces that can be completed in one iteration

3. The PMO wants to have a copy of your project plan. What will you send them?

    A. A copy of your product roadmap
    B. The team's requirements hierarchy
    C. A photo of the team's Kanban board
    D. Your user story backlog

4. The product owner has told the team how much work will need to be completed in the next iteration. In this scenario:

    A. The iteration planning process is proceeding smoothly.
    B. The product owner is overstepping their role.
    C. The product owner is taking over the ScrumMaster's responsibilities for planning.
    D. The team should claim more responsibility for planning in their next retrospective.

5. Your team has decided they need an Iteration 0 before starting the development work. Why?

    A. To set up the build server for the project
    B. To practice working together and get through the Storming stage before the real work starts
    C. To hold planning poker sessions to estimate the user stories
    D. To minimize as many of the project risks as possible before development begins

6. The team believes that it will take 15 hours of effort to write the user guide for the new product they are building. What should their estimate be for that task?

    A. 15 hours
    B. 18 hours, to add a buffer for distractions and availability issues
    C. 12 hours, since teams tend to overestimate how long a task will take
    D. 12 to 18 hours

7. Your team members are estimating their tasks. What process are they engaged in?

    A. Release planning
    B. Progressive elaboration
    C. Iteration planning
    D. High-level visioning

8. In the daily stand-ups, the team coach should:

    A. Schedule and facilitate the meeting.
    B. Let the team members resolve their own conflicts.
    C. Listen and note any problems for immediate follow-up.
    D. Ask questions to determine the root cause of any problems that are raised.

9. We can say that an iteration demo is successful if _____.

    A. The product got shipped and the team would work the same way again.
    B. The customer accepts the minimal viable product that has been built.
    C. A gulf of evaluation is cleared up.
    D. The product owner says the product increment is done.

10. Your lead engineer just came down with the measles in the middle of a sprint. As team coach, what should you do?

    A. Call his functional manager and request a new lead engineer for your team.
    B. Ask the team how much of the planned work can be done.
    C. Ask everyone else to work overtime.
    D. Postpone the release date.

11. Your sponsor has asked for clarification on when releases of your product will ship and what those releases will contain. Which agile deliverable would best address this need?

    A. Product demo
    B. Product roadmap
    C. Product backlog
    D. Product owner

12. Your team committed to delivering 10 story points this iteration, but it looks like you will only complete 8. You should:

    A. Extend the iteration.
    B. Add more resources to the team.
    C. Complete 8 points, and put 2 back in the backlog.
    D. Adjust the iteration plan from 10 points down to 8.

13. When agile teams use the term "timeboxed," what do they mean?

    A. Work shall take a minimum amount of time.
    B. Work can take no more than a maximum amount of time.
    C. Work must be done by a given time, plus or minus 20 percent.
    D. Work must happen at a set time.

14. The project management office is auditing your agile project and asks to see your iteration plans. They notice that only the next couple of iterations have plans. As a result, they give the project a "red flag" for having incomplete plans. The most responsible thing to do is:

    A. Explain the agile principles of progressive elaboration and rolling wave planning.
    B. Create detailed iteration plans for the remainder of the project.
    C. Ignore them, since they clearly have no right to be reviewing your project.
    D. Ask the team to create detailed plans for the remaining iterations in the release.

© 2015 RMC Publications, Inc • 952.846.4484 • info@rmcls.com • www.rmcls.com

15. Which of the following statements correctly describes agile planning?

    A. Plan at multiple levels, and have managers create iteration plans.
    B. Use appropriate estimate ranges, and exclude diversions/outside work.
    C. Plan at multiple levels, and have team members create iteration plans.
    D. Use fixed-point estimates, and base projections on completion rates.

16. You are leading a team with an average velocity of 50 points per iteration. Another team of the same size in your organization is working on a project with similar complexity. The other team's velocity is averaging 75 points per month. Your team should:

    A. Perform affinity estimating to check their estimates, since something is off.
    B. Work longer hours.
    C. Ignore the difference.
    D. Request additional resources to get more work done.

17. Estimates should be presented as ranges to:

    A. Allow for change requests.
    B. Keep the sponsors flexible.
    C. Allow for scope creep.
    D. Reflect the level of uncertainty in the estimates.

18. Your team is averaging 40 story points per two-week iteration. They have 200 points' worth of functionality left in the user story backlog. How many weeks can we expect it will take until development is completed?

    A. 2.5
    B. 5
    C. 10
    D. 20

19. Affinity estimating allows a team to:

    A. Average the over- and under-estimations in our estimate ranges.
    B. Confirm that stories that have been estimated as the same size are of equivalent magnitude.
    C. Check that stories within the same functional areas are of similar proportion.
    D. Estimate the most important stories first to deliver value early and reduce risk.

20. You are a full-time ScrumMaster on an agile team. A team member becomes ill partway through an iteration in which the team committed to deliver 25 story points. Which action is most appropriate?

    A. Work the remaining team longer hours.
    B. Send work home to the sick team member.
    C. Start development yourself to assist the team.
    D. Deliver what you can within the sprint.

Answers

1. **Answer**: C

   Explanation: Depending on the circumstances, it's possible that any of these options could be a helpful response. However, based on the information provided, the BEST thing to do next would be to use affinity estimating to compare the estimates that have been made so far. A team normally wouldn't re-estimate their entire backlog, since much of that work might not have been estimated yet. They also probably wouldn't stop to analyze the root cause. Over the course of a large project, it's perfectly normal for the size of a story point to drift; so it's more helpful to compare and adjust the estimates than to figure out what caused the problem. For the same reason, it's also unlikely that such a normal occurrence would require postponing the next sprint to perform a spike.

2. **Answer**: D

   Explanation: Slicing user stories is the process of dividing stories into pieces that can be completed in one iteration. The other options are incorrect.

3. **Answer**: A

   Explanation: Agile's closest equivalent to a traditional project plan is the product roadmap. The team's requirements hierarchy isn't a plan; it simply outlines how they will break down the requirements into progressively smaller pieces. Their Kanban board isn't a plan, either—it just shows the work moving through the development process at the moment. Finally, the user story backlog is a prioritized master list of the work that still needs to be done; it doesn't include the planned release dates or the features that will be included in the releases, the way a product roadmap does.

4. **Answer**: B

   Explanation: During iteration planning, the product owner's role is to prioritize the backlog items. The team then decides how many of the top-priority items in the backlog can be completed in the next iteration timebox. So this product owner is overstepping their role, since the amount of work that can be completed in the next iteration is decided by the team, not the product owner or the ScrumMaster. While it's true that in this scenario the team isn't doing their own planning, based on the information provided, it isn't clear that there is a problem on their side that should be addressed in a retrospective. The product owner probably just needs to be educated about agile and encouraged to allow the team do their own planning.

5. **Answer**: A

   Explanation: Iteration 0 is an optional iteration that the team can use to set the stage for their development efforts. It isn't used for estimating or for working together, since those activities are done in the development iterations. Although agile teams do try to minimize risk early in the project, they usually do this by prioritizing risk mitigation stories or by scheduling a risk-based spike. So although risk mitigation could theoretically be part of Iteration 0, the answer that BEST fits the definition of Iteration 0 is to set up the build server for the project.

6. **Answer**: A

   Explanation: Agile teams estimate tasks in ideal time—how long it will take if there are no interruptions or distractions. In this case, they have decided that the effort will take 15 hours, so their estimate should also be 15 hours. If they think the work will take 15 hours, then they wouldn't estimate less than that (and in general, teams tend to underestimate how long a job will take, not overestimate it). Although agile teams do rely on estimate ranges to convey the uncertainty of larger estimates to stakeholders, tasks are typically given single-point estimates. That's because those estimates are only for the team, and also at this point (shortly before the work is done) they should have enough information to agree upon a single-point estimate rather than using a wide range such as 12 to 18 hours.

7. **Answer**: C

   Explanation: Agile work units are progressively broken down from large to small and estimated at the last responsible moment. Since tasks are the smallest agile unit of work, we can deduce logically that these work items wouldn't be estimated until the last planning step just before the work is done, which is iteration planning. Although it's true that all agile estimating is progressively elaborated over time, that isn't the BEST answer to this question, since it is too general.

8. **Answer**: C

   Explanation: In a daily stand-up, the role of the ScrumMaster or team coach is to listen and note any impediments to the team's progress for quick follow-up. This meeting is generally held at the same time and place every day, so scheduling usually isn't required—and because the discussion is run by the team members, no facilitation is necessary either. Since daily stand-ups are strictly limited to answering three questions, team conflicts shouldn't be an issue in these meetings. Root cause analysis of the problems would be done in a separate meeting, rather than during the stand-up itself.

9. **Answer**: D

   Explanation: This question requires you to think through the options carefully, since each of them is applicable to some aspect of an agile project. "The product got shipped and the team would work the same way again" describes methodology success criteria—these are signs that our methodology is working well. However, these criteria don't apply to iteration demos since only one increment is built in an iteration, not the entire product. We can also rule out "The customer accepts the minimal viable product that has been built" because the minimal viable product isn't built in one iteration. Although we do want to identify and clear up any gulfs of evaluation that may exist during the iteration demo, that isn't what determines if the meeting was successful. The success of the demo is based on whether the product owner accepts the product increment built in the iteration as "done."

10. **Answer**: B

    Explanation: This question tests your grasp of the agile principle of timeboxing. The correct answer is to discuss with the team how much of the planned work they will be able to complete within the timebox. We wouldn't request a new lead engineer, either temporarily or permanently, because swapping people in and out of the team would be likely to throw the team back to the Storming stage, lowering its productivity. The option of asking everyone to work overtime isn't consistent with the agile principle of sustainable development. Although postponing the release date might be necessary in some cases, we aren't given enough information to support the conclusion that this is the BEST answer.

11. **Answer**: B

    Explanation: The product roadmap shows release dates and the high-level contents of releases, so it would be the best deliverable for answering these questions. A product demo might be good for showing the sponsor what has already been built, but demos are not targeted at communicating the release schedule for upcoming features, and neither is the backlog. Although the product owner will likely know the answers to these questions, the question asked "Which agile deliverable"—and the product owner is generally a person, not a deliverable.

12. **Answer**: C

    Explanation: Since iterations are timeboxed, the duration won't be changed. You also wouldn't change the iteration plan or expand the team. Instead, work that isn't completed within the iteration is returned to the backlog. Therefore, the choice of completing 8 points and returning 2 points to the backlog is the correct option.

13. **Answer**: B

    Explanation: If an activity is timeboxed, that means a maximum duration has been assigned to it. This ensures that the team spends an appropriate amount of time on the work without allowing waste.

14. **Answer**: A

    Explanation: An incomplete set of iteration plans may be a surprise to a PMO that is not familiar with agile methods. When faced with this situation, you should explain the benefits of agile planning and how an agile approach ties into the concepts of progressive elaboration and rolling wave planning, which are discussed in the *PMBOK® Guide*. Making up plans too early is a poor use of time on an agile project and could mislead stakeholders. The choice of ignoring the request is incorrect, because it is counter to the "Respect" principle defined in PMI's Code of Ethics and Professional Conduct (see chapter 7).

15. **Answer**: C

    Explanation: The only correct combination is to plan at multiple levels and have team members create iteration plans. All of the other choices contain incorrect elements. Managers do not create iteration plans—teams do. Diversions and outside work are included when determining availability, and we use range estimates on agile projects, not fixed-point estimates.

16. **Answer**: C

    Explanation: Velocity is team-specific and unique to that team. In other words, a story point for one team probably wouldn't have the same value as a story point for another team. Therefore, it is not appropriate to compare velocities between teams. The best choice would be to ignore the difference.

17. **Answer**: D

    Explanation: We present estimates as ranges to show the level of uncertainty in the estimates and to manage stakeholder expectations.

18. **Answer**: C

    Explanation: Since this question doesn't indicate there will be any differences in the team's availability or known distractions going forward, we can do a fairly straightforward calculation to get the answer. If we average 40 points per iteration, we should get through a 200-point backlog in 5 iterations (200 / 40 = 5). Each iteration is 2 weeks long, so 5 iterations is equivalent to 10 weeks (5 × 2 = 10). The key to answering this question correctly is noticing that the question asked for the number of weeks, not the number of iterations.

19. **Answer**: B

    Explanation: Affinity estimation involves a process of triangulation to confirm that stories estimated to be the same size are of equivalent magnitude.

20. **Answer**: D

    Explanation: In such a situation, we deliver what we can within the sprint. Sending work home and working the team longer are counterproductive in the long term and don't adhere to the agile principle of maintaining a sustainable pace. Starting development work yourself would leave nobody to do the ScrumMaster role. It is better to just deliver what is possible within this iteration and explain the variance.

© 2015 RMC Publications, Inc • 952.846.4484 • info@rmcls.com • www.rmcls.com

CHAPTER 6

Problem Detection and Resolution

Domain VI Summary

This chapter discusses domain VI in the exam content outline, which is 10 percent of the exam, or about 12 exam questions. This domain deals with the agile practices used to prevent, identify, and resolve threats and issues, including catching problems early, tracking defects, managing risk, and engaging the team in solving problems.

Key Topics

- » Control limits
- » Cost of change
- » Cycle time
- » Defect rate
- » Escaped defects
- » Expected monetary value
- » Failure and success modes
- » Lead time

- » Problem solving
  - – As continuous improvement
  - – Team-based
- » Risk-adjusted backlog
- » Risk burndown graphs
- » Risk severity
- » Technical debt
- » Throughput/productivity

- » Trend analysis
  - – Lagging metrics
  - – Leading metrics
- » Variance analysis
  - – Common cause
  - – Special cause

Tasks

1. Create a safe and open environment to surface problems.
2. Engage team in resolving threats and issues.
3. Resolve issues or reset expectations.
4. Maintain a visible list of threats and issues.
5. Maintain a threat list and add threat remediation efforts to the backlog.

All project teams need to be proficient at managing problems, threats, and issues. Since problems will always arise on a project, our effectiveness in preventing, detecting, and resolving them is likely to determine whether our project succeeds or fails. The topic of problem solving isn't mentioned in the Agile Manifesto, since it isn't unique to agile methods. However, agile teams do have some preferred ways of managing problems that reflect the values and principles laid out in the Manifesto. This chapter will discuss how agile teams manage threats and identify and solve problems.

This chapter is broken into four themes: understanding problems, detecting problems, managing threats and issues, and solving problems. The tools discussed in the managing threats and issues section tie back to our discussion of risk management in chapter 2, because these tools are how agile teams manage risk. (Recall that managing risk is essential to value delivery since risk is negative value on a project.) It's also worth noting that as a rule, we will be using the concept of a "problem" very broadly here, to encompass any threat or risk to the project.

Understanding Problems

When a problem occurs on a project, it can be tempting to ignore the issue and just continue pushing forward, in the hope that it will somehow go away or resolve itself. But even a single problem can result in delay, waste, and rework that can bring our progress to a halt and even reverse it, completely derailing our project schedule. And the longer the problem is left unaddressed, the more its effects will be compounded.

How Problems Impact a Project

Let's use an example to see how even a single problem can quickly escalate and impact a project. We'll start with a simple scenario—we have been assigned the project of driving ten miles in ten hours. The diagram below shows the ten miles that we need to cover, beginning at mile 0. (Yes, this would be a very slow, boring project, but it keeps the math simple.) At hour zero, our task looks simple enough, and the way ahead is clear:

Figure 6.1: Our Plan Is to Drive 10 Miles in 10 Hours

At mile 2, everything is still going well. We have been driving for 2 hours, and we have traveled 2 miles. This means our plan and our actual performance results are the same.

Figure 6.2: At Mile 2 the Plan and the Actual Results Are the Same

In fact, things go exactly according to plan for the first 4 hours, in which we cover 4 miles—but then we encounter an unexpected problem that we didn't plan for.

Figure 6.3: At Mile 4 We Encounter a Problem

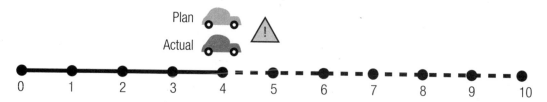

When we encounter a problem, it takes time to diagnose what the issues are and then determine what we should do about them. The time spent diagnosing the problem is time that is not spent on the execution of our project, and so we fall behind the plan. In this example, let's say that it takes us one hour to diagnose the problem.

Figure 6.4: Problem Diagnosis Takes One Hour

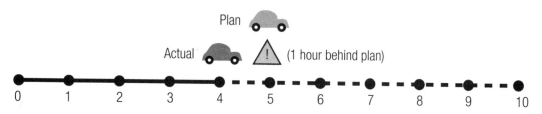

Solving a problem often involves undoing flawed work. Perhaps we thought our original approach was appropriate for the project, but now we've found out it doesn't work. So we need to backtrack and try a new approach. This is a double whammy—it takes time to remove or undo the bad work, and then we also have to redo that work. Since our completed progress has moved backward to an earlier point, we fall even further behind our plan.

In our example, let's assume it takes two hours to undo the bad work and that this takes us two steps backward in the completed scope on the project. Including the hour we spent diagnosing the problem, the total time we've lost is now 1 + 2 + 2 = 5 hours.

Figure 6.5: Work Has to Be Undone,
Taking Two Hours and Moving Two Steps Backward

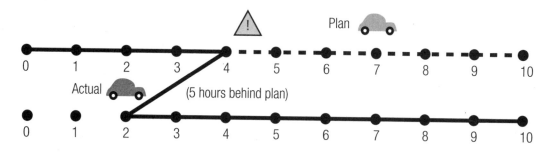

Given this unplanned impact on the project, pretty soon we will consume all our scheduled time, even if we encounter no other problems. (And if our budget is driven by effort, then we will also use up all of our budget.) In fact, if there are no further delays, when we reach our 10-hour deadline for the project, we will be only 50 percent complete—not a good outcome.

Figure 6.6: The Project Should Be Finished Now but Is Only 50% Done

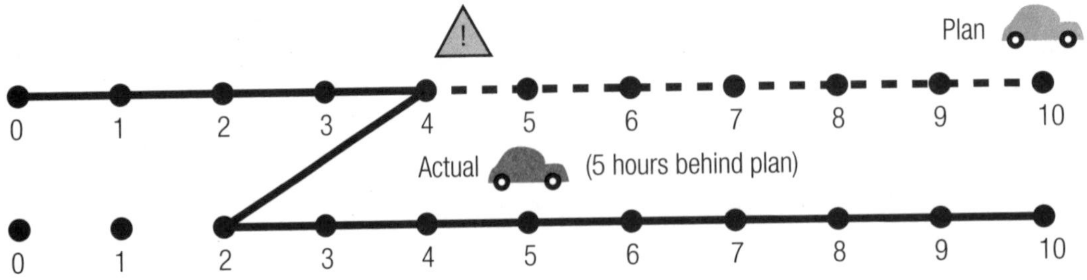

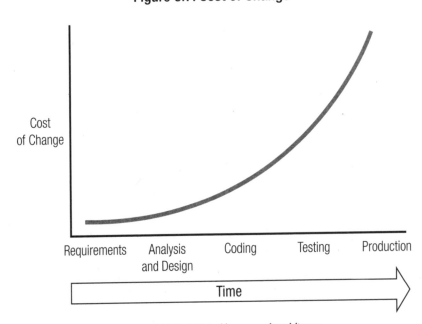

The secret to minimizing the impact of problems is to identify them as soon as possible, since early detection reduces the potential for rework. Once a problem is detected, we also need to diagnose and solve it as quickly as possible so we don't consume any more unplanned time than necessary. This may all sound like pretty basic stuff, but many projects languish in the face of problems, seemingly unaware of the double hit that each day of dithering brings.

The Cost of Change

The reason it's so costly to procrastinate in dealing with problems is because of the cost of change curve. We briefly mentioned this idea in chapter 2 when discussing incremental delivery. To jog your memory, here is the graph again. The curve of this graph shows that the longer a defect is left unaddressed, the more expensive it will be to fix.

Figure 6.7: Cost of Change

Image copyright © Scott W. Ambler, www.agilemodeling.com

There are a number of reasons why the cost of fixing a problem increases over time, but here are two of the key factors:

» Over time, more work may have been built on top of the error or problem, so that more work will need to be undone to fix it (and then have to be redone).

» The later we are in the development cycle, the more stakeholders will be impacted by the defect, making it that much more expensive to fix.

As we've seen, agile methods emphasize frequent verification and validation—both through active stakeholder participation (such as modeling, demonstrations, and reviews) and through software development practices (such as pair programming, continuous integration, and test-driven development). These practices are all intended to find defects and problems as soon as possible, before the costs escalate too far up the cost of change curve.

The figure below illustrates the impact of this approach, by showing how much sooner issues can be identified with agile techniques than with traditional techniques. On this diagram, the black arrows higher up the curve show where issues are found by traditional methods, and the green arrows lower on the curve show where issues are found with agile methods.

Figure 6.8: Where Issues Are Found Using Agile and Traditional Methods

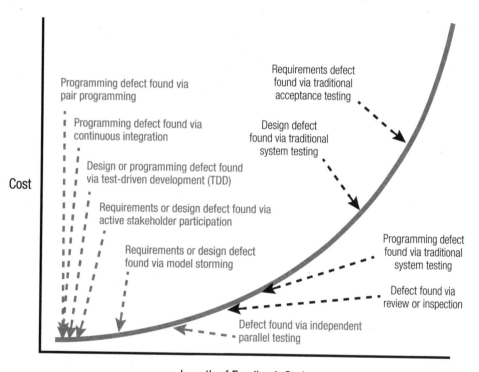

Image copyright © Scott W. Ambler, www.agilemodeling.com

Technical Debt

Technical debt is the backlog of work that is caused by not doing regular cleanup, maintenance, and standardization while the product is being built. It is a backlog of things that should be done to make work easier in the future, but aren't done because of a push to deliver features. Technical debt can be thought of as an inflated cost of change curve. It increases the cost of development and making changes in the future because we'll have to do all the standardization and clean-up work that has been put off—or if we don't do that, development and changes will take longer because they'll have to be done piecemeal, since the solution we are building hasn't been standardized.

Figure 6.9: Technical Debt

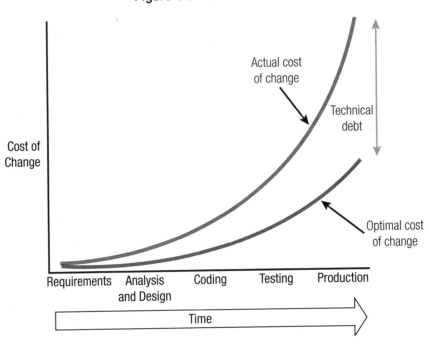

In software projects the solution to technical debt is *refactoring*. (We defined refactoring as one of XP's core practices in chapter 1 and discussed it as part of the "Red, Green, Refactor" process in chapter 2.) Refactoring is the process of taking time to simplify and standardize the code to make it easier to work on in the future. For example, if one of the developers comes up with a better way to validate user-entered text, then that approach should be applied everywhere, and the existing code will need to be updated to include it. We don't want some parts of the system to use one way of validating text and other parts of the system to use another way. If multiple approaches are used, then future upgrades and changes will need to be made in several places. This will take longer, and since it is easy to miss one of the discrepancies, this can easily lead to errors.

All of this extra effort is technical debt. So we need to do frequent refactoring to keep the code clean and reduce technical debt.

When asking the team for estimates, always ask them to include time for refactoring, since this should be part of their regular work routine. The idea that we aren't done until the code has been refactored is baked into many agile practices. For example, in test-driven development we first write tests that fail, and then write code until the tests pass. However, we don't stop there—we still need to do the last step, refactoring, which is why this process is referred to as "Red, Green, Refactor."

To keep technical debt as low as possible, refactoring should be done frequently. There's a saying that "Refactoring should be like daily hygiene, not spring cleaning." This means that we don't save it up for a refactoring sprint, instead we incorporate it into all our regular work. Creating space for the team to refactor and reduce technical debt can take commitment and courage, since the business is likely to be pushing for more and more features. So we need to explain that it is in their best interest to take the time to refactor as we go, to reduce technical debt and streamline future changes.

One analogy we can use for technical debt is preventative maintenance. We want to drive our car, but we also need to maintain it. If all we do is drive the car, and we never do any maintenance, it will eventually break down; then we won't be able to drive it at all. In the same way, we want the team to be productive, but we also need to protect and expand their production capability. Taking time to refactor the work, sort things out, and standardize and simplify it will enhance the team's ability to produce more features in the future.

Outside of software development, lean manufacturing uses an approach to workplace organization called 5S, based on five Japanese words—*seiri, seiton, seiso, seiketsu,* and *shitsuke*—that can be translated as "Set in order, Sort, Shine, Standardize, and Sustain." This describes how to organize a work space for efficiency and effectiveness by identifying and storing the necessary tools, maintaining the area and the tools, and sustaining the new order. If you've ever done a DIY project and spent more time looking for your tools and trying to find that tape measure you just had five minutes ago, you'll understand the benefits of this approach.

In a similar way, reducing technical debt involves keeping the work environment organized and the product we are building organized, streamlined, and standardized. Only when this is in place can we proceed effectively with the project.

Create a Safe and Open Environment

The PMI-ACP exam content outline highlights the importance of creating a safe and open environment in several of its domains. For example, we've already discussed this topic in relation to agile leadership (in chapter 1) and agile teams (in chapter 4). This concept is also important for the problem-solving domain. We want people to feel comfortable not just to experiment, but also to admit their problems, failures, and mistakes and ask for help so that the project can recover as quickly as possible.

This isn't important just because it makes people happier; it also has hard economic benefits. In a work environment where a problem is treated as a personal failure or lack of competence, people will try to fix their own problems before admitting they have a problem and asking for help. That's understandable; if you're supposed to be an expert, you might think twice about admitting you're stuck on a small issue on day 10 of a new project. However, that's exactly what we want people to do—when someone gets stuck, we want them to share it with their teammates quickly. This will give the team more options to fix the issue and try different approaches before things fall hopelessly behind (as occurred in our example of trying to drive ten miles in ten hours).

So creating a safe and open environment is as much about avoiding catastrophic delays as it is about protecting people's feelings. The team leader can look for opportunities to reinforce the importance of asking for help. For example, if a demo shows that the team has made little progress in a particular area, maybe they are facing a problem that they haven't shared yet. This can be a coaching opportunity to remind people to share issues early.

Failure Modes

Problems often arise for reasons that can be prevented, and understanding the human factors that contribute to problems can help us minimize problems and handle them more effectively. Alistair Cockburn describes some "failure and success modes" that can help us understand this human side of performance. They explain why, even though we know what we should be doing, we often behave differently. Although these modes aren't mentioned in the exam content outline, it's helpful to understand them—and they might be referenced in an exam question or two, if only as an option that you need to rule out. There are five failure modes that Cockburn describes in his book *Agile Software Development: The Cooperative Game*.[1] Let's review them to understand the issues involved.

» **We make mistakes**. It's no surprise that people make mistakes. This is one of the main reasons why iterative and incremental development was created—this approach recognizes that mistakes happen and provides us with mechanisms to quickly recover from them so they don't overwhelm the project.

» **We prefer to fail conservatively**. When faced with uncertainty, people tend to revert back to what they know, even if they are aware that it might not be the optimal approach. (For example, when an agile project begins to go off track or encounters multiple issues, the leader may be tempted to revert back to a command-and-control way of running the project.) Of course, there are exceptions to this failure mode. Some people feel they have nothing to lose with a new approach. There are ways to move forward with either mindset—reverting to the familiar or embracing a new approach—but when there are people with opposing mindsets on the same project, it can create challenges.

» **We prefer to invent rather than research**. This describes a tendency that many people have (especially engineers) to rely on inventing new ways of doing things, rather than researching options that have already been invented and reusing them. This tendency may be a product of education systems that reward individual thinking and scientific experimentation, combined with the intellectual satisfaction we get from solving problems. However, while it may be more fun to invent rather than research, it is also usually more costly, time-consuming, and error-prone.

» **We are creatures of habit**. People are creatures of habit, so getting us to change how we do things is always going to be difficult. Often, we know there are better approaches, but we don't adopt them because at some level (consciously or unconsciously) we don't want to change our ways.

» **We are inconsistent**. Most people are very inconsistent at following a process. So the challenge is not just finding a better way of doing things; it is getting people to accept the new way, change their habits, and then apply the new approach consistently. As Karl Wiegers has said, "We are not short on *practices*, we are short on *practice*."

Success Modes

So do Cockburn's failure modes mean that we are doomed to fail? No; fortunately he has also identified some common success modes—common traits that help counter the failure modes. It can be helpful to share the following success modes with our team members, and try to leverage them for the good of the project:[2]

» **We are good at looking around**. This refers to our ability to observe, review, and notice when things are not right.

» **We are able to learn**. After seeing what's wrong, we are able to find ways to fix it and expand our skills and knowledge along the way.

» **We are malleable**. Despite the common resistance to change, we do have the ability to change and accept new ideas and approaches.

» **We take pride in our work**. We are able to step outside of our job descriptions to repair or report an issue, because it is the right thing to do for the project.

Success Strategies

Based on his success modes, Cockburn has come up with ten strategies for overcoming the failure modes.[3] These strategies are equally human, although—like common sense—they are often not applied. Like the failure and success modes, the following strategies aren't likely to be directly tested on the exam, but they can help you understand the outlook that the exam questions are based on.

» **Balance discipline with tolerance**. This strategy involves establishing a standard way of doing things and encouraging people to adopt it, but also building some capacity for tolerance and forgiveness into the approach. Then, if people diverge from the standard, we focus our energy on re-establishing the practice, providing it doesn't need modification.

» **Start with something concrete and tangible**. We solve a problem in our mind first, and then turn the solution into reality. To help people overcome their resistance to a change or solution, we can start with concrete and tangible tools that represent the final solution. For example, we can create low-fidelity mock-ups that not only show how the screens in a new computer system will flow, but also invite touch and interaction by allowing stakeholders to reorder and annotate the screens. This interaction can transform an abstract model into a tangible workflow that people can understand.

» **Copy and alter**. It is often easier to modify a working design to fit our needs than to create something from nothing. When creating a brand new solution from scratch, we're likely to procrastinate due to "blank page syndrome"—the feeling of paralysis engendered by a blank page. If we instead take the approach of adapting an existing solution, we'll have a working framework to start from.

» **Watch and listen**. We learn by watching others and listening to them. Simply putting junior team members close enough to see and hear the more experienced team members has been found to improve the junior team members' performance, even with no other training. This is referred to as the "expert in earshot" approach.

» **Support both concentration and communication**. Knowledge work requires both concentration and communication. For example, many knowledge workers need a certain amount of time to get into the quiet and productive mode known as "flow." They may spend 20 minutes getting there, only to have it interrupted by questions. However, communication and conversation are also essential for knowledge work and should be encouraged, since they help surface gaps in understanding and provide solutions to questions. The key is in balancing these needs. As we saw in chapter 4, agile teams use the "caves and common" model—they have both quiet areas where people can retreat to and a common work room where ideas are exchanged. In their book *Peopleware*, Tom Demarco and Anthony Lister also suggest establishing a "quiet work period" every day for a set period of time (such as 2 hours). During this quiet work period, meetings and phone calls are banned so that people can focus and get work done.

» **Match work assignments with the person**. We should try to match work assignments to personality types and look for mismatches. For example, a developer who would prefer to just rewrite someone else's bad code rather than explain what is wrong and how to improve it probably won't make a good coach. When assigning team roles, we should try to match them to people's personality traits—detail-oriented, social, analytical, supportive, etc.

» **Retain the best talent**. With skilled professionals such as knowledge workers, the difference between worst in class, average, and best in class is very large. We need to recognize the huge difference that talent makes and find ways to attract and retain the best talent. If this sounds like an obvious statement, then why do so many organizations ignore it? Maybe because it's easier to measure the short-term impact of a 5 percent drop in salaries on the bottom line than it is to measure the unknown, longer-term (and likely much larger) impact of losing the best people.

» **Use rewards that preserve joy**. Reward structures are tricky to set up, because once people start to expect a reward, it can have a huge demotivational impact to later remove that reward. For example, rewarding a child with a gift for reading may seem like a good idea at first to get her started. But this tactic could actually backfire and cause her to read less after the gift is phased out or she no longer sees it as valuable. We need to give rewards that tap into long-term, self-esteem-based motivators, such as pride-in-work, pride-in-accomplishment, and pride-in-contribution. Such rewards have a long-term appeal that doesn't fade over time.

» **Combine rewards**. The best reward structures combine elements to make not only a compelling package, but a truly supportive, caring work environment where personal and company objectives are aligned. And since no single reward system will work for everyone, we must build in many different kinds of rewards.

» **Get feedback**. A little bit of feedback can replace a lot of analytical work. We've already seen that agile practices such as pair programming, continuous integration, and iteration reviews are designed to "bake in" frequent feedback. The benefits of checking work before going on to something new are universal, which is why agile methods encourage so much feedback.

EXERCISE: CATEGORIZE THE PROBLEMS

Read the scenarios in the following table, and determine which of the five common failure modes they relate to. The answer for the first scenario is provided for you as an example.

To jog your memory, here is the list of common failure modes with numbers added for easy reference.

1. Making mistakes
2. Preferring to fail conservatively
3. Inventing rather than researching
4. Being creatures of habit
5. Being inconsistent

	Scenario	Failure Mode(s)
1	Tim's five-year financial forecasts did not include inflation, although including inflation is a standard company practice.	1
2	Rather than writing requirements in user story format, Bill continued to craft detailed use case descriptions. While these descriptions were useful, they took him four times longer to produce than the user stories created by other business analysts on the project, and they still needed to be confirmed with the business.	4

	Scenario	Failure Mode(s)
3	Jim fared little better than Bill. While most of Jim's requirements were in user story format, whenever he encountered something tricky, he reverted back to use case formats.	4, 2, 5
4	Pete and Kim paired on designing a new sorting algorithm that, while late, did improve reporting performance.	3
5	Mary's estimates omitted any allowance for remediation work after the testing of the user interface, but she did include the allowance in her estimates for testing the reports.	1, 5
6	After the first stakeholder demo of the real-time language-translator robot that swore at the CEO, Tom switched back to playing prerecorded samples in subsequent demos, though the team agreed it was not nearly as much fun.	2, 4

ANSWERS

	Scenario	Failure Mode(s)
1	Tim's five-year financial forecasts did not include inflation, although including inflation is a standard company practice.	Making mistakes (1)
2	Rather than writing requirements in user story format, Bill continued to craft detailed use case descriptions. While these descriptions were useful, they took him four times longer to produce than the user stories created by other business analysts on the project, and they still needed to be confirmed with the business.	Being creatures of habit (4)
3	Jim fared little better than Bill. While most of Jim's requirements were in user story format, whenever he encountered something tricky, he reverted back to use case formats.	Being creatures of habit (4) Maybe some level of preferring to fail conservatively (2) Being inconsistent (5)
4	Pete and Kim paired on designing a new sorting algorithm that, while late, did improve reporting performance.	Inventing rather than researching (3)
5	Mary's estimates omitted any allowance for remediation work after the testing of the user interface, but she did include the allowance in her estimates for testing the reports.	Making mistakes (1) Being inconsistent (5)
6	After the first stakeholder demo of the real-time-language-translator robot that swore at the CEO, Tom switched back to playing prerecorded samples in subsequent demos, though the team agreed it was not nearly as much fun.	Preferring to fail conservatively (2) Being creatures of habit (4)

Detecting Problems

The concepts we'll discuss in this section are focused on finding problems and defects. Diagnostic tools such as cycle time, trend analysis, and control limits can point to potential problems before they occur or help us identify problems as soon as possible after they have occurred.

The daily stand-up meeting is also an important mechanism for identifying problems. Recall that the third question in this meeting asks whether there are any problems or impediments to the team member's progress. The purpose of this question is to bring potential issues and problems to the surface early—rather than waiting until the team member can't continue working, the customer complains, or the team holds its next retrospective. When a concern is mentioned in a daily stand-up meeting, it is the responsibility of the team's coach or ScrumMaster to further investigate the issue and resolve it.

T&T Lead Time and Cycle Time

Lead time is a diagnostic tool that can be used to help identify and diagnose problems. This concept measures how long something takes to go through the entire process, for example, from design to shipping, or from requirements gathering through development to deployment.

Cycle time is a subset of lead time that measures how long something takes to go through part of the process, such as from product assembly to painting, or from coding to testing. For example, the cycle time for building a user story begins when the team starts working on it and ends when that item is finished, accepted, and begins delivering business value.

Here's a diagram that illustrates the difference between lead time and cycle time on a Kanban board:

Figure 6.10: Lead Time and Cycle Time Illustrated on a Kanban Board

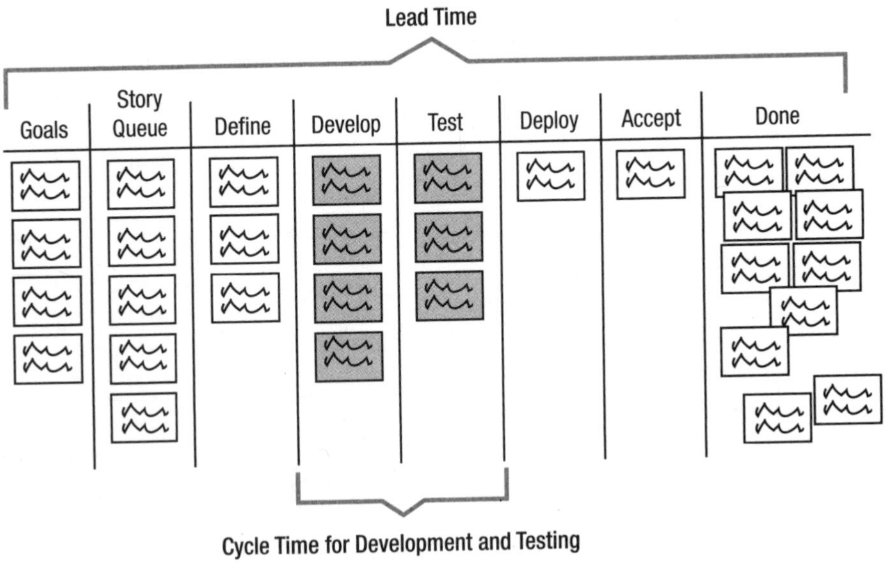

Cycle Time for Development and Testing

In the diagram above we can see that the lead time for a new feature in a software project spans the entire life cycle, from goal-setting through development and acceptance, until it is done. We can also see the cycle time for its development and testing, which is a component of its lead time. So lead time is how long it takes for the entire process, and cycle time is how long it takes for one component of that process. The lead time for a new system will likely have multiple cycle time elements like this.

Here's another example of these concepts. Lean-based approaches tend to look for global rather than local optimization. Although we may be able to make development and testing go faster by adding resources, if this doesn't speed up the overall process it's unlikely to be of much use to the business. So in lean terms, instead of trying to optimize the local throughput of development and testing (the cycle time) we want to reduce the overall time from inception to fruition (the lead time). Nevertheless, cycle time is likely to be a more important topic for the PMI-ACP exam than lead time, because of the close relationship between cycle time, WIP, and throughput.

Cycle Time, WIP, and Throughput

In mathematical terms, cycle time is a function of WIP and throughput and can be calculated by using the following formula:

$$\text{Cycle time} = \frac{\text{WIP}}{\text{Throughput}}$$

We've already defined WIP (work in progress) in chapter 2. Throughput, the other variable in this equation, is the amount of work that can be processed through a system in a given amount of time—such as the amount of work the team can get done in one iteration. (This can also be expressed as their average completion time.)

Knowing the team's throughput allows us to forecast future capability without specifically needing to know what the team might be asked to do. Knowing their cycle time allows us to make reliable commitments to the customer or organization about how long it will take to deliver work. WIP measures how much work we have "in the hopper" and gives us insight into issues, bottlenecks in the process, and rework-related risks.

Exam tip

Typically, the PMI-ACP exam contains few, if any, calculations or formulas. However, if there is any formula that is likely to appear on the exam, it is the one shown above for cycle time. And even if you don't encounter the formula itself, you may see questions that require you to understand the relationship between cycle time, WIP, and throughput. For example, if a team's WIP or throughput go up or down, how will that impact their cycle time?

Let's look more closely at the relationship between cycle time and work in progress (WIP) to see why this link is so important for agile. During the cycle when an item has been started but not yet completed, it is considered to be work in progress. As explained in chapter 2, excessive WIP is associated with a number of problems:

» It represents money that has been invested but isn't producing any return yet.
» It hides bottlenecks and masks efficiency issues.
» It carries the risk of potential rework.

It is because of these risks that agile and lean methods place a big emphasis on limiting WIP. And in order to do so, we must pay close attention to cycle time, since long cycle times lead to increased amounts of WIP. This is why agile approaches break the project work down into small batches and focus on finishing items and getting them accepted by the customer as soon as possible. For example, Scrum projects have small user stories, work in one- or two-week sprints, and get the product owner's acceptance of user stories at the end of each sprint.

These types of practices, which are common (in some form) to all agile methods, result in a reduction in cycle time on agile projects compared to non-agile approaches. For example, on a traditional project, the analysis, design, code, and test steps for a given piece of functionality might take months—or even years. This represents a very long cycle time, which leads to high levels of WIP and all the risks and issues associated with it. With agile methods, on the other hand, there is a constant effort to minimize cycle time. For example, to keep their focus on minimizing cycle time, agile teams may record the "Date Started" and "Date Done" on their task board cards, as shown below.

Figure 6.11: Using Task Boards to Track Cycle Time

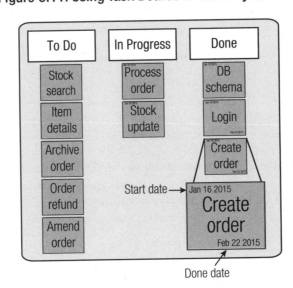

Project Cycle Time

We've said that the cycle time for a work item is the elapsed time between its start and finish. So the project cycle time is how long the entire project will take, from start to finish. This may seem a strange way of describing the project duration, but in lean-speak, the project duration is the cycle time for the entire project.

T&T Throughput and Productivity

It's important to understand the difference between throughput and productivity, since the exam questions may use both terms. We've said that throughput is the average amount of work the team can get done in a time period (or their average completion time). Productivity, on the other hand, is the rate of efficiency at which the work is done—such as the amount of work done per team member.

For example, if a team's throughput goes up, it might be because their productivity (output per person) went up. But not necessarily. They might have gotten more work done because they added another person to the team. In that case, their individual productivity might have actually gone down in the process of getting the new team member oriented to the project and the team. Yet they still might have been able to get more work done since there were more hands on deck.

EXERCISE: CYCLE TIME

Here's a chance to test your understanding of cycle time, WIP, and throughput with an exercise.

Imagine a bicycle factory that produces 25 bikes per day and typically works on 100 bikes at any given time. Calculate the average length of time it takes to make a bike.

$$\text{cycle time} = \frac{WIP}{Throughput} = \frac{100}{25} = 4 \text{ days}$$

Improvements to the assembly process reduce the cycle time to 3 days and the WIP to 90 bikes in progress. What is the percentage of improvement in throughput?

$$\text{Throughput} = \frac{WIP}{\text{cycle time}} = \frac{90}{3}$$

$$\frac{5 \, (diff)}{25 \, (old)} = 20\% \text{ improvement} \qquad \text{Throughput} = 30 / \text{dg}$$

ANSWER

The first question is a pretty basic one that we should be able to answer if we understand the definitions of WIP, throughput, and cycle time. The WIP, or the measure of work in progress, is the 100 bikes being worked on at any given time. Throughput is how many things go through the process per time interval; in this case, the throughput is 25 bikes per day. We are being asked to calculate how long it takes on average to make a bike, which is the cycle time (Ha!). Since cycle time = WIP / throughput, we can calculate our cycle time as 100/25 = 4 days.

The second question is trickier, because we now need to rearrange the equation to find throughput. The rearranged equation is throughput = WIP/cycle time. With our new, improved process, WIP = 90 and cycle time = 3 days, so throughput = 90/3 = 30 bikes per day.

Once we've done this calculation, we still have to find the percentage of improvement over the old throughput of 25 bikes per day. So we subtract the old throughput from the new measurement (30 – 25 = 5), and then divide the difference by the old throughput (5/25 = 0.2, or 20%). Therefore, increasing our throughput from 25 to 30 bikes per day is a 20 percent improvement.

Defects

We've seen that because of the cost of change curve, it is advantageous to catch and fix defects as quickly as possible to minimize rework and reduce costs. This is where the idea of "defect cycle time" comes in. So far we have been talking about cycle time in terms of developing new work, but cycle time is also useful for finding and fixing defects. "Defect cycle time" is the period between the time the defect was introduced and the time it was fixed. The length of the defect cycle time dictates how far up the cost of change graph the defect will go.

To help minimize the cost of fixing defects, some project teams actively track their average defect cycle time and set goals for the quick resolution of defects, as shown below.

Figure 6.12: Average Defect Cycle Time (Hours)

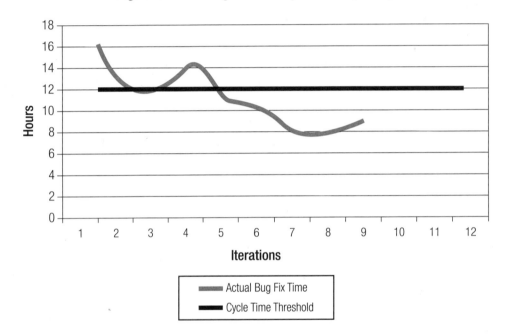

In this example, the team's cycle time threshold is 12 hours. This means that, on average, their goal is to have all reported defects fixed within 1.5 business days. So if a defect is found at noon on Tuesday, it should be fixed by the end of the workday on Wednesday. We can see that for the first five iterations, the team's "bug fix" time was usually above 12 hours. But as the project continued they were able to improve their defect cycle time—and in later iterations, the average dropped below that threshold, eventually going as low as 8 hours.

By tracking both their defect cycle time and their cycle time for creating new work, agile teams can minimize both the potential for rework and the cost of any rework that is required.

T&T   Defect Rates

Unfortunately, no matter how hard we try to identify and prevent defects, there may be an occasional defect that makes it through all our tests and quality control processes and ends up in the final product. (In software development, these are called *escaped defects*.) Defects that are missed by testing are the most costly kinds of defects to fix, since they are on the top right end of the cost of change curve. By the time we learn about such a defect, it is obviously too late to prevent it, but we can use this information to improve our processes going forward and hopefully reduce the number of defects of that nature in the future.

A project's defect rate measures the frequency of defects found, such as "one defective feature per 50 successful features delivered." To assess and improve the effectiveness of their testing and quality control processes, agile teams often track their defect rates so they can monitor the trend over time. These graphs are often posted as information radiators that show the number of defects found per time period, like this:

Figure 6.13: Defects Found per Month

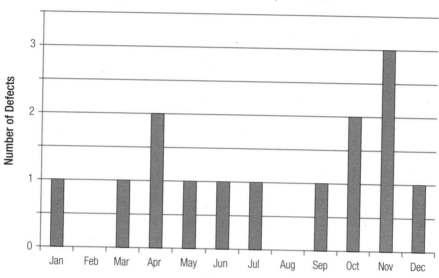

Software development teams commonly track their escaped defects back to the release they originated in ("escaped from"), rather than by iteration or time period. Such a graph might look like this:

Figure 6.14: Defects Found per Release

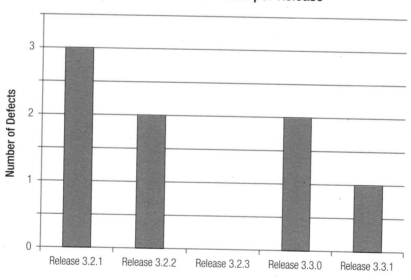

Defect rates can be used to understand how well a process is working. An increasing defect rate should be a cue to ask why the process seems to be getting worse. Is a new type of work being undertaken? Is the team being pushed too hard for productivity at the expense of quality? Defect rates and defect counts can be used to give insight into these issues.

T&T **Variance Analysis**

Variance is the measure of how far apart things are, or how much they vary from each other. For example, if you ask multiple people to estimate the same job, there will be some variance (differences) between their estimates. And once the work has been executed, there will be a variance (difference) between the closest estimate and the actual results. Much variance is normal and should be expected—we just want to keep it within acceptable limits. If it fluctuates beyond the acceptable limits, then we need to know so we can take action.

When evaluating performance or tracking results, we should understand that there will always be a certain amount of variance due to normal fluctuation. Even highly engineered processes have to take into account some degree of normal variation. We don't want to mistake this normal variation for a significant outcome or trend, or vice versa. We usually assume that normal variation is like background noise—too small to be mistaken for the performance outcomes we are trying to measure—but this isn't always the case.

Causes of Variation

Quality expert W. Edwards Deming classifies variance into common cause variation and special cause variation. Common cause variation refers to the average day-to-day differences of doing work, and special cause variation refers to the greater degrees of variance that are caused by special or new factors.[4]

For example, let's say we are given the job of driving nails into wood all day. We can expect that not all the nails will be perfectly straight—some of them will go in straight and others will go in at a slight angle; this is common cause variation. However, now let's say that someone turns off the lights while we are nailing. If we keep nailing in the dark, the variance in our nails is likely to be much larger, since the environment has changed (special cause variation). Deming goes on to say that there are two classic mistakes that managers make:[5]

> Mistake 1: Reacting to an outcome as if it came from a special cause, when it actually came from common causes of variation.

An example of this kind of mistake would be stopping work to investigate why the team estimated a user story as four days of work when it actually took five days to complete.

> Mistake 2: Treating an outcome as if it came from common causes of variation, when it actually came from a special cause.

An example of this kind of mistake would be continuing our nailing project after the lights go off and expecting all the nails to be as straight as before. Or, an agile team's estimates seem to be increasingly off, but they decide that must be normal variation rather than stopping to investigate whether their story point unit has become inconsistent.

Variance Example: Training Pilots

Lean experts Mary and Tom Poppendieck tell a story about variance in pilot training.[6] Pilots were evaluated on their performance over several training flights, and their results were tracked and plotted as shown below. On this diagram, the pilots' worst performances are circled in black, and their best performances are circled in green.

Figure 6.15: Training Flight Performance

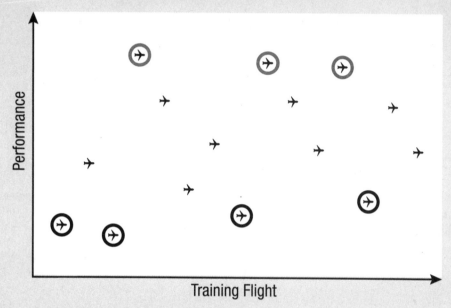

When the pilots performed poorly, they were yelled at and told to improve. When pilots did well, they were congratulated and praised. The chart shows us that the poorly performing pilots improved in their subsequent flights, while the top-performing pilots got worse in subsequent flights.

So one conclusion we could draw from these results is that yelling at people is a very effective way of improving performance and that we should do it more often—while praise isn't effective and should be stopped. But this would be an incorrect conclusion. What is really shown here is normal variation. Some flights go well, while others do not. Within any variable data set, extreme lows are usually followed by a higher value, just as extreme highs are usually followed by a lower value. So what we should learn from this scenario is that processes simply vary, and we need to accept a certain degree of variance in any real-world system.

Accept the Variance or Take Action?

The main point is that we should simply accept common cause variation on our projects; we only need to investigate or take action in the case of special cause variation. In other words, we want to avoid micromanaging the project and instead focus on removing true bottlenecks and impediments. Asking our developers why they only coded four features this week when they completed five features last week is an example of failing to accept common cause variation; this type of stuff just varies a bit and isn't perfectly predictable.

Figure 6.16: Know When to Intervene

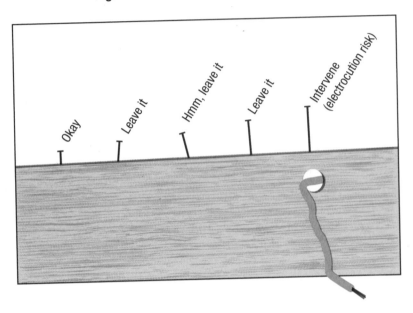

So on agile projects in particular, focusing a lot of effort on tracking conformance to a rigid plan isn't the best use of a leader's time. Instead, we should look to external indicators and the daily stand-up meetings where the team reports any issues or impediments to their work to see if there are special issues that need to be resolved.

EXERCISE: COMMON CAUSE OR SPECIAL CAUSE— LEAVE ALONE OR INTERVENE?

Read the following team comments and determine whether the issue being discussed is related to a common cause or special cause. The first answer is provided for you as an example.

	Comment	Common Cause or Special Cause?
1	"Our initial estimates for product testing times were on average 5 percent too high."	Common cause (based on this limited description)
2	"I am still working on the tether for the screw-top fastener. I know it should have been done yesterday, but I'll get it done today."	Common
3	"I read a press release today saying our competitor's product now does single-pass printing. Do we want to move that feature up in our backlog?"	Special

	Comment	Common Cause or Special Cause?
4	"The paint shop is running a day behind again."	Common
5	"Bill, our electrical safety inspector, caught malaria on vacation."	Special
6	"That's nothing. Helen in accounting got married in Cuba over the weekend!"	Common
7	"I am waiting on Ted again to finish the last of the renderings."	Common

ANSWER

	Comment	Common Cause or Special Cause?
1	"Our initial estimates for product testing times were on average 5 percent too high."	Common cause
2	"I am still working on the tether for the screw-top fastener. I know it should have been done yesterday, but I'll get it done today."	Common cause—it sounds like a one-off task took a little longer than anticipated.
3	"I read a press release today saying our competitor's product now does single-pass printing. Do we want to move that feature up in our backlog?"	Special cause—an external change may trigger a project reprioritization.
4	"The paint shop is running a day behind again."	This is most likely common cause—hopefully it's not a big deal unless some important paint job is waiting.
5	"Bill, our electrical safety inspector, caught malaria on vacation."	This may be special cause—will the loss of the safety inspector while he recovers from malaria impede the overall project progress?
6	"That's nothing. Helen in accounting got married in Cuba over the weekend!"	Gossip/common cause—it's difficult to see how this will impact the team too much.
7	"I am waiting on Ted again to finish the last of the renderings."	This is most likely common cause—the comment does not include commentary to indicate it is a big issue.

T&T **Trend Analysis**

Trend analysis is a particularly important tool for detecting problems because it provides insights into future issues before they have occurred. Although measurements like the amount of budget consumed, for example, are still important, such measurements are lagging metrics; in other words, they provide a view of something that has already happened. While lagging metrics that provide a perfect view of the past might be exciting to accountants, leading metrics that provide a view into the future are more exciting to agile leaders. This is because leading metrics provide information about what is occurring now or may be starting to happen on the project. This early indication of a potential problem is more useful than lagging metrics, since it helps the team adapt and replan appropriately.

For example, the figures below show the defects, change requests (CR), and clarifications (questions such as, "How do I use the product to do X?") that were logged for a team's first release during the month of March. For each category, the number of items opened, the number closed (resolved), and the ongoing tally of remaining items in that category are tracked.

Figure 6.17: Trends in Defects, Change Requests, and Clarifications

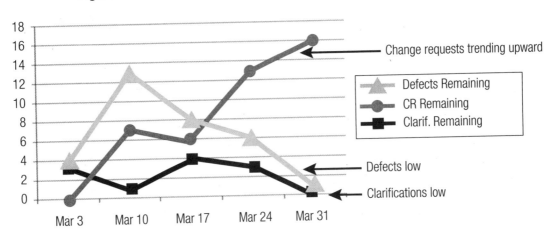

Here we can see that the number of defects remaining spiked on March 10, but that the number has gone down since then, and the number of clarifications bubbled along with low numbers. However, change requests are escalating, and there is a trend emerging. For the five periods tracked, there are more change requests being opened every month than being closed.

This trend is useful not only because it tells us we are receiving change requests faster than we can process them, which is increasing our WIP, but also because it indicates that we may not be spending enough time validating feature requirements before developing them. So this metric provides insight into the project and should prompt a discussion with the team about the issues and possible solutions.

Another point to keep in mind here is that identifying trends is more important than analyzing the actual data values. So for this example, the valuable information is that we have an escalating number of changes, rather than the specific number of change requests that were opened in a given week.

T&T Control Limits

Control limits are a concept from manufacturing process control, where statistically calculated upper and lower limits are used to help determine the acceptable variation in a process. However, in an agile context, control limits have a much looser interpretation that includes tolerance levels and warning signs. Such limits can help us diagnose issues before they occur or provide guidelines for us to operate within. Some of the agile recommendations or rules of thumb, such as limiting teams to 12 or fewer members, could be interpreted as control limits.

One way we can use control limits on an agile project is to monitor our velocity to gauge how likely it is that we will be able to complete the agreed-upon work by the release date. For example, if we have 600 points' worth of functionality left in the backlog and 10 months until deployment preparations begin, we should set our lower control limit at 60 points per month (600 / 10 = 60), since that is the minimum velocity required to meet this goal.

Figure 6.18: Velocity Control Limits

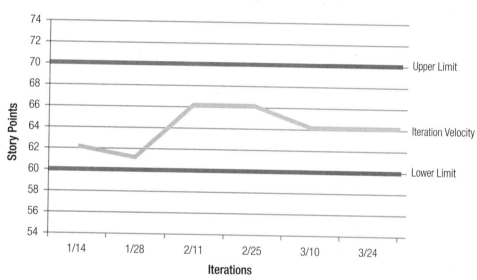

This diagram shows that our lower control limit is set at 60 points per iteration. If our velocity dips below this limit, there is a chance that we won't finish the agreed-upon functionality by the release date. Obviously, a quick dip below 60 points per month wouldn't be as bad as a sustained period—so it is really our rolling average that we are most concerned with. However, this kind of graph is easy to interpret and provides a visual control tool.

Kanban and task boards that limit WIP are also a form of control limits on agile projects. These tools prevent too much work from being included in an activity, and they help the team control the amount of work in progress.

Managing Threats and Issues

In chapter 2 we explained that since risk is essentially anti-value, managing risk is critical for value-driven delivery. As a result, agile teams need to balance the goals of delivering business value and reducing risk each time they select a new batch of features or stories to work on. As the project continues, they also need to continually assess the severity of the project threats and monitor their overall project risk profile. All of these practices are included in the problem detection and resolution domain, so in this section we'll finally get into the nitty-gritty of how agile teams manage threats and issues, by examining three tools—the risk-adjusted backlog, risk severity, and risk burndown graphs.

✓ Exam tip

For the exam, bear in mind that in agile, "risk" generally refers to threats and issues that could negatively impact the project, as opposed to the *PMBOK® Guide* definition, which includes positive risks, or "opportunities." Therefore, in this discussion we will be using the terms "risks" and "threats and issues" interchangeably.

T&T Risk-Adjusted Backlog

In planning each iteration, agile teams seek to balance delivering the highest-value features and mitigating the biggest risks that remain on the project. They do this by moving the items with the greatest value and risk to the top of their backlog. The backlog might start out as just a list of the business features involved in the project, divided into practical bundles of work—but once the risk response activities are added and prioritized (based on their anti-value), it can be referred to as a "risk-adjusted backlog." This single prioritized list is what allows agile teams to focus simultaneously on both value delivery and risk reduction activities. We'll look at the specifics of how this is done shortly, but the general process is illustrated below.

Figure 6.19: Using the Backlog to Prioritize both Features and Risk Responses

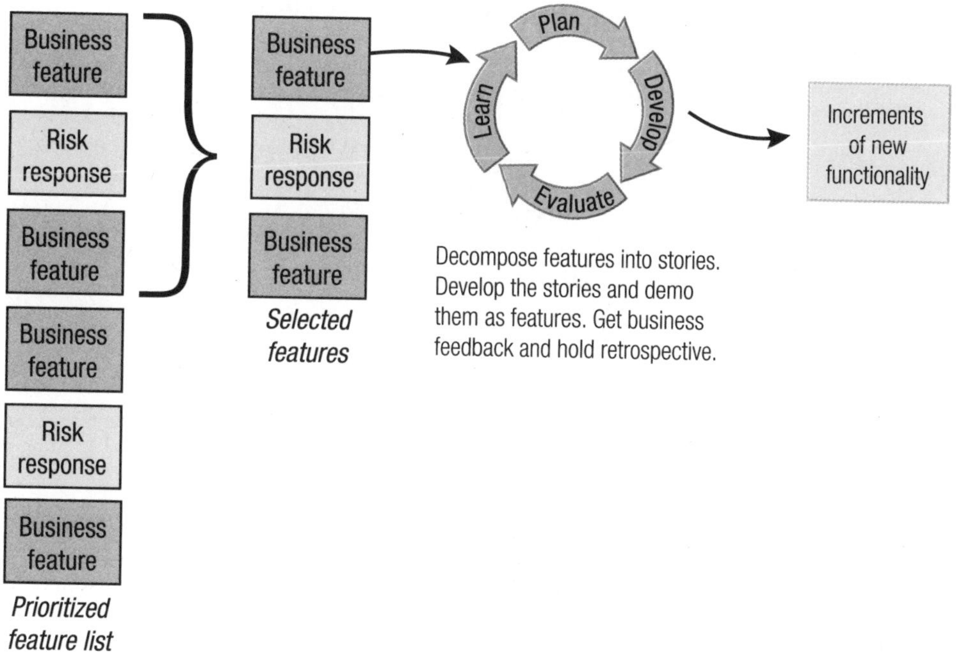

Business feature

Risk response

Business feature

Business feature

Risk response

Business feature

Prioritized feature list

Business feature

Risk response

Business feature

Selected features

Plan — Develop — Evaluate — Learn

Decompose features into stories. Develop the stories and demo them as features. Get business feedback and hold retrospective.

Increments of new functionality

Creating the Risk-Adjusted Backlog

Most project teams are comfortable with ranking customer requirements (or in agile terms—stories, features, and use cases) on the basis of business value and risk level. This ranking is often subjective, based on a customer's gut feeling or preference. However, we can get much more scientific about building a risk-adjusted backlog by using the return on investment per feature. This process starts with the financial return expected from the project as a whole. (For regulatory compliance or maintenance projects, we would instead use the financial impact of not doing the project—fines, lost business, equipment failure, etc.)

For example, let's say that before approving a $2 million project, a company does a cost-benefit analysis to determine whether the revenue or savings to be gained from it justify the cost of undertaking it. In the process, it determines that the expected return for the project is $4 million in three years.

Once we have a dollar figure like this, the next step is for the business representatives (not the development team) to distribute or prorate this amount across the project features. Business representatives often push back at this, with comments like, "I can't put a dollar value on a sales report; that would be too subjective!" It can be helpful to remind them that someone came up with a projected return for this project. If that figure cannot be divided across the product's features, then where did it come from?

After the business representatives have (perhaps somewhat arbitrarily) attributed a dollar value to each of the product features, we can prioritize those features based on business value, as shown below.

Figure 6.20: Features Prioritized by Business Value

Must	$5,000
Must	$4,000
Must	$3,000
Should	$2,000
Should	$1,000
Should	$500
Could	$100

Next, we need some way to monetize the risk avoidance and risk mitigation activities. To get this figure, we can calculate the expected monetary value of each risk. For a sample calculation, let's assume that using an in-house reporting tool has no cost to the project, but buying a high-performance reporting engine will cost $10,000. There is a 50 percent chance that we will end up needing the high-performance tool. We can use the following formula to calculate the expected monetary value of this risk:

Expected Monetary Value (EMV) = Risk Impact (in dollars) × Risk Probability (as a percentage)

So in our example, EMV = $10,000 × 50% = $5,000. This figure allows us to say to the project sponsors that we believe the economic value to the organization of mitigating this risk is $5,000. Therefore, the priority of the response action for this risk should be on par with a functional feature valued at $5,000.

This EMV calculation can be done for most risks. Technical risks usually have an associated purchase cost (e.g., $10,000 for a high-performance reporting engine) or a time penalty that can be translated into a dollar amount (e.g., it will take two developers an additional three weeks at x dollars). Human resource and business risks can likewise be estimated in monetary terms. We tend to think that we cannot assign a monetary value to certain things, but remember that insurance companies are able to determine a value for a lost finger or emotional distress for their clients. The same concept applies to assigning value to project risks.

However, it is important to keep in mind that we are looking for relative amounts rather than precise numbers. In many cases, the initial return that the business established to justify the project is suspect anyway, so we should not get hung up on the level of accuracy of the risk values or try to create a perfect balance sheet of risk. Instead, we should focus on coming up with general, justifiable numbers that have consensus from the project stakeholders to use as a basis for prioritization.

Using this approach, we can rank the project risks to produce a prioritized list of threats and issues, ordered by expected monetary value, as shown below.

Figure 6.21: Threats and Issues Prioritized by EMV

Risk 1 ($9,000 × 50% = $4,500)
Risk 2 ($8,000 × 50% = $4,000)
Risk 3 ($3,000 × 50% = $1,500)
Risk 4 ($6,000 × 25% = $1,500)
Risk 5 ($2,500 × 25% = $625)
Risk 6 ($500 × 25% = $125)
Risk 7 ($500 × 20%= $100)

Of course not all risks will have avoidance or mitigation steps that we can schedule into the project. Some risks may have to be accepted (e.g., "We are waiting for service pack 2") or transferred (e.g., "We have taken out insurance"). But for the risks that can be proactively tackled, the next step is to prioritize the response actions along with the functional features to get the risk-adjusted backlog, as illustrated below.

Figure 6.22: Risk-Adjusted Backlog
(with Features and Risk Response Actions)

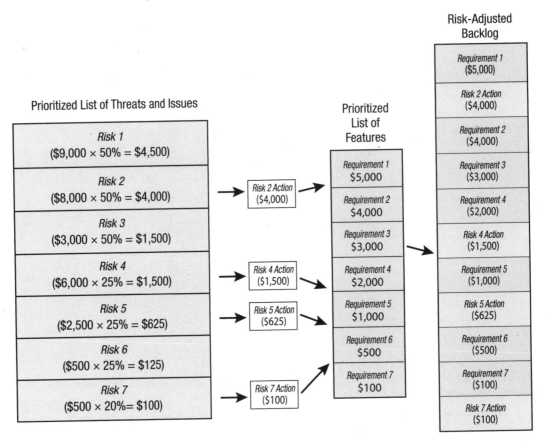

Attributing a dollar figure to both functional and nonfunctional requirements allows the team to have a more meaningful discussion with the sponsor and customer. For example, in this diagram, we can see that Risk 2 has an expected monetary value of $4,000; as a result, when it is time to select the work for an upcoming iteration, the response action associated with Risk 2 now has a similar ranking to Requirement 2. In other words, a risk response action that did not seem to have a compelling business value now makes sense because we can show that it has equal value to the organization as a high-priority feature.

Again our focus here is not really on the precision of the numbers. A guess multiplied by a guess is unlikely to be highly accurate. Instead, the purpose of this exercise is to facilitate better discussions with business representatives about how to sequence the work items. If people start rigging the numbers to serve their personal agendas, the purpose and true power of the process will be lost, and technical dependencies and sponsor mandates will take precedence over delivering value.

We should think of the risk-adjusted backlog as a tool that uses calculations only to get at what is truly important—the priority of the work items that need to be done. The true benefit of this tool is that it helps the product owner and the development team bridge the communication gap and have meaningful discussions about schedule and scope trade-offs. So although we assign numerical values to help level the playing field, creating a risk-adjusted backlog is really more of a qualitative practice than a quantitative practice.

Risk Severity

We've seen that on agile projects, risk management should be a driver for work scheduling, as the team moves high-risk activities into earlier iterations of the project and incorporates risk mitigation actions into the backlog. As the project progresses, the team needs to continue to monitor risks and track the effectiveness of their risk reduction efforts. So let's look at how that is done, using the concept of risk severity. To explain it, we have to start by backtracking a bit.

As we've seen, risks are generally assessed via two measures—risk probability (a measure of how likely a risk is to occur) and risk impact (a measure of the consequence to the project should the risk actually occur). When the probability is indicated as a percentage and the impact is defined in monetary terms, we can multiply these two figures to calculate the risk's expected monetary value (EMV). For example, if a risk is estimated to have a 25 percent probability of occurring and its financial impact is estimated at $8,000, its EMV will be 0.25 × $8,000 = $2,000.

But as we've seen, there is a drawback to using EMV. It can be tempting to focus too much on the exact dollar amounts rather than the relative value of the risks. To help us avoid that problem, there is another metric we can use to rank risks and determine risk response priorities—risk severity.

To calculate risk severity, instead of using a risk's probability percentage and dollar impact, we instead rank its probability and impact on a simple scale—such as low (1), medium (2), and high (3). Then, we multiply those probability and impact rankings to calculate the risk's severity:

$$\text{Risk Severity} = \text{Risk Probability} \times \text{Risk Impact}$$

Now, this formula might look just like the one we used for EMV—however, its output will be different because the inputs are rankings. For example, let's say we decide to rank the probability and impact of our risks on a three-point scale, as low (1), medium (2), or high (3). Using this scale, a risk that has a high probability and a high impact will have a risk severity of 3 × 3 = 9. On the other hand, a risk that has a high probability and a low impact will have a risk severity of 3 × 1 = 3. You can see how this would help shift our attention away from specific dollar amounts so that we can focus on the bigger picture.

Of course, some projects might require an in-depth risk analysis based on expected monetary value. However, for most agile projects, we are interested in relative risk profiles and trends. For that, all we really need is the abstract value of risk severity based on a three-point scale, which will generate severity scores ranging from 1 through 9.

We start by doing an analysis of the risks so that we can assign a probability and impact score to each one. Then we use those scores to calculate the severity of each risk. The figure below shows the risks identified for an example project with their probability, impact, and severity scores.

Figure 6.23: Identified Risks with Impact, Probability, and Severity Scores

ID	Short Risk Name	Impact	Prob.	Sev.
1	JDBC driver performance	3	2	6
2	Calling Oracle stored procs. via web service	2	2	4
3	Remote app. distribution to PDAs	3	2	6
4	Oracle warehouse builder stability	2	2	4
5	Legacy system stability	2	1	2
6	Access to user community	2	1	2
7	Availability of architect	2	2	4
8	Server upgrade necessary	1	2	2
9	Oracle handheld warehouse browser launch	3	1	3
10	PST changes for British Columbia	0	0	0
				33

As the project progresses, the team can expand this table to record how their attempts to manage the project risks are working. The next chart shows the progress of the risks over the first four months of the project.

Figure 6.24: Progress of Risks

ID	Short Risk Name	January Imp.	Prob.	Sev.	February Imp.	Prob.	Sev.	March Imp.	Prob.	Sev.	April Imp.	Prob.	Sev.
1	JDBC driver performance	3	2	6	3	0	0	3	0	0	3	0	0
2	Calling Oracle stored procs. via web service	2	2	4	2	0	0	2	0	0	2	0	0
3	Remote app. distribution to PDAs	3	2	6	3	1	3	3	0	0	3	0	0
4	Oracle warehouse builder stability	2	2	4	2	3	6	2	2	4	2	0	0
5	Legacy system stability	2	1	2	2	1	2	2	0	0	2	0	0
6	Access to user community	2	1	2	2	2	4	2	1	2	2	1	2
7	Availability of architect	2	2	4	2	3	6	2	2	4	2	0	0
8	Server upgrade necessary	1	2	2	1	1	1	1	0	0	1	0	0
9	Oracle handheld warehouse browser launch	3	1	3	3	1	3	3	3	9	3	1	3
10	PST changes for British Columbia	0	0	0	0	0	0	2	2	4	2	1	2
				33			25			23			7

Image originally published in "The Game of Risk" by Mike Griffiths on gantthead.com on September 20, 2011, copyright © 2011 gantthead.com. Reproduced by permission of gantthead.com.

During these four months, many of the project risks were mitigated or avoided. For example, the first risk, "JDBC driver performance," ended up not being a problem because database features and performance testing were deliberately included in the first (one-month) iteration. As a result, the probability of this risk occurring was reduced from 2 (medium) to 0, and its risk severity went from 6 to 0.

T&T Risk Burndown Graphs

Most projects will have a longer list of threats and issues than shown in the example we just saw. But even in these tables, all the numbers make it difficult to detect trends and understand what is really happening. To make sense of all this data, we have to put in some effort and take time to study the details. This is where another key agile tool comes into the picture, reflecting the agile emphasis on fast, visual communication. To make the data in the above risk severity table easier to grasp at a glance, we can convert the numbers into a risk burndown graph, like this:

Figure 6.25: Risk Burndown Graph

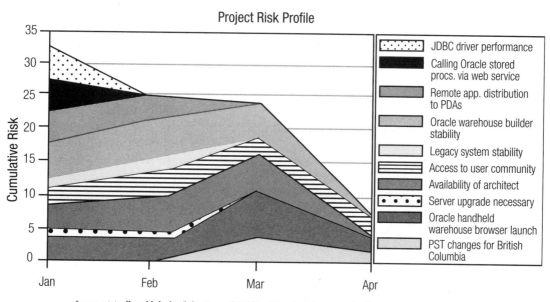

Image originally published in "The Game of Risk" by Mike Griffiths on gantthead.com on September 20, 2011, copyright © 2011 gantthead.com. Reproduced by permission of gantthead.com.

Risk burndown graphs are essentially stacked area graphs of cumulative project risk severity. The severity scores for each risk are plotted one on top of another to show the project's cumulative severity profile. When risks and their history of severity are displayed in this format, it is much easier to interpret the overall risk status and trends of the project.

For instance, we can tell from the general downward trend of this graph that the project risks are being reduced. This is an excellent way to demonstrate the value of "Iteration 0" activities—iterations that may not directly deliver much (or any) business value, but are used to reduce risks or prove whether a particular approach will work.

New and escalating risks are also easy to spot on a risk burndown graph. In our example, Risk 9, "Oracle handheld warehouse browser launch," escalated from a severity 3 risk to a severity 9 risk in March. Risk 10, "PST changes for British Columbia," is a new risk that was assessed as a severity 4 risk in March and went down to a severity 2 risk in April.

Risk burndown graphs quickly inform stakeholders whether the risks are moving in the right direction (downward), or if they are escalating. These graphs provide an easy-to-interpret way to check the health of a project and can be quickly produced in tools like Microsoft Excel. Although project risk is not a core metric like features delivered, it is useful to track. Agile teams need to actively control their risks, since the effectiveness of their risk reduction efforts ultimately impacts whether they are successfully maximizing value.

K&S Solving Problems

The final topic we'll discuss in this chapter is problem solving. Although we should take steps to find indications of problems early and then take action to avoid them before they occur, it is inevitable that our projects will still encounter problems. Therefore, we need to understand how to resolve them.

Problem Solving as Continuous Improvement

Agile methods don't rely on a reactive approach of "fix the problem after it arises." Instead, they include focused efforts to identify potential issues during the iteration reviews and retrospectives that are done at the end of each iteration. This minimizes the need for ad hoc problem solving during the iterations.

Another way of saying this is that agile methods consider the team's lessons learned to be too important to be reserved until the end of the project. Therefore, they include iterative opportunities for the team to capture those lessons, probe the strengths and weaknesses of their approach, and ensure that their mistakes won't be repeated. These team-based "problem solving" efforts are actually part of the continuous improvement process. Reflecting that emphasis, in this chapter we have been discussing agile problem solving conceptually, but we'll leave the details of how it is actually done for the next chapter ("Continuous Improvement"), where we'll examine the problem-solving steps that are done in the retrospective process.

Engage the Team

Another important difference from traditional approaches is that agile problem-solving methods are team-based and inclusive; they aren't just relegated to the customer or managers. Agile methods engage the team in identifying, diagnosing, and solving threats and issues. Rather than leaving problem solving and correction to the project manager or external "fixers," this is very much a whole-team activity.

Why is it so important to engage the team in this effort? When attempting to resolve threats and issues, organizations too often overlook the best source of solutions—the project team. The team usually has the best practical solutions, even if they may not be the best theoretical solutions. Also, there are many benefits to involving the team in solving problems. One of the most important is that you gain the team's buy-in from the start; you don't have to sell your solution to the team.

Background information

Gaining Team Buy-In

A story from one of my projects illustrates how the team's practical solution may be preferable to the best theoretical solution. While managing a government project during an IT vendor change, we needed to quickly build rapport with the business users of the system and the subject matter experts (SMEs) in order to maintain the previous team's development pace. The problem we faced was how to get to know the business folks better and connect the team members and the SMEs. Instead of contriving some project social event and trying to get buy-in from the business and the team, I presented the problem to the team in a planning meeting.

After some back and forth, a team member suggested that since quite a few of the business people went bowling, perhaps we should arrange a bowling social event. Other team members concurred, and we brainstormed ideas for mixing up the teams (so they weren't "us versus them").

We ran the event. It went really well and led to other activities, all of which helped to build strong relationships between stakeholders and a successful project. As the project manager, I could have chosen to consult the PMO, the human resources department, or social psychology experts to design the optimal team-building event—but in that case, I would have had to sell the event to the team.

The Benefits of Team Engagement

Let's look more closely at the benefits of engaging the team in finding solutions to problems:

» **By asking the team for a solution, we inherit consensus for the proposal.** In the bowling story, the fact that the solution came from the team and not from me meant I didn't have to sell it to them—it already had their support. It is easier to guide a suboptimal solution that has good support to a successful outcome than it is to build support for an optimal solution and make sure it is successfully executed. Once challenges arise, people will start thinking, "Whose bright idea was this?" If they have to answer, "Oh yeah, it was ours," they will be more motivated to continue working to find a solution.

» **Engaging the team accesses a broader knowledge base.** Asking the group for ideas taps the collective knowledge of the team. Since team members are closer to the details of the project, they bring additional insights to bear on problems and their potential solutions. (In my story, I didn't know that the business folks liked to bowl. This information was collective knowledge; one team member, Chris, knew that three users bowled, and Julia, another team member, knew that two SMEs did, as well. So together we found a new piece of useful information.)

» **Team solutions are practical.** Anyone who has worked hard to craft a solution only to be told "That won't work here because . . ." will know how frustrating these words are. Team-sourced solutions have already been vetted for practicality, and because they are created internally, they also include solutions for implementation issues.

» **When consulted, people work hard to generate good ideas.** People don't want to be treated as work drones. When asked for their input, they generally work hard to create innovative and effective solutions that are supported by those who need to implement them. The simple act of asking for suggestions engages team members beyond narrow roles such as "coder," "tester," or "engineer." (Treating workers as interchangeable resources is a poor model inherited from the command-and-control methods of the industrial revolution. Leading companies such as Toyota and 3M recognize that their best ideas come from inside their companies, and that they need to make use of this intellect. It is partly due to these methods that these companies innovate better, produce higher-quality products, and have better labor relations.)

» **Asking for help shows confidence, not weakness.** Asking for ideas and solutions to problems is not a sign of incompetence or an inability to manage. Just because a leader asks for input does not mean he or she doesn't know what to do. Instead, it demonstrates that the person is thoughtful and values the opinions of others.

» **Seeking others' ideas models desired behavior.** In agile, the leader's role includes modeling desired behavior—we need to behave as we wish others to behave. So by asking for the team's input, a leader demonstrates how all problems should be tackled. If instead the leader stays silent, makes decisions without knowing all the facts, or doesn't ask for help that the team could provide, what message does that send? Whether we realize it or not, the message is obvious to team members that we expect them to behave the same way and work in isolation. Any time and money we spend on team-building activities is wasted if we take a management-in-a-vacuum approach. In contrast, when we model collective problem-solving behavior, team members are encouraged to solve their problems in this way, too. Teams that can effectively solve problems and build support for their solutions are the real powerhouse of successful projects. That's why it's important for agile leaders to support and mirror these best practices.

In summary, the team is best suited for coming up with solutions to project problems. Not only do the team members provide the best solutions; including them in the problem-solving process improves their performance and motivation as well. People relish having their input valued, and they appreciate opportunities to help solve problems. Agile methods take advantage of these benefits by unleashing the team's creative power on project problems.

Considerations and Cautions for Engaging the Team

Despite the benefits of engaging the team in problem solving, this is not a silver bullet or a cure-all approach. There are some points that we need to keep in mind about using this approach, including some cautions:

» **Involve the team where it can be most helpful.** The team engagement approach works especially well on complex, embedded process problems that would take days to explain to an expert. In cases like this, it is a huge benefit that the team already has first-hand knowledge of the problem and, more importantly, what types of actions are practical to solve it.

» **Solve real problems.** We should use the team to help solve real problems only, not to make decisions like which brand of printer toner to buy. Remember that agile leaders are always setting an example and modeling behavior for the team. If a leader goes too far and consults the team for every little decision (such as what color dialogue box they should create), no work will get done. Engaging the team in solving a problem is a tool to use when you are stuck and the problem is important.

» **Team cohesion is necessary.** If the team is fragmented and there are opposing groups, then resentment that the other group is "fixing their problems" is likely to undermine the process. We need to get the team members aligned with each other for team problem solving to be most effective.

» **Check in after team or project changes.** If a significant portion of the team changes, we need to recanvas the team to make sure they are still on board with the approach that was chosen. Having to implement the bright ideas of other people is nearly as bad as not being consulted in the first place. So if the team changes significantly, we need to check in to find out whether people still agree that the solution is a good approach. Likewise, if the project changes significantly, we need to perform a checkpoint and have the team review their approach in light of the new circumstances.

» **Be sure to follow through.** Once you ask the team for solutions, make sure you follow through on executing them. It is pretty demoralizing to be asked to work on a solution and then see that solution wither. It is fine to go back to the team with implementation problems that need to be solved, but we shouldn't waste people's time by asking for their input if we are simply going to ignore it.

Some Problems Can't Be Solved

Sometimes, even when we engage the team and try our best, we end up deciding that certain problems cannot be solved. If there is a way to work around those problem, then the best approach is accept them and move on to delivering value where we can. This is preferable to spending more and more time trying to solve a problem. It all comes back to the saying, "Where is our next dollar best spent?"

In other words, we have to know when to let a problem go, and not throw good money after bad. We try to solve all the problems we encounter, but sometimes the smartest thing to do is just get out of the situation and reset our expectations for the project instead.

Chapter Review

1. A new risk has been discovered halfway through the project. What should your team do first?

    A. Add it to the top of the backlog.
    B. Ask a subject matter expert to assess its probability and impact.
    C. Evaluate its root cause.
    D. Schedule a risk-based spike to resolve or minimize it.

2. What would be most helpful for improving a team's problem-solving proficiency?

    A. Focus on keeping arguments and disagreements to a minimum.
    B. Encourage them to share their mistakes and problems with each other.
    C. Ask more experienced team members to mentor their peers.
    D. Score their suggestions and post a leaderboard in the team space to encourage competition.

3. Which tool or metric would allow a team to find problems most quickly?

    A. Variance analysis
    B. Technical debt
    C. Risk burnup chart
    D. Defect cycle time

4. As the coach of an agile team, you expect the team members to:

    A. Come to you whenever they encounter a problem
    B. Report all their problems in the daily stand-up meeting
    C. Solve most problems collectively as the work proceeds
    D. Figure out the best solution on their own

5. Ideally, who will catch and fix a coding error?

    A. The customer will spot it in the demo.
    B. The developers will find it during unit testing.
    C. The reviewer will catch it during pair programming.
    D. The testers will find it in testing.

6. Two team members are having a difference of opinion about how to build the next user story. What should be done?

    A. The team coach should assess the level of conflict and intervene appropriately.
    B. The ScrumMaster should decide the issue, since it is becoming an impediment to progress.
    C. The product owner should be consulted.
    D. The team should gather to discuss the issue and come up with a collective solution.

7. Ideally, what does your team want to see on the top line of your risk burndown graph?

    A. A steady, consistent upward trend
    B. A sharp upward trend as early as possible in the project
    C. A steady, consistent downward trend
    D. A sharp downward trend as early as possible in the project

8. As the lead tester on an XP team, you discover a problem. What should you do?

    A. Discuss the issue with the developer.
    B. Try to fix the problem yourself.
    C. Tell the customer.
    D. Alert the other coders to the problem.

9. An agile team is refactoring their code. Why are they doing this?

    A. To check the unit tests for errors
    B. To make sure the tests are ready before the code is written
    C. To make the code easier to update and maintain
    D. To get a consistent level of technical debt that will make it easier to forecast velocity

10. We put risk mitigation stories in the backlog to:

    A. Avoid having to keep a separate list of threats and issues.
    B. Keep the team focused on risks.
    C. Ensure that risk reduction efforts are done in the early iterations.
    D. Make sure the team doesn't forget to do something about the risks.

11. Which scenario is an example of treating a change as coming from common causes of variation, when it actually comes from a special cause?

    A. Gathering the team to investigate why their last retrospective ran 15 minutes longer than the 2-hour timebox
    B. Sending a mass e-mail to stakeholders to remind them not to interrupt the team members during an iteration
    C. Asking the product owner to delay the release deadline next month because the team won't have finished all the release scope by then
    D. Assuring the sponsor that the team will finish the release on time next month based on their historical average velocity, even though the senior developer just broke her leg and will be out for six weeks

12. As the team lead, you have been asked to explain to the business what the project team has accomplished during Iteration 0. Which of the following tools would you use for this presentation?

    A. Technical debt burnup chart
    B. Risk burndown graph
    C. Risk-adjusted backlog
    D. Average defects per release

13. As the team's agile coach, you measure small amounts of variance in task durations. What should you do?

    A. Undertake root cause analysis to eliminate it.
    B. Engage the team in diagnosing the problem.
    C. Diagnose the issue as part of your leadership role.
    D. Accept some variance as inevitable.

14. What is the advantage of using risk severity instead of expected monetary value (EMV) to rank risks?

    A. Risk severity helps us focus on the relative value of the risks.
    B. Risk severity is more accurate than EMV.
    C. EMV is less realistic than risk severity.
    D. EMV can't be ranked on a simple scale (low, medium, high).

15. What is one cause of technical debt?

    A. Trying to keep all the code perfectly neat and standardized
    B. Refactoring that hasn't been done on a regular basis
    C. Building a stripped-down minimal viable product
    D. Pair programming stress and burnout

16. What is the agile approach to problem solving?

    A. Fix the problem after it arises, at the last responsible moment.
    B. To keep velocity consistent, only fix problems that are posing impediments to progress.
    C. When a problem arises, either fix it on the spot or add it to the backlog.
    D. Capture impediments, problems, and lessons learned daily in the stand-up meetings.

17. As a team member, if you encounter a tricky problem during a development iteration, agile recommends that you:

    A. Stop what you're doing until you figure out a solution, using your individual expertise and ingenuity.
    B. Tell the ScrumMaster about the problem and let them decide what to do about it, since it's their job to remove impediments to progress.
    C. Just keep moving ahead so your velocity isn't disrupted, since most problems eventually take care of themselves.
    D. Quickly bring the problem to your team members and ask for their help in solving it, since many heads are better than one.

18. Stories and features are prioritized in the risk-adjusted backlog based on their:

    A. Risk impact or risk probability
    B. Risk mitigation impact or user impact
    C. Expected monetary value or business value
    D. Cost-benefit ratio or customer value

19. When a problem arises, what steps are needed?

    A. Remove the bad work and redo it; then look for other problems that might be related to that issue.
    B. Revise the plan since we have fallen behind.
    C. Diagnose the problem, decide what to do about it, and fix the problem (remove the bad work and redo it).
    D. Decide what to do and add the new work to the backlog.

20. What is the least likely reason why changes found later in the project are more costly to fix?

    A. More rework might be needed to fix the problem.
    B. More stakeholders might be affected by the problem.
    C. More code might have to be refactored.
    D. More features might have to be supported.

Answers

1. **Answer**: B

   Explanation: To assign a new risk to the right place in the backlog, we need to assess its probability and impact so we can calculate its expected monetary value. Although the team members might be able to assess some risks themselves, that option isn't offered here. Other risks might require the specialized knowledge of a subject matter expert, such as the customer—so this is the correct answer here. We wouldn't add the risk directly to the top of the backlog without analyzing its probability and impact. Also, in most cases it wouldn't be worthwhile to stop the project to evaluate its root cause. Although we could certainly schedule a risk-based spike, before doing that we would want to know the priority of the risk in comparison to the other work, based on its expected monetary value.

2. **Answer**: B

   Explanation: The best answer here is to create an environment in which people are encouraged to openly share their mistakes and problems with each other. This will not only allow problems to be solved more quickly, it will also lead to better solutions by drawing upon a wider range of viewpoints. Minimizing arguments and disagreements stifles the healthy debate that is necessary for finding the best solution to a problem. Asking senior team members to mentor their peers would probably be misinterpreted and lead to problems—also, on an agile team, mentoring is done by the team coach or ScrumMaster, not the team members (who have their own work to do). Posting scores on a leaderboard to encourage competition wouldn't be consistent with the agile principles of respect and team empowerment.

3. **Answer**: A

   Explanation: The phrasing of is question is a bit misleading since only one of these options (variance analysis) is a method for detecting problems. Because Option A is the only viable answer, the speed of finding problems isn't relevant for selecting the answer. Although our level of technical debt and defect cycle time may be useful metrics for diagnosing or analyzing problems, they don't help us identify the problems in the first place. Risk burnup chart is a made-up term.

4. **Answer**: C

   Explanation: Agile team members are expected to solve most of their technical problems collectively as the work proceeds. They don't try to figure out solutions on their own or bring their problems to their coach, since those approaches wouldn't draw upon the team's collective technical expertise and diverse viewpoints. They also don't report all their problems in the daily stand-up meeting; the issues mentioned in that meeting are those that the team members can't resolve themselves and that pose impediments to their further progress.

5. **Answer**: C

   Explanation: Although this question uses the word who, it is actually asking for the IDEAL way to catch and fix an error. Because of the cost of change curve, the sooner an error can be found, the better. So we are looking for the fastest way to identify an error, not who will do it—and the quickest way to find a coding error is to catch it in pair programming.

6. **Answer**: D

   Explanation: On an empowered agile team, it is up to the team members to resolve their technical disputes collectively. The coach, ScrumMaster, or product owner is unlikely to have the knowledge required to make such decisions.

7. **Answer**: D

   Explanation: A risk burndown graph is a stacked area chart that shows the project's cumulative risk severity. When the top line of this graph moves downward, it means that the project's cumulative risk is being reduced. So we ideally want to see a sharp downward trend as early as possible in the project, to show that the project risks are being quickly resolved. Minimizing risk early is one of the ways that agile teams maximize value, as reflected in the agile value proposition.

8. **Answer**: A

   Explanation: This question isn't really about XP teams—although it does require some understanding of team roles, it's really checking whether you understand the agile approach to problem solving. On an agile team (regardless of the methodology), problems are approached and resolved collectively, rather than individually. Telling the customer and alerting the other coders wouldn't help to solve the problem and might even stir up unnecessary trouble.

9. **Answer**: C

   Explanation: This question tests whether you know that refactoring is the process of streamlining and standardizing the code to make it easier to update and maintain. The unit tests don't need to be checked for errors since they are themselves error-checking tools. Agile software teams do write the tests before the code, but that isn't why they refactor their code; refactoring is simply the last step in that process. The remaining option that refers to technical debt is made-up nonsense.

10. **Answer**: C

    Explanation: Here we have a few reasonable choices to consider, but only one option is the main reason. Putting risks in the backlog does prevent us from having to maintain a separate list, but that is not the reason we do it. It also helps keep the team focused on the risk, but again, that is not the main reason we do it. Keeping the team from forgetting about the risk is a lot like the previous idea, and is also not the real reason we put risks into the backlog. Instead, the main reason we do that is to ensure that the risk mitigation work is done early in the project, to rapidly address any risks that can reduce value.

11. **Answer**: D

    Explanation: Although all the scenarios listed are mistakes in one way or another, only option D is an example of treating a change as if it came from common causes of variation, when it actually came from a special cause. In this scenario, the absence of the team's most experienced developer during the final project push will most certainly affect their velocity, even if someone else is brought on board to fill in. (The addition of a new person to the team, even temporarily, would be likely to send them back to the Storming stage.)

© 2015 RMC Publications, Inc • 952.846.4484 • info@rmcls.com • www.rmcls.com

12. **Answer**: B

    Explanation: Since Iteration 0 is concerned with establishing tools and environments and proving approaches, a risk burndown graph would be a good way to show what the team has been working on. While they may not have built any business functionality, hopefully they will have reduced some technical risks and proven some of the key approaches that will be used. The other tools are either made-up terms (technical debt burnup chart) or not relevant to the task at hand (risk-adjusted backlog, average defects per release).

13. **Answer**: D

    Explanation: The question states that the variance amounts are small. Since some amount of variance is inevitable, we should simply accept it. It would be inappropriate to undertake root cause analysis or take more time to try to diagnose this variation since it is due to common causes.

14. **Answer**: A

    Explanation: The advantage of using risk severity instead of EMV to rank risks is that risk severity helps us focus on the relative value of the project risks rather than their exact dollar amounts. The other options are incorrect statements.

15. **Answer**: B

    Explanation: If you know that technical debt is code that hasn't been refactored or standardized, then this answer should be clear. Building a minimal viable product shouldn't lead to more technical debt than any other kind of product design. Also, pair programming shouldn't lead to burnout because the pairs change roles regularly. (If you and your teammate just don't get along, the best approach would be to raise and address that problem before the quality of your work slips and starts to affect the project, rather than afterward.)

16. **Answer**: C

    Explanation: When a problem is found, we first try to fix it immediately (as we can do in pair programming or unit testing). If we can't fix it right away, the remediation effort is added to the backlog to be implemented as soon as possible. In regard to option A, agile teams prefer to prevent rather than fix problems—and once a problem arises, they try to address it quickly (not "at the last responsible moment") to keep technical debt to a minimum. The other two options are incorrect.

17. **Answer**: D

    Explanation: Agile teams rely on collective problem solving rather than individual ingenuity because problems are solved more quickly and effectively when diverse viewpoints are brought to bear, rather than when team members try to push through on their own. And although it is the ScrumMaster's role to remove impediments to progress, that refers to external roadblocks. When it comes to development issues, in many cases only the team members have the expertise needed to resolve the issue, so those kinds of problems can't be delegated to the ScrumMaster. Also, one thing we definitely don't want to do is to ignore a problem and hope it will go away; that's a surefire recipe for technical debt, if not project failure.

18. **Answer**: C

    Explanation: We prioritize risk stories based on their expected monetary value and feature stories based on their business value. Option A, "risk impact or risk probability" is partly correct, because that's how we calculate expected monetary value; however, that doesn't account for feature stories, and the question doesn't specify only risk stories. The other two options are made-up.

19. **Answer**: C

    Explanation: This question is a bit tricky, since there are certain correct aspects to all the answer options. However, the question asks for the steps needed in response to a problem, and option C provides the most complete list of actions to take. Option A misses the important steps of diagnosing the problem and deciding what to do about it. Option B is least likely to be correct, since agile plans are re-evaluated anyway in planning each iteration. Option D is correct as far as it goes, but often we might be able to fix the problem right away rather than adding it to the backlog.

20. **Answer**: D

    Explanation: The first three options are all classic reasons why the cost of change goes up over time. That leaves option D, "more features might have to be supported." Now, this statement could be interpreted as still another reason why changes would be more costly—but the need to support more features isn't necessarily related to the cost of change. So this is the option least likely to affect the cost of change, and therefore the correct answer.

© 2015 RMC Publications, Inc • 952.846.4484 • info@rmcls.com • www.rmcls.com

CHAPTER 7

Continuous Improvement

Domain VII Summary

This chapter discusses domain VII in the exam content outline, which is 9 percent of the exam, or about 11 exam questions. This domain focuses on continuous improvement in the areas of product, process, and people, including process analysis and tailoring, product feedback methods, reviews, and retrospectives.

Key Topics

- » Agile hybrid models
- » Approved iterations
- » Continuous improvement
- » Feedback methods
- » Fishbone Analysis
- » Five Whys
- » Kaizen
- » Learning cycle
- » PMI's Code of Ethics and Professional Conduct
- » Process analysis
  - – Anti-patterns
  - – Success criteria
  - – Success patterns
- » Process tailoring
  - – Risks
  - – Recommendations
- » Product feedback loop
- » Project pre-mortems
- » Retrospectives (intraspectives)
  - – Five-step process
  - – Three problem-solving steps
- » Reviews
- » Self-assessment tools and techniques
- » Systems thinking
- » Value stream mapping
  - – Nonvalue-added time
  - – Process cycle efficiency
  - – Total cycle time
  - – Value-added time

Tasks

1. Periodically review and tailor the process.
2. Improve team processes through retrospectives.
3. Seek product feedback via frequent demonstrations.
4. Create an environment for continued learning.
5. Use value stream analysis to improve processes.
6. Spread improvements to other groups in the organization.

This last chapter is about continuous improvement on agile projects, which the exam content outline defines as spanning the areas of process, product, and people. In this chapter we'll build on our discussion of problem solving in chapter 6 by describing the problem-solving process that takes place during retrospectives. We'll also cover additional practices that agile teams use to identify and make improvements during the project.

Most traditional projects capture the majority of their lessons learned at the end of the project. The intent is to allow the organization to apply these lessons to future projects that are in a similar business or technical domain or have similar team dynamics.

This approach, frankly, is too little, too late. We need to apply the benefits of learning as we go—on our current project, and as soon as possible. The immediate application of lessons learned is especially critical for projects in a quickly changing environment or those with a high degree of uncertainty and risk, where staying the course could be fatal to the project if circumstances change.

This is why agile methodologies incorporate continuous improvement activities into their iterative approach. The agile approach to lessons learned is deliberate and frequent, and it helps ensure that we regularly consider adaptation and improvement to the point where they become a habitual part of our normal way of working. As shown in the diagram below, the "Learn" step of inspecting, adapting, and improving is built into every agile iteration.

Figure 7.1: Capturing Lessons Learned in Each Iteration, While They Are Still Actionable

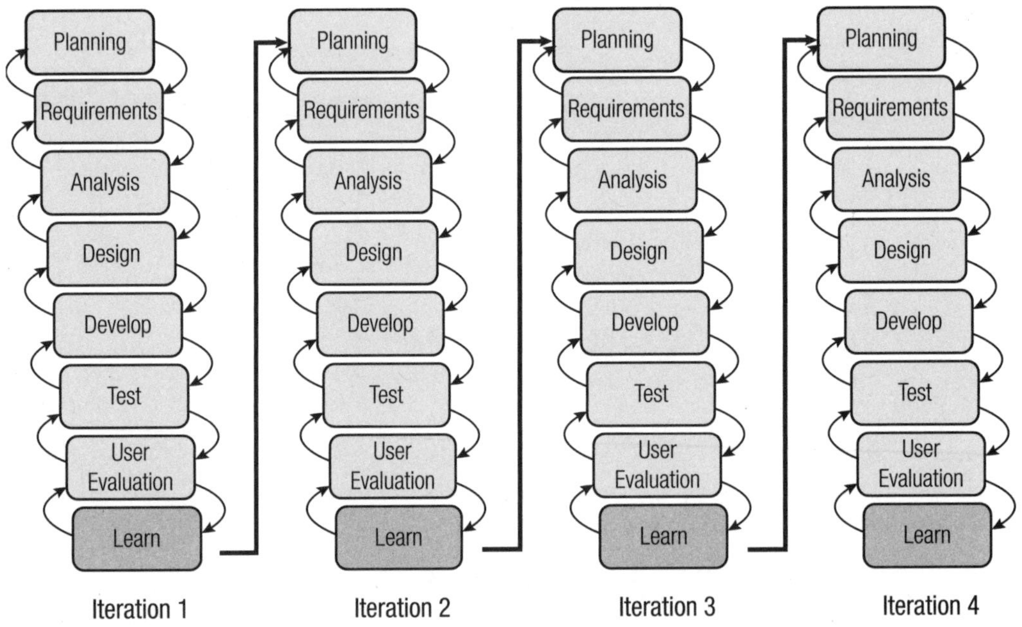

The "Learn" step shown here includes the team's iteration reviews and retrospectives, but isn't limited to them. This is an ongoing process of enhancing our process, product, and people that is never completed—it continues throughout the project, in much the same way as stakeholder communication. So continuous improvement is more like a journey than a destination, since it is an ongoing part of the iterative life cycle that drives agile methods.

In this chapter, we'll discuss the concepts, tools, and practices involved in this learning effort. We'll organize these topics into the three themes outlined in the exam content outline—continuous improvement of our process, product, and people—and then we'll conclude with a final section on PMI's Code of Ethics and Professional Conduct. However, the divisions between tools for improving process, product, and people are somewhat arbitrary since there isn't always a sharp dividing line between them. Some of these tools, such as retrospectives, could lead to improvements in any of these areas.

T&T Kaizen

Kaizen is a process for continuous improvement that is named after the Japanese word *kaizen*, which means "change for the better." The kaizen approach is the basis for agile's way of doing continuous improvement. (The software experts who originally developed agile didn't so much invent a new way of running projects as repurpose Japanese design and manufacturing ideas from the 1950s.)

Kaizen focuses on encouraging the team—the people who are doing the work—to frequently initiate and implement small, incremental improvements. This approach is based on a completely different mindset than the top-down, management-led re-engineering initiatives that are so common in the West, which are typically large, one-time efforts.

At Toyota, home of the lean production system, kaizen usually follows the Plan-Do-Check-Act (PDCA) cycle developed by W. Edwards Deming—which closely mirrors agile's Plan-Develop-Evaluate-Learn cycle:[1]

Figure 7.2: PDCA Cycle versus Agile Cycle

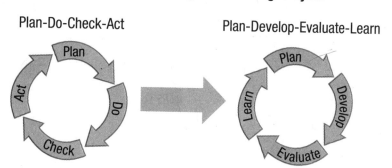

Since agile is based on lean, we would expect agile to have its own kaizen process, and it does—the team's retrospectives. It's true that agile retrospectives aren't exactly the same as kaizen, which looks for continuous daily improvements rather than learning after each iteration. However, retrospectives are the closest thing that agile has to kaizen, and the techniques used in these meetings include kaizen tools such as Five Whys and Fishbone Analysis (which will be explained later in this chapter). Nevertheless, kaizen isn't really a set of practices; it's basically an attitude or mindset that can help us understand why agile methods approach improvement as a continuous team-based, iterative process.

Agile Cycle versus PDCA Cycle

As explained above, agile projects follow a continuous cycle of Plan-Develop-Evaluate-Learn that is similar to Deming's Plan-Do-Check-Act cycle. We can see how this cycle fits into the agile life cycle in this diagram.

Figure 7.3: Plan-Develop-Evaluate-Learn within the Agile Life Cycle

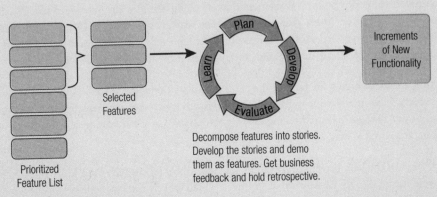

Prioritized Feature List

Selected Features

Decompose features into stories. Develop the stories and demo them as features. Get business feedback and hold retrospective.

Increments of New Functionality

The "Plan" and "Develop" steps of the agile cycle are equivalent to Deming's "Plan" and "Do" steps. In agile's "Evaluate" step, we do testing and get customer feedback, which parallels the "Check" step in Deming's cycle. In the "Learn" step, we reflect on our process and immediately apply the lessons we've learned; this is much like the "Act" step in Deming's cycle. The PDCA cycle is also mirrored on a smaller scale within agile demonstrations, reviews, and retrospectives.

The similarities between these two cycles are not coincidental. Lean is derived from the work of Deming, and agile is basically an instance of lean—so these parallels are simply revealing the common core of both approaches.

Multiple Levels of Improvement

Agile continuous improvement efforts are layered like an onion—they occur on multiple levels within the project. For example, on IT projects continuous improvement starts at the code level with pair programming. While one person writes the code, the other person reviews it, critiques it, and suggests improvements in real time. On a daily basis, the team members share any impediments they're facing in their stand-up meetings so that the ScrumMaster can quickly remove them. Finally, each iteration ends with a review, in which the team demonstrates their work to the customer, and a retrospective, in which the team members examine their process and decide on improvements to make in the next iteration.

Figure 7.4: Multiple Layers of Continuous Improvement on Agile Projects

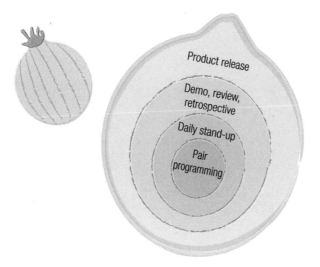

K&S **Continuous Improvement—Process**

Just as our plan will inevitably evolve as we learn more about the project, the environment, and the stakeholders, our approach to doing the work may also need to evolve. The main triggers for driving process changes on agile projects are the iteration retrospectives, which will be discussed later in this chapter. To evaluate our process and identify areas to improve, we can employ analytical tools such as systems thinking, process analysis, and value stream mapping. We can also use project pre-mortems to identify and mitigate potential failure points in our process. However, before we start making any changes to our process, we need to have a solid understanding of how process tailoring works and the best practices for implementing it.

T&T **Process Tailoring**

"Process tailoring" refers to adapting our implementation of agile to better fit our project environment. However, like do-it-yourself electrical work, it can be dangerous to tailor agile processes if we don't fully understand why things are the way they are in the first place. As a general recommendation, teams that are new to agile should use their methodology "out-of-the-box" for a few projects before attempting to change it. That's because the problems a new team encounters with a standard technique or practice may be due to their lack of skill or experience using that technique, rather than issues with the technique itself. By discarding or changing a practice before its value is recognized, we risk losing the benefit the practice was originally designed to bring to the project.

Another consideration is that all the techniques and practices in an agile methodology are designed to work in balance with each other. The balance and interrelationships of agile practices can be complicated, as illustrated by Kent Beck's view of XP practices shown below.

Figure 7.5: Relationships between XP Practices

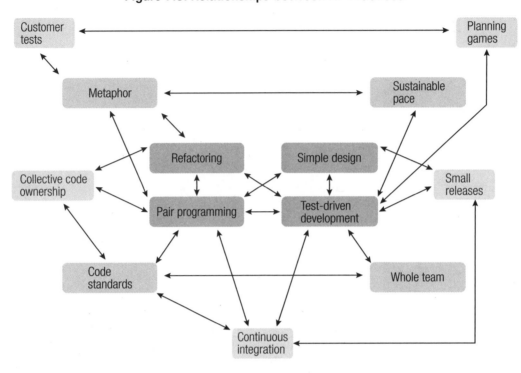

Removing or augmenting any of these elements without understanding the relationships among them can lead to problems. For example, ruthless testing allows for courageous refactoring, and having frequent user conversations allows the project to have light requirements. If you remove one practice without understanding its counterbalance, your project may be headed for trouble.

Having said that, some agile methodologies are quite tailoring-friendly. For example, in his book on Kanban, David Anderson notes:

> *Kanban is giving permission . . . to create a tailored process optimized to a specific context . . . You have permission to try Kanban. You have permission to modify your process. You have permission to be different. Your situation is unique and you deserve to develop a unique process definition tailored and optimized to your domain, your value stream, the risks that you manage, the skills of your team, and the demands of your customers.[2]*

In contrast, methods like Scrum are less keen on tailoring. Instead, the Scrum methodology makes recommendations about how to transform the enterprise into an agile environment that will better support a Scrum approach, since changing the methodology to suit the environment could potentially damage or weaken Scrum with nonstandard practices.

Clearly there are benefits and risks with both extremes. When considering the question of whether process tailoring is appropriate, we should first recognize that all projects are different—they solve different problems, face different challenges, include different people, and operate in different organizations with different cultures and norms. The unique nature of projects is why luminaries like Jim Highsmith and Alistair Cockburn recommend a process-per-project approach—in other words, creating processes that are situationally specific for the job at hand.

Others take a different view. Ron Jefferies asserts that "Agile ain't just any damn thing"—he aims to maintain an adherence to agile values and practices. His concern is that agile methodologies may be transformed beyond recognition—and then, when projects start to fail, those failures will be blamed on agile.

Mitigating the Risks of Process Tailoring

So we have two different perspectives on process tailoring, which both have valid points to consider. Where does this leave us? As I see it, the balance lies in mitigating the risks involved in tailoring a process. The risk of failure will be high when people who are inexperienced in agile modify the methods. A team that hasn't run through several projects using a by-the-book agile approach will not have firsthand experience of how the practices balance each other.

Personally I'm very happy to tailor processes; I view agile methods as useful tools, not as some sacred procedure that must not be touched. I look beyond the method to effective project delivery and happy sponsors, customers, and teams. Frankly, if calling it "spinach" instead of "agile" and wearing silly hats would help with a project's implementation and execution, then I would do it. However, we need to make sure we are changing things for good reasons, and not just for the sake of change.

Sometimes it's tempting to avoid a practice just because it is hard or is being met with resistance. This could be a sign that the practice is a poor fit for the organization, but it's equally likely that the practice is highlighting an underlying issue that needs to be resolved. For example, if an IT team is pushing back on two-week iterations because it takes them three days to build and release to a test environment, we shouldn't automatically lengthen the iterations to three or four weeks; instead, we should investigate why the build and release process is taking so long, and improve it.

Exam tip

For the exam, you should understand that process tailoring can be effective and productive, but that teams should be aware of the risks involved in this practice and mitigate these risks with the following best practices:

» Get used to normal, out-of-the-box agile before attempting to change it. Agile methods were created based on the collective wisdom of many experienced practitioners, so don't be too hasty to change them.

» Carefully examine the motivation to drop, amend, or append a practice. Is the change a cop-out to avoid a more fundamental problem, or will it truly address a gap or add value in this project's unique environment?

K&S Hybrid Models

One approach for customizing our process is to use elements from different models—to splice together processes from different methodologies to make a hybrid model. A hybrid is simply a combination of two different types of things—for example, a hybrid car might be powered by gasoline and battery, diesel and battery, or hydrogen and battery. Likewise, a hybrid agile model might be a combination of two (or more) agile methods, or an agile methodology and (heaven forbid!) a traditional approach.

Of course, I'm just joking about being shocked by mixing agile and traditional approaches—in some situations, that can make perfect sense. In fact, "The 2015 State of Scrum Report" noted that 63 percent of the respondents practiced Scrum alongside a traditional approach to project management.[3]

Let's look at examples of both an agile-agile and an agile-traditional hybrid model.

Agile-Agile Hybrid: Scrum-XP

One very popular hybrid—in fact, so popular that some people wouldn't even consider it a hybrid—is the use of both Scrum and XP on the same project. The reason this combination works so well is that these two approaches are quite complementary, since they each focus on different aspects of a project:

» XP provides great technical guidance but not much in the way of project governance guidance.
» Scrum provides a project governance model but not much in the way of how to do the work.

Because there is so little overlap between these two methodologies, this hybrid can offer well-formed, cohesive coverage for both managing the project and building the solution.

Figure 7.6: Scrum and XP Are Complementary

Scrum
Helps us organize and
manage the project

XP
Helps us build the
software or solution

Agile-Traditional Hybrids

Sometimes certain portions of a project are best suited to an agile approach and other parts to a traditional approach. For example, we can see that it would be helpful to use agile to validate understanding, iterate to find the best solution, drive out risks, and engage the team in a more respectful, rewarding way—but we see few benefits from using an iterative, incremental approach to build the solution. In a case like this, why use it?

Even if the plan for the execution of the project is linear rather than iterative, I think a team can always benefit from using certain components of agile—such as daily stand-ups, retrospectives, and local decision making. And on an agile project, it might be best to use a traditional approach for certain portions, such as a procurement effort that has to follow a strict purchasing workflow—we simply plan, specify, and order the equipment in good time, and then wait for it to arrive. In a similar way, for a software installation project we might start with ten trial installations, gathering our lessons learned incrementally, and then plan a traditional rollout of the system to the sixty remaining data centers. There are many different ways to combine agile and traditional methods, depending on the needs of the project.

Here's a more in-depth example. Years ago, I worked on a project for the British government to update how they calculated export credit insurance for companies trading with foreign entities. (If you are exporting an oil rig across the North Sea to Norway or building a hydroelectric dam in Africa, you can buy export credit insurance to protect that deal against nonpayment, loss of the asset mid-ocean, or swings in the currency rate.) The way countries calculate export credit insurance rates is regulated by the OECD (Organisation for Economic Co-operation and Development), which had just changed their rules. So we had a big specification document of nonnegotiable calculations from the OECD.

For this project, we used a hybrid approach. One team used agile methods to build all the underwriting screens and reports. Another, more traditional team converted the specification document into the executable code that would calculate the new rates. That made sense because that team wouldn't have much to demo while they were working on the calculations—and since the OECD specifications were fixed, even if someone didn't like the resulting numbers, we couldn't change them.

Once the calculation team was done, its results were added to the agile-developed underwriting system to complete the project. These two teams were coordinated under a single hybrid plan that contained the agile and traditional synchronization points as shown below:

Figure 7.7: Agile-Traditional Hybrid Example

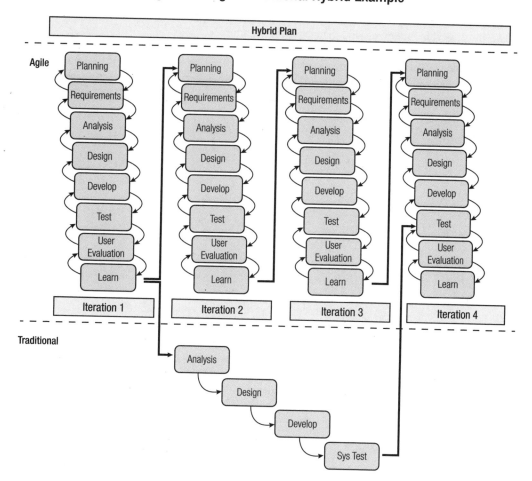

As it turned out, the project went well. The United Kingdom was the first country to implement the new OECD rules, and the large consulting company I was working for had the opportunity to repeat the project for multiple European governments.

Agile purists might argue that this could all have been done in an agile way—and yes, it probably could have. But I'm a pragmatist, not a purist; I value finding the lowest-risk route to project success over using the purest process. (You can also probably knit shoes as well as socks, but most people are happy to wear conventional shoes.)

This often seems to be the way—we might value, promote, and earnestly practice a pure way of working, but real projects have a way of throwing challenges at us. We can refuse to compromise our values, and not get the work—or we can develop a hybrid model that will allow us to complete the work as well as we can within awkward constraints. The latter method often works out just fine, and turns out to be better than many of the alternatives.

K&S Systems Thinking

When a team is considering changing their process, it can be helpful to understand the systems-level environment for the project. This type of analysis is called systems thinking. One part of the systems-thinking approach involves classifying projects in terms of their complexity (i.e., level of uncertainty) in two areas: the project requirements and the technological approach.

So far we have been saying that knowledge work projects are characterized by high levels of uncertainty—and compared to industrial work projects, this is true. But it turns out that, when you look at the full spectrum of possibilities, most knowledge work projects actually occupy the middle ground between total chaos and absolute certainty, in terms of both their requirements and their technology. This middle ground is where agile is most effective and useful, as shown below.

Figure 7.8: Different Levels of Complexity on Projects

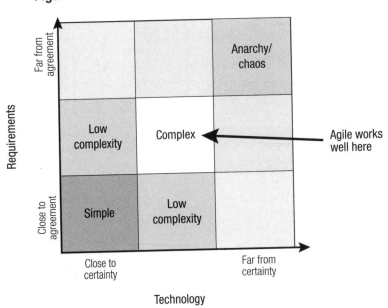

Agile works best in the middle cell of this matrix because these "complex" projects have some uncertainty around both requirements and technology, but not so much that they are chaotic or impossible to get our hands around. Of course, agile methods can also be used on simpler projects to reap the benefits of increased collaboration, communication, and visibility—but those kinds of projects can also be run just fine with a traditional approach. Complex projects, on the other hand, become a struggle if the team tries to use traditional methods.

So when we think about modifying our methods or tailoring our process, it's helpful to consider the project environment from this systems-level perspective. Agile is no silver bullet or panacea, however. If there is no clear agreement on what we are supposed to be building or what approach and tools we should be using, then our project will be in a state of chaos—and neither agile nor any other approach can be successful.

K&S **Process Analysis**

Process analysis is closely related to process tailoring and systems thinking. Process analysis involves reviewing and diagnosing issues with a team's agile methods (or, more often—their home-grown add-ons and replacements for agile methods). This analysis can help us reach a decision about whether to tailor the process.

Methodology Anti-Patterns

Alistair Cockburn provides the following list of anti-patterns (bad or unhelpful attributes) to watch out for in our methodologies:[4]

» **One size for all projects**: It isn't possible to create one optimal methodology for all types of projects, all technologies, and all team sizes. Therefore, be wary of claims of a one-size-fits-all approach.

» **Intolerant**: A methodology can be like a straightjacket, in the sense that it is a set of conventions and policies that we agree to follow and use. The size and shape of the straightjacket should be chosen by the team and should not be made any tighter than necessary, to give people a little wiggle room in their choices.

» **Heavy**: There is a common but incorrect belief that the heavier a methodology is (the more artifacts, procedures, and practices it prescribes), the safer it is. However, adding weight to a methodology is not likely to improve our chance of delivering the project successfully. Instead, it simply diverts our attention from the real goal of the project.

» **Embellished**: All methods tend to get embellished over time. Teams tend to add things that they think they "should" be doing—but that way of thinking usually just leads to potentially expensive, error-prone add-ons to the process. To find the embellishments in a process, gather together the people who are directly affected by the methodology, and ask them to review it—then watch their nonverbal signals and body language closely for signals of which things they don't do, but are afraid to admit they don't do. These are the embellishments.

» **Untried**: Many methodologies are untried. They are proposals created from nothing and are full-blown "shoulds" in action. For example, how often have you heard statements like, "Well, this really looks like it should work"? Instead of creating a complicated new theoretical methodology, it is better to reuse, adjust, tune, and create just what is needed. See what actually works on the project and use that, not something untried that someone believes "should work."

» **Used once**: A methodology that is used once is a little better than one that is untried, but it is still no recipe for success. The reality is that different projects need different approaches, and just because an approach worked under one set of circumstances doesn't guarantee it will work under another.

Success Criteria

If the above characteristics are things to watch out for, are there any signs that tell us we are doing something right? Yes—our approach is probably working well if our project meets these three success criteria:

» **The project got shipped**. The product went out the door.
» **The leadership remained intact**. They didn't get fired for what they were doing (or not doing).
» **The team would work the same way again**. They found the approach to be effective and enjoyable.

If the above success criteria seem too simple, consider how many projects fail to create high-quality products that satisfy stakeholders, are delivered on time, and keep both the leadership and the team happy.

Methodology Success Patterns

In addition to his methodology anti-patterns, Cockburn also lists seven patterns of successful methodologies:[5]

» **Interactive, face-to-face communication is the cheapest and fastest channel for exchanging information.** We should make face-to-face communication the default communication approach, and structure the team space and the project meetings to leverage this approach.

» **Excess methodology weight is costly.** Written documentation is slow to produce and takes time away from completing the project, so we want to minimize such documentation to a barely sufficient level.

» **Larger teams need heavier methodologies.** As team sizes grow, osmotic communication and tacit knowledge become harder to maintain, so more knowledge needs to be committed to written documentation.

» **Projects with greater criticality require greater ceremony.** As the penalty for errors in our product escalates, so too should the ceremony and care we take in developing it. For example, failure in developing a video game simply wastes the users' leisure time and may result in poor game sales. But failure in developing life-saving medical equipment may result in loss of life. As the criticality of a product increases, the rigor associated with developing it should also increase.

» **Feedback and communication reduce the need for intermediate deliverables.** We could write a long document to demonstrate that we understand the customer's requirements, but that would be an intermediate deliverable that takes a lot of time away from actually building the product. So instead, we build a prototype that shows we understand the requirements; since this is actually part of building the product, it moves the development work along.

» **Discipline, skills, and understanding counter process, formality, and documentation.** Jim Highsmith cautions us, "Don't confuse documentation for knowledge," since knowledge can be tacit (unwritten). We should also recognize that "Process is not discipline." Discipline is choosing to do something a certain way, while process is just following some instructions. And finally, "Don't confuse formality with skill." In terms of formality and skill, one does not relate to the other. We should be looking for smart people who can perform exploratory work and apply their skills and understanding, rather than people who simply follow a formal process or create documentation.

© 2015 RMC Publications, Inc • 952.846.4484 • info@rmcls.com • www.rmcls.com

» **Efficiency is expendable in nonbottleneck activities**. According to the Theory of Constraints, the improvements we make won't improve the project's total output unless they address a constraint in the system. For example, improving our requirements-gathering or analysis activities will not benefit the project if the bottleneck in our process is in coding or testing. Unless the coding or testing processes are improved, we will not see any increases in efficiency. So we need to look carefully for the constraints on our projects, and make our improvements there.

Exam tip

As explained in chapter 1, you don't need to memorize the above lists for the exam. Once you have absorbed the agile mindset, simply read through these lists a few times (preferably spaced out in time), and think about how each item makes sense and fits naturally into the agile way of doing things. That way, if you are asked about any of these ideas on the exam, you will be able to figure out the answer logically.

EXERCISE: PROCESS ANALYSIS KEY POINTS

See how many of the anti-patterns (items to watch out for), success criteria, and success patterns you can remember by listing them in the following table.

Anti-Patterns	Success Criteria	Success Patterns

Anti-Patterns	Success Criteria	Success Patterns

ANSWER

Anti-Patterns	Success Criteria	Success Patterns
One size for all projects	The project got shipped.	Face-to-face communications
Intolerant	The leadership remained intact.	Excess methodology weight is costly
Heavy	The team would work in the same way again.	Larger teams need heavier methodologies
Embellished		More ceremony for more critical projects
Untried		Increasing feedback and communications reduces the need for intermediate deliverables
Used once		Discipline, skills, and understanding over process, formality, and documentation
		Efficiency is expendable in nonbottleneck activities

T&T Value Stream Mapping

Value stream mapping is a lean manufacturing technique that has been adopted by agile methods. The goal of this technique is to optimize the flow of information or materials required to complete a process, thereby reducing the time it takes to create value and eliminating wasteful or unnecessary work. In value stream mapping, we create a visual map of a process flow, so that we can identify where delays, waste, and constraints are occurring. Once we identify the areas that could be improved in the process, we can then look for ways to remove those problems and make the process more efficient.

The value stream mapping process involves the following steps:

1. Identify the product or service to be analyzed.
2. Create a value stream map of the current process, identifying steps, queues, delays, and information flows.
3. Review the map to find delays, waste, and constraints.

4. Create a new value stream map of the desired future state of the process, optimized to remove or reduce delays, waste, and constraints.
5. Develop a roadmap for creating the optimized state.
6. Plan to revisit the process in the future to continually refine and optimize it.

Value stream mapping involves the concept of cycle time—as you may recall from chapter 6, this is a measure of how long it takes to get something done. For example, the cycle time for developing a user story begins when the team starts working on the story, and ends when that feature is accepted by the customer and available to deliver business value. However, to understand value stream mapping, we need to break down the concept of cycle time into more detail—there are four more terms you'll need to know:

» **Total cycle time**: In value stream mapping, cycle time is referred to as "total cycle time" because we are breaking cycle time into two parts—value-added time and nonvalue-added time. So, the total cycle time for a process is the sum of its value-added time and its nonvalue-added time.

» **Value-added time**: This is the time during the cycle where value is being added to the process.

» **Nonvalue-added time**: This is all the rest of the time in the cycle, where we will find the delays, waste, and constraints that we want to remove or reduce.

» **Process cycle efficiency**: This is the value-added time divided by the total cycle time. Once we have made improvements to the system, we can use this ratio to compare the efficiency of the new process compared to its previous state.

Value Stream Mapping Example

Next, we'll create a value stream map to show how this technique works. The first step in the process is identifying the activity we want to map—we'll use this scenario:

> *You're buying a cake so that you and a friend can have a little party to celebrate your passing the PMI-ACP exam.*

For our purposes, we'll say that this process involves choosing a cake, waiting at the bakery counter to get the cake, paying for the cake at the checkout register, and walking home with the cake. We then unpack and slice the cake before enjoying the benefit of the process—eating the cake!

In step 2, we begin mapping the value stream. First, we identify the starting point of the process (who initiates it) and the end point (who gets the end result).

Figure 7.9: Value Stream Map—Process Starting and End Points

Next, we identify the high-level steps, inventories, and queues in the process, focusing on the primary flow.

Figure 7.10: Value Stream Map—Primary Flow

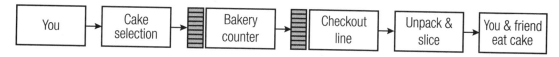

We then add any supporting groups (such as bakers and sales clerks) and alternative flows (such as selecting another cake if the bakery counter doesn't have the one you want).

Figure 7.11: Value Stream Map—Supporting Groups and Alternative Flows

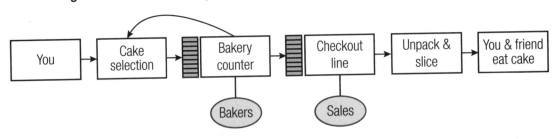

With our value stream map now complete, we add up all the value-added and nonvalue-added activities and calculate the efficiency of the process.

Figure 7.12: Value Stream Map—Calculations and Analysis

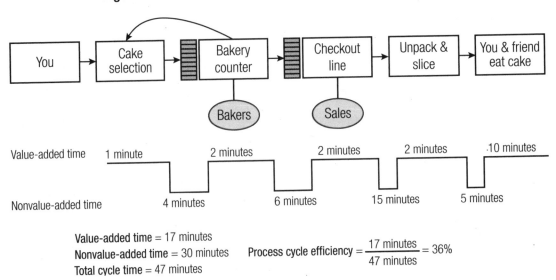

In step 3 of the process, we review the value stream map to identify waste, bottlenecks, and process inefficiencies. In this example, we can see that there are waiting delays at the bakery counter and the checkout counter. We can also see that there are motion inefficiencies, with lots of time spent between paying for the cake and being able to eat it, as we travel from the store back home.

The next steps are to remove the waste and create a new value stream map of the future optimized state, along with a roadmap of how to get to that state.

To design a more efficient process, we could consider phoning a specialty cake catering service and having the cake express-delivered to our house. In this new, more efficient process, let's say that our time waiting to place the order is reduced from 4 minutes to 2 minutes. Our 6-minute wait to pay for the cake is reduced to 1 minute, and the 15-minute walk home is replaced by a 10-minute delivery period.

Figure 7.13: Value Stream Map—Improved Process

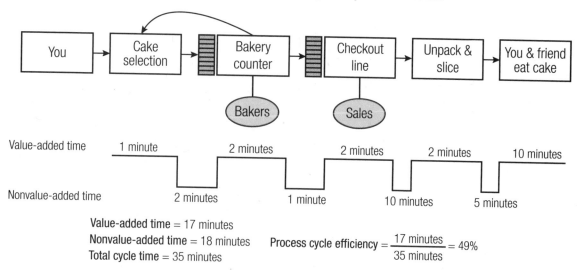

Value-added time = 17 minutes
Nonvalue-added time = 18 minutes
Total cycle time = 35 minutes

$$\text{Process cycle efficiency} = \frac{17 \text{ minutes}}{35 \text{ minutes}} = 49\%$$

With these changes, the value-added time remains at 17 minutes, but the nonvalue-added time is reduced from 30 minutes to 18 minutes, and total cycle time is reduced from 47 minutes to 35 minutes. This means that the process efficiency goes up from 36 percent to 49 percent. Of course, process efficiency is not the whole story—if the express delivery of the cake triples the cost, we might think twice about using it. However, this design does maximize the efficiency of the transaction.

In the final step, we complete the process by planning in the points where we will revisit the new process in order to keep refining it.

Exam tip

The exam is unlikely to ask you to create or analyze a detailed value stream map like this—we're using this example simply to explain the concept. However, you should understand what value stream mapping is and how an agile team might use it. Although you most likely won't be asked to do any calculations, the concepts of total cycle time, nonvalue-added time, value-added time, or process cycle efficiency might appear in one or two questions.

EXERCISE: PROCESS CYCLE EFFICIENCY—CAR WASH

Calculate the process cycle efficiency of waiting in line for 10 minutes to use a car wash for 5 minutes.

Total cycle = 15
PCE = 5/15 = 33%

ANSWER

To do this calculation, we first need to know what the total cycle time is: 10 minutes in line + 5 minutes washing the car = 15 minutes of total cycle time.

Now, let's calculate the process cycle efficiency, using the following formula:

Process cycle efficiency = Value-added time / Total cycle time

So process cycle efficiency = 5 / 15 = 33%.

Our perception of which aspects of a process add value affects our ability to evaluate a value stream. Test yourself with this next exercise to see how well you understand the concept of process cycle efficiency.

EXERCISE: PROCESS CYCLE EFFICIENCY—CUPCAKES

On Saturday morning, Tom and his young son Tim make cupcakes together. Although Tom does not like cupcakes, he values the 30 minutes with his son that it takes to make them. Tim thinks that making the cupcakes is a chore, but he really values the 5 minutes that he spends eating them. What are the process cycle efficiencies for Tom and Tim?

5/35 = Tom
30/35 = Tim

© 2015 RMC Publications, Inc • 952.846.4484 • info@rmcls.com • www.rmcls.com

ANSWER

This question involves two separate calculations of process cycle efficiency—one for Tom, who considers the baking process to be the value-added time, and one for Tim, who considers the eating process to be the value-added time. To work out the process cycle efficiency, we need to know both the total cycle time and the value-added time. The total cycle time is the same for both Tom and Tim: 30 + 5 = 35 minutes. The formula for calculating the process cycle efficiency is:

Process cycle efficiency = Value-added time / Total cycle time

For Tom, who values the baking time with his son, the process cycle efficiency is 30/35, or 86%.

For Tim, who values actually eating the cupcakes, the process cycle efficiency is 5/35, or 14%.

T&T Project Pre-Mortems

A project pre-mortem is a facilitated team exercise that aims to identify the possible failure points on a project *before* they happen, so that we can avoid or minimize those risks. (This technique originally came from Gary Klein, and was popularized by Daniel Kahneman's book *Thinking, Fast and Slow*.) Pre-mortems are especially valuable on long-running projects that are likely to experience more change than short projects, simply because they are exposed to a longer horizon of risk.

We can think of pre-mortems as a pessimistic version of the "Remember the Future" collaboration game described in chapter 3. As in that game, we ask the team to imagine they are in the future—but this time, instead of telling them that the project is a success, we tell them that it has failed—and their task is to tell us why that happened. After generating a list of potential failure points, the team looks for ways to adapt the project plan to avoid or mitigate these issues.

The participants in this failure analysis exercise include the development team and the product owner. The product owner's participation is key, because the team will be proposing risk avoidance or mitigation actions that the product owner will need to agree to add to the product backlog.

A pre-mortem exercise typically includes four working steps: imagine the failure, generate the reasons for the failure, consolidate the list, and revisit the plan. Let's examine each of these steps in more detail.

1. Imagine the Failure

The facilitator asks the participants to imagine that they are in the future after the project has failed completely, and they have been tasked to report the reasons for this failure.

Imagining what problems might befall the project might feel strange at first, since we are trained from childhood to have a "positive can-do attitude" and "look on the bright side." However, envisioning that the worst has already happened actually enhances our creativity. Painting a vivid picture of a future calamity frees the mind and gives us permission to explore negative thoughts rather than suppressing them.

2. Generate the Reasons for Failure

In this step, the participants work independently for 3 to 5 minutes, coming up with a list of reasons why the project has "failed." It's important that these ideas be generated independently to avoid the effects of cognitive and psychological bias (as discussed under wideband Delphi in chapter 5). Each person will have a unique perspective, concerns, and difficult memories of previous project challenges—we want to cast a broad net to capture all of this input without any self-filtering.

If the participants seem to be having trouble coming up with ideas, the facilitator can suggest scenarios to help them generate fresh insights, such as:

» The team members are no longer talking to each other.
» All funding for this and similar projects has been withdrawn.
» Stakeholders are questioning the project governance and decision making.

3. Consolidate the List

Once the lists are complete, the facilitator asks each person, in turn, to read out one item from their list and writes it on a whiteboard. This round-robin process continues until everyone has gone through their list. This approach works better than going through everyone's list in one pass because it keeps the group more engaged and helps prevent long monologues that could stall progress.

When the list of failure points is complete, the team prioritizes the list, usually by dot voting.

4. Revisit the Plan

In this step, the team reviews their plan to see what they might do to avoid or mitigate the issues they've identified. Recall that on agile projects, the plans consist of the release roadmaps and the backlog. So as the participants come up with risk mitigation ideas, they propose new candidate user stories to the product owner for approval. Once approved, the stories are added to the backlog.

If the team has generated a large number of issues, another meeting may be necessary to discuss them all. But if possible, I prefer to get through at least all the high priority issues before closing the session, to reduce the likelihood that after multiple meetings the group will develop a habit of negativity and start dwelling on potential problems.

Although pre-mortems are typically done near the start of a project, on a long project it's a good idea to repeat them over the course of the project—this might involve gathering the team to review the list every few months, to see if anything has changed. As the work proceeds, they might be able to identify new root causes of problems or propose better avoidance and mitigation strategies as candidate stories for the backlog.

K&S Continuous Improvement—Product

Just as we continuously refine our processes, so do we continuously improve the evolving product. We've said that agile methods rely on iterative and incremental development—and these are both forms of continuous improvement, in which customer feedback steers us toward the ultimate solution.

When we build in small increments and get feedback, the product evolves toward the true business requirements—and sometimes the true business requirements may be quite different from the originally stated requirements, as the creative process illuminates better options. By this cycle of developing in small increments, reviewing, discussing how to improve, and then doing some more development and maybe enhancing a few things, the product or solution is incrementally built through a process of continuous improvement.

Figure 7.14: Iterative and Incremental Product Improvement

The image above visualizes the progression of a product from the first sketch to the final outcome over five iterations, using the analogy of a superhero. In the first sketch, both of the superhero's arms are extended backward, but in their review the stakeholders decided to change the position of the right arm to show more movement. Subsequent iterations incrementally added texture, color, and detail to reach the final version. This is the same process used by agile teams to test approaches and adapt and iterate through increments to the final fit-for-business-purpose solution.

T&T Reviews

Although reviews are a key component of all agile methods and have been mentioned frequently in this book, we haven't really stopped to explain what they are, or why they are so important. So far we've mostly discussed iteration (or sprint) reviews, where the team demonstrates the product increment to the customer. But that is only one type of review; if we think of agile reviews in a larger context, some of the other topics in this chapter—such as product feedback and retrospectives—are also types of reviews. So to begin this section on product improvement, we'll take a moment to examine why agile teams do so many reviews, and how they fit into the overall agile process.

Exam tip

This high-level explanation of the role of reviews is intended to deepen your understanding of agile; it is unlikely to be tested on the exam. For the exam, simply make sure you understand the description of sprint reviews in the Scrum section of chapter 1, since the iteration reviews that are done in generic agile are essentially the same as sprint reviews, other than the name.

The Scientific Method

Agile methods and their lean cousins are based on an approach called the scientific method. Let's review this concept and then examine how it applies to agile.

The scientific method is a process for investigating things and learning new knowledge or correcting previous knowledge. This process involves making observations, thinking of a hypothesis to explain our observations, conducting experiments to test our hypothesis, and then confirming, adapting, or rejecting our hypothesis based on the data we have gathered. The results of this method are defendable theories that are backed up by repeatable experiments. A more detailed explanation of this process is shown below.

Figure 7.15: The Scientific Method

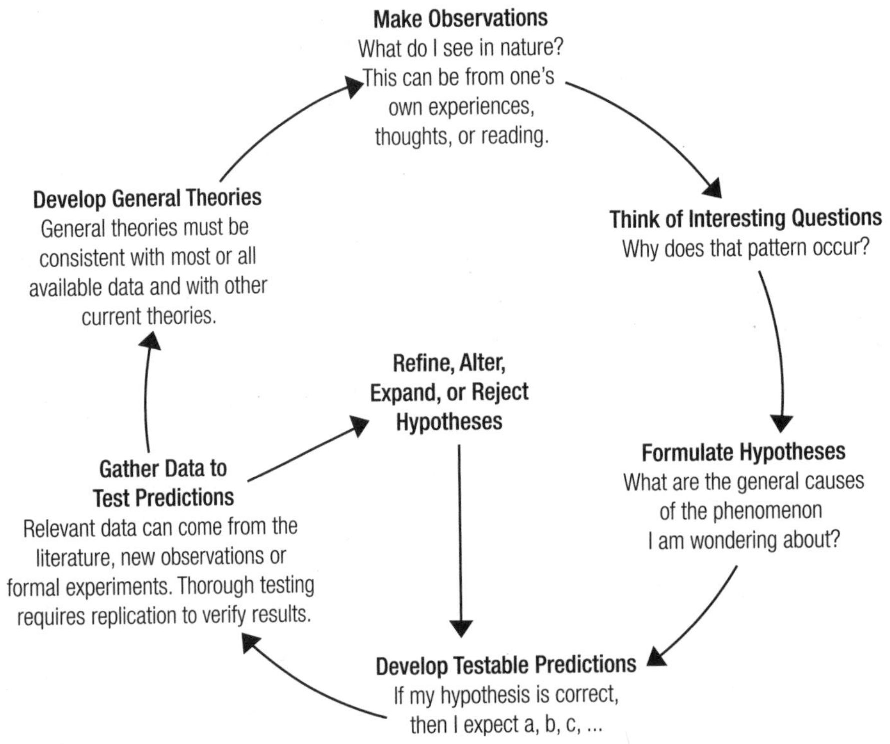

Make Observations
What do I see in nature? This can be from one's own experiences, thoughts, or reading.

Think of Interesting Questions
Why does that pattern occur?

Formulate Hypotheses
What are the general causes of the phenomenon I am wondering about?

Develop Testable Predictions
If my hypothesis is correct, then I expect a, b, c, ...

Gather Data to Test Predictions
Relevant data can come from the literature, new observations or formal experiments. Thorough testing requires replication to verify results.

Refine, Alter, Expand, or Reject Hypotheses

Develop General Theories
General theories must be consistent with most or all available data and with other current theories.

Lean and agile teams follow the scientific method by systematically and continuously making small experiments to improve their processes. Once we understand the scientific model, we can recognize that many agile practices are either an experiment (i.e., sprints and testing) or a review (i.e., product demos and retrospectives).

Deming's PDCA cycle—which we've said is the basis of both lean and agile processes—can be viewed as an experimentation framework that reflects the scientific method. In this cycle, we plan some work, do the work, check if the product is progressing correctly or if the process can be improved, and then act on these observations.

Figure 7.16: PDCA Cycle (Plan-Do-Check-Act)

Each iteration around this cycle can be viewed as a set of experiments: Can we make the product enhancements? How are our new process suggestions working? Where can we improve next? And since we are continuously running experiments, we also need to be continuously reviewing the results—that is what reviews are for.

Agile reviews follow some basic ground rules or guidelines:

» **Let the data speak for itself.** We don't try to prejudge the results of our experiments or filter out unlikely suggestions. Instead, we run the experiments that we agree are the most promising, and then examine the results.

» **Respect individuals.** We value everyone's view equally; we don't judge suggestions based on the person's role or seniority.

» **Diverge then converge.** We encourage diverse suggestions to increase the likelihood of generating valuable insights and ideas. To generate these ideas, we *diverge*—we generate our ideas individually so that they aren't influenced or inhibited by the views of our colleagues. Then we *converge*—we gather together to review the ideas, agree on the most likely root causes of issues, or identify the best suggestions for further experiments. This converging is usually done by collectively categorizing, ranking, and prioritizing the ideas that have emerged.

T&T Product Feedback Loops and Learning Cycles

You may recall the figure below from our discussion of frequent verification and validation in chapter 2. We can build on that earlier explanation now by adding that these cycles are also reviews that are based on feedback loops.

Figure 7.17: Frequent Verification and Validation on Agile Projects (aka Feedback Loops)

Each of these feedback loops—whether short and informal, or longer and more formal—reveals new information about the product or service we are building and the effectiveness of our knowledge and process. Agile teams need to continually try to confirm their understanding and the correctness of their work like this, since these projects are often unprecedented, and designs are largely invisible until the end product is completed. Product feedback loops help ensure that we are building the right thing in a robust and sustainable way. We continually ask ourselves, at increasingly more granular and significant levels, questions such as:

» "Does it meet the customer's needs and expectations?"
» "Does it work in all conditions?"
» "Did we break anything while building this?"

The answers to these questions will either solidify our understanding or prompt further analysis and change. Each review or product feedback loop is also a learning cycle where we ask questions such as:

» "How can we improve efficiency?"
» "How can we improve quality?"
» "How can we share the lessons learned with other groups?"

© 2015 RMC Publications, Inc • 952.846.4484 • info@rmcls.com • www.rmcls.com

Agile product feedback loops and learning cycles go hand in hand. Defects or issues that are found during a review lead to efforts to find a solution and build the right product—here a review issue leads to a learning cycle. However, positive feedback should also prompt learning. For example, if the business representative really, really likes the site fly-by video that we developed, maybe we should show it to other teams or the PMO, so that it can become a standard for other projects.

T&T **Feedback Methods**

Product feedback methods such as prototypes, simulations, and demonstrations of functionality are critical for success on agile projects. We've seen that knowledge work products are intangible and difficult to reference and that companies rarely build the same system twice. This means that the team is usually developing a product that, for the most part, is new to the users. Therefore, the customer needs to be able to look at something and try it out before they can confirm whether its functionality meets their needs. The term IKIWISI ("I'll Know It When I See It!") is the agile catchphrase for this process, reminding us that the customer's true requirements will only emerge once we demonstrate the product in a tangible way.

In addition to helping us clarify requirements, product feedback methods can uncover the need for new features. Often, it isn't until the customer actually sees the functionality that they will realize additional elements are required. For example, "The order entry screen looks great, but after trying it, we realize we also need a duplicate order function."

So when we demonstrate functionality to the product owner, it serves two purposes. First, we learn about any differences between what was asked for and what we interpreted and built (the gulf of evaluation). And second, we learn about any new or adjusted functionality that is required (IKIWISI).

In other words, requirements evolve with prototypes, simulations, and demonstrations. When users have a chance to evaluate and use the emerging product, it helps uncover the true business requirements. In the process, the solution we have built converges toward the emerging requirements, and the gulf of evaluation grows smaller, as illustrated below.

Figure 7.18: Requirements Evolve with Feedback over Iterations

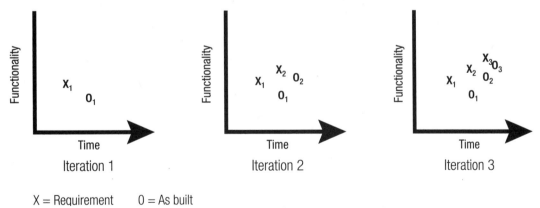

X = Requirement   0 = As built

Here, X_1, X_2, and X_3 are the emerging requirements, and O_1, O_2, and O_3 represent the product iterations that we have built and demonstrated to the customer. Over these three iterations, the ultimate solution takes form as the true requirements and what we are building move closer together.

T&T Approved Iterations

The term "approved iterations" in the exam content outline is related to iteration (or sprint) reviews. At the end of each iteration, the team holds a review meeting with the stakeholders to demonstrate the new increment built in the iteration. If the product owner is satisfied that the increment has met the iteration goal—the items selected from the backlog for that iteration—then they will "approve" that iteration.

When working in an outsourced environment, approved iterations can be used to control the release of incremental funding payments to a vendor. For example, "Since the work for this iteration has been approved, here's your payment for this portion of the project."

K&S Continuous Improvement—People

In this section, we'll get into the nitty-gritty of how agile teams evaluate themselves and identify areas for improvement by examining the retrospective process and team self-assessments. Of these two topics, agile retrospectives are much more important for the exam, since these are the primary events for learning, reflection, and readjustment on an agile project. Retrospectives are common to all agile methods, and serve as the main trigger for driving changes in all three areas—process, product, and people.

In fact, we could have covered retrospectives in either of the earlier sections of this chapter, since they can lead to improvements in process or product, as well as people. However, retrospectives are fundamentally about people, since they are meetings held for and by the team members. (Although these meetings are usually defined as being only for the development team, on some projects other key stakeholders are also invited to participate.)

T&T Retrospectives

A retrospective, or intraspective, is a specialized meeting that may be held after a release, or even the entire project. However, this term most often refers to the meeting that is held at the end of each iteration, after the iteration review. Regardless of which part of the project is being reviewed, this meeting is an opportunity for the members of the development team to inspect and improve their methods and teamwork. (Retrospectives can also be scheduled on a regular basis or when required on lean and Kanban projects that may not use iterations.) During the retrospective, we grapple with the following questions:

» What is going well?
» What areas could use improvement?
» What should we be doing differently?

As problems are identified, we brainstorm solutions, and then we commit to trying the selected solutions for one or two iterations before meeting again to discuss (in another retrospective) whether the situation has improved. If the changes helped—great! We adopt them as part of our processes on the project. If they did not help, we consider the effort a learning opportunity and decide whether to try something else or revert to our earlier process.

The Benefits of Retrospectives

Since retrospectives happen during the project, the lessons and improvements that result from them are highly applicable and pertinent to upcoming work; after all, that work will have the same business domain, technical domain, and team dynamics as the iteration we are assessing. In other words, the retrospective offers immediate value to the current project, rather than just documenting good advice in the hopes that a project with a similar business domain or team dynamics will come along.

I started out my career as a project manager using traditional project management methods. When I would read other project managers' lessons learned reports, I would be dismissive of the risks and problems they encountered. I would tell myself that I would never be so foolish as to fall into those traps or make such basic mistakes, so those problems couldn't happen on my projects. As a result, I never really took the lessons learned as seriously as I could have.

Now, however, when my agile team identifies issues, those issues are very real and applicable. This is our project, and this is my team reporting these problems right now—so I'd better help them create some solutions or get ready to experience the ongoing impact of those problems. Reviewing lessons learned throughout the project makes the issues and lessons very real and pressing. Like getting bad news sooner, this is actually a good thing, even if the advantages can be hard to see at the time.

Retrospectives offer a number of benefits for teams, including the following types of improvements:

» **Improved productivity**: By applying lessons learned and reducing rework, the team can get more productive work done.

» **Improved capability**: Retrospectives provide a venue for spreading scarce knowledge, and as the number of people who have the scarce knowledge increases, so does the number of people who can perform tasks associated with the knowledge.

» **Improved quality**: We can improve quality on our projects by finding the circumstances that have led to defects and removing the causes.

» **Improved capacity**: Retrospectives focus on finding process efficiency improvements, which can improve the team's capacity to do work.

The Retrospective Process

So what do we have to do to get these benefits? According to Esther Derby and Diana Larsen, the retrospective process includes five steps—an initial opening step, a final closing step, and three middle steps that outline a process for team-based problem solving, as shown in the next figure.

Figure 7.19: Retrospective Process

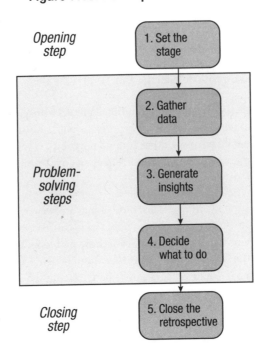

The duration of a retrospective should reflect the length of the iteration. For a 1- or 2-week iteration, a two-hour retrospective is a good fit; for a 3- or 4-week iteration, we may want a three-hour retrospective. To timebox the five steps, Derby and Larsen suggest breaking down the total time by percentages. For a two-week retrospective, they suggest that might look something like this:[6]

Figure 7.20: Retrospective Process with Typical Timings

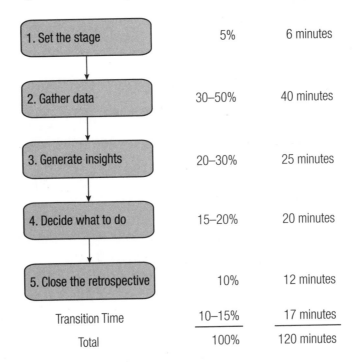

Step	Percentage	Time
1. Set the stage	5%	6 minutes
2. Gather data	30–50%	40 minutes
3. Generate insights	20–30%	25 minutes
4. Decide what to do	15–20%	20 minutes
5. Close the retrospective	10%	12 minutes
Transition Time	10–15%	17 minutes
Total	100%	120 minutes

The steps in the retrospective process are part of an ongoing cycle in which retrospectives alternate with iterations:

Figure 7.21: Retrospectives and Other Iteration Activities Feed into Each Other

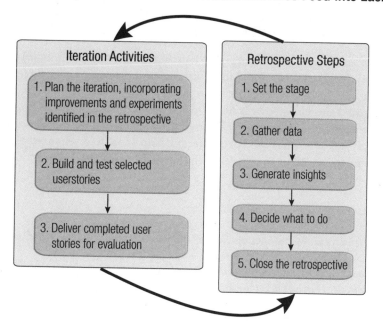

The cycle begins with the iteration activities (on the left)—then we complete the retrospective steps for that iteration (on the right). The outputs from that meeting feed back into planning the activities for the next iteration, as we execute the improvements and experiments agreed upon in the retrospective. In that iteration, we build and deliver a product increment, which generates another opportunity for review and adaptation—and so the cycle continues.

In the remainder of this section, we'll take a closer look at the five steps in the retrospective process.

Step 1: Set the Stage

At the start of the retrospective, we need to set the stage to help people focus on the task at hand of reflecting on how things went during the period that is being reviewed. We also want to prepare the participants for the next steps in the retrospective, gathering data and generating insights. In setting the stage, we aim to create an atmosphere where people feel comfortable speaking about things that may not have gone so well.

One technique we can use to encourage participation is to make sure we get people talking early in the retrospective process. The theory behind this approach is that if we allow people to remain silent at the start of the retrospective, we are establishing an understanding that it's okay to not participate. The idea that silence is acceptable is counterproductive, since the whole point of a retrospective is to get people talking about how things went and what they want to do in the future. One way to get people talking early is to ask them to introduce themselves in the first retrospective session. We can also ask participants to outline what they hope to get from the retrospective, or to say one or two words that describe how they felt about the iteration and the team's progress. The goal is to get them used to speaking in the group setting early in the process, and then encourage them to continue contributing throughout the meeting.

The next part of setting the stage is outlining the retrospective's approach and the topics for discussion. Providing such an outline establishes a clear purpose and agenda and prevents people from regarding this as "another aimless meeting." We also need to establish some team values and working agreements about how to run the retrospective—also known as "rule setting." With these rules or agreements, the team lays out what behavior is acceptable (e.g., talking about problem areas) and what is not acceptable (e.g., personal criticism or unsubstantiated complaints) during the retrospective.

The final element of setting the stage is to get people into the right mood for contributing information and ideas. Participants may be feeling uneasy about bringing up problematic issues, for fear of conflict. They might feel that criticisms about the process may reflect poorly on them, their peers, or management. However, the real goal of the retrospective is simply to find ways to improve. The fact that we've made mistakes before is of little consequence if we can move on and not make those mistakes again. Getting people to understand and believe this concept is easier said than done, however, so we need to gauge the participants' willingness to share and speak openly during the retrospective, and try to increase their comfort level.

The activities we can use to help set the stage include:

» **Check-In**: Participants answer a series of check-in questions with one or two words or a short phrase.

» **Focus On/Focus Off**: The team members discuss productive and unproductive ways of participating and agree to stay in the "Focus On" column.

» **ESVP**: Participants anonymously identify their attitude toward the retrospective as Explorer, Shopper, Vacationer, or Prisoner.

» **Working Agreements**: The team brainstorms and then defines the working agreements they would like to put in place for the retrospective.

To see how this step is done, let's take a closer look at each of these four activities.

Check-In

We can use this exercise to help people put aside their concerns and focus on the retrospective. In a round-robin format, we ask people to summarize in one or two words what they hope to get from the retrospective, the main thing on their mind, or how they are feeling about the retrospective.

Focus On/Focus Off

We can use this activity to establish a mindset for productive communication in the retrospective. For this exercise, we use the diagram shown below, which summarizes the most productive emphasis for a retrospective meeting.[7]

Figure 7.22: Focus On/Focus Off

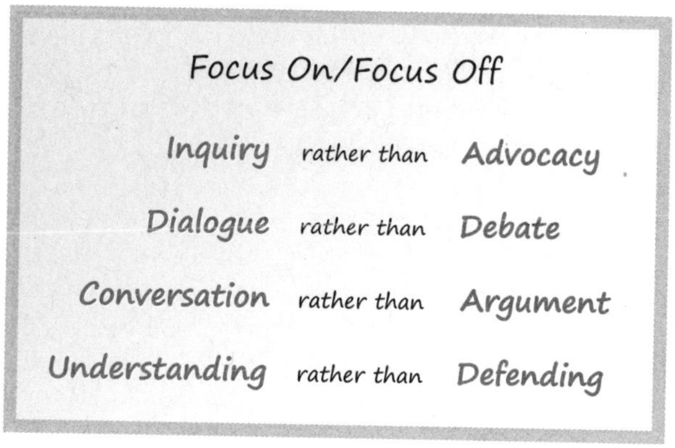

> Focus On/Focus Off
>
> Inquiry *rather than* Advocacy
> Dialogue *rather than* Debate
> Conversation *rather than* Argument
> Understanding *rather than* Defending

During the exercise, we ask participants to discuss what each of these pairs of words means to them, inviting them to provide examples. Then we ask people if they are willing to stay in the left "Focus On" column. If there is disagreement, we can ask them to discuss the impact of the behavior in the right-hand column, and then try for consensus again.

ESVP

In this exercise, participants anonymously identify their attitude toward the retrospective as one of the following roles, recording their choice on a slip of paper:

» **Explorers**: Explorers are eager to discover new ideas and insights, and they want to learn everything they can.

» **Shoppers**: Shoppers will look over all available information and will happily go home with one useful new idea.

» **Vacationers**: Vacationers aren't interested in the work of the retrospective, but they are happy to be away from their regular job.

» **Prisoners**: People who classify themselves as prisoners feel like they are being forced to attend the retrospective and would rather be doing something else.

The anonymous results are collected and tallied for the group to see, as shown below, to gauge the participants' level of energy and commitment.[8]

Figure 7.23: ESVP Exercise

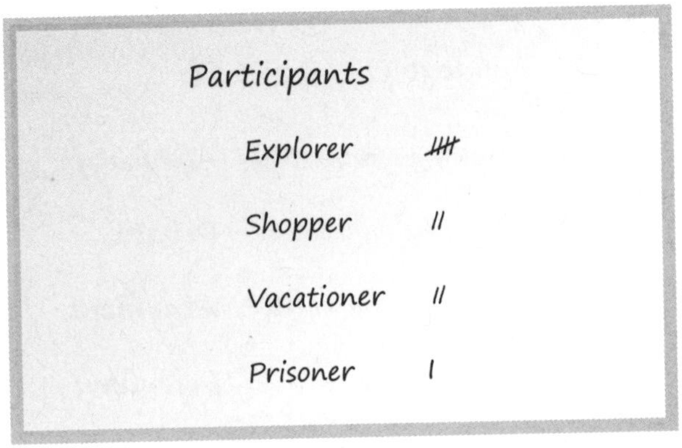

Participants

Explorer	ⅢⅠ
Shopper	Ⅱ
Vacationer	Ⅱ
Prisoner	Ⅰ

After the results have been tallied, the facilitator should conspicuously tear up and discard the slips of paper so no one will worry about their answer being traced back to them via handwriting analysis. Then the facilitator will ask the participants how they feel about those scores and what the scores mean for the retrospective.

Working Agreements

For this activity, the participants are formed into small groups and asked to develop candidate working agreements. The entire group reviews the suggestions and selects three to seven topics to develop further. Then each small group is given a topic to work on. Finally, the participants gather together again as one group and spend some time clarifying and refining the ideas they have generated, building a single master list to work from.

EXERCISE: SET THE STAGE

Test your understanding. What do you think the goals of the "set the stage" step in the retrospective process are?

ANSWER

In setting the stage, we aim to:

» Explain why we are doing the retrospective.
» Get people talking so they are comfortable contributing throughout the retrospective.
» Outline the approach and topics of the retrospective.
» Establish ground rules.
» Determine if people feel comfortable enough to contribute to the retrospective.

Step 2: Gather Data

In the data-gathering stage, we are trying to create a shared picture of what happened during the iteration. Without a common vision of what occurred, we would be simply speculating on what changes or improvements should be made—and we might actually end up addressing different issues or concerns, without realizing it. So in this step, we begin the problem-solving process by collecting and integrating the pieces of the puzzle we are trying to solve. When we are finished with this step, we should have a shared understanding of a comprehensive collection of observations, facts, and findings.

There are several team-based facilitation techniques that we can use to gather data, including:

» **Timeline:** The team members create a timeline to track the progression of the iteration.

» **Triple Nickels:** The team is divided into five groups that spend five minutes gathering or building on five ideas, five times.

» **Color Code Dots:** The team members identify where their energy was high and low during the course of the iteration.

» **Mad, Sad, Glad:** The participants identify their emotional reactions throughout the iteration.

» **Locate Strengths:** The team identifies what went well during the project.

» **Satisfaction Histogram:** The team members create a graph that shows how satisfied they feel about a particular area or issue.

» **Team Radar:** The team assesses how they have performed against their previous process improvement goals.

» **Like to Like:** The team compares their reactions to events that occurred over the course of the iteration.

Next, we'll examine the first two techniques (Timeline and Triple Nickels) more closely.

Timeline

The team can use the Timeline technique to either diagnose the origin and progression of a single problem or a number of problems. To start, the facilitator draws a timeline for the review period—which might be the iteration, or the timeline of the problem. The facilitator then asks team members to recall good, problematic, and significant events that occurred during the timeline.

Working individually at first, team members write events on colored sticky notes. The color of the notes indicates how they categorize the event—as good, problematic, or significant. When everyone has a collection of notes, they are invited to place them on the timeline and review what other people have posted.

By recalling the timeline like this, the team members get better insight into the events and what contributed to the issues. As a result, they may get clues about how to avoid similar problems in the future. The graphical representation of cause and effect and the opportunity to see other people's interpretations of the same events can help the participants recall additional details and build on each other's insights.

After the notes are posted, the team discusses the timeline from left to right, adding new notes as required. Below the timeline, they also record their emotional response to the events and draw a trend line to reflect those feelings. Expressing their emotions can help surface additional information that might be useful in generating insights (the next step of the retrospective process).

Figure 7.24: Timeline Exercise

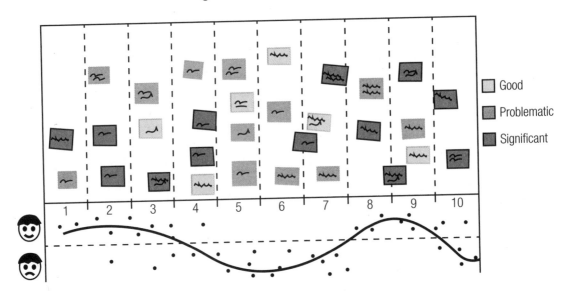

Triple Nickels

Triple Nickels is a data-gathering exercise in which participants spend five minutes gathering data on at least five ideas related to a specified topic. (This technique gets its name from a shooting competition that engages five targets from five yards in five seconds.) The team is divided into groups of five (or if the entire group is smaller than seven people, the exercise is done as one group), and the groups conduct five rounds of expansion.

This exercise asks people to work individually at first and think of at least five issues that occurred during an iteration. The participants then record the issues on sticky notes or index cards. After the first five minutes are up, everyone passes their cards to the person on their right. Then each person spends five minutes building on the ideas on the cards they have received. This process repeats five times so that everyone gets to contribute their own ideas and has an opportunity to think about and add to the ideas of the other group members.

The goal of this technique is to create an environment where participants have time to do both personal reflection and to expand on other people's ideas.

Figure 7.25: Triple Nickels

Step 3: Generate Insights

This part of the retrospective process allows us to evaluate the data we gathered in the previous step and derive meaningful insights from it. The goal of generating insights is to help us understand the implications of our findings and discussions. So this step involves collaborative exercises that are aimed at analyzing the data from the previous step and making sense out of it. These activities help us interpret and understand the implications of the problems we are faced with, before we move on to solving them.

There are several team-based activities that we can use to generate insights, including:

» **Brainstorming**: The team focuses on generating a high volume of ideas that will be filtered afterward. Common approaches include Free-for-All, Round-Robin, and Quiet Writing.

» **Five Whys**: The participants analyze the underlying cause of a problem by asking "Why?" five times to move beyond their automatic answers and identify the actual root cause of the issue.

» **Fishbone Analysis**: The team uses this diagramming tool—often along with the Five Whys exercise—to display their root cause analysis of a problem.

» **Prioritize with Dots**: To determine their priorities, the team members use the dot voting technique described in chapter 2.

» **Identify Themes**: The participants identify recurring patterns in the strengths they identified in Step 1.

Let's look at the first three techniques (Brainstorming, Five Whys, and Fishbone Analysis) in more detail.

Brainstorming

As explained in chapter 3, brainstorming exercises aim to generate a large number of ideas that are then filtered into a select list of ideas that will move forward in the process. The high volume of ideas generated during such exercises helps counter the common phenomenon that a team's best ideas rarely emerge first. There are many different approaches to brainstorming, including the three methods described in chapter 3—Quiet Writing, Round-Robin, and Free-for-All.

Once the ideas have been generated, they are filtered based on criteria that have been generated by the team. For example, if we are brainstorming ideas about problems with the availability of business representatives, we might decide to filter for items we are "able to influence" and items we are "unable to influence." We then review each idea in reference to these filters and decide whether to include or exclude the idea for the next step in the retrospective process, "Decide what to do."

T&T Five Whys

The aim of the Five Whys exercise is to discover the cause-and-effect relationships involved in a particular problem and get to the root causes of the problem. This technique originated with Toyota and is routinely used in the lean approach.

When using this exercise in a team setting, people work in pairs or small groups. Within these pairs or groups, they ask "Why?" five times to move beyond their automatic, habitual answers and to try to get to the root cause of the problem. Here is an example:

Question 1: Why did we get that system crash in the iteration demo?
Answer 1: We tried to access sales data for a store with no sales.

Question 2: Why does accessing a store with no sales cause a problem?
Answer 2: The fetch routine returns a null value that is not handled by the system.

Question 3: Why don't we catch null values and display a more meaningful error message?
Answer 3: We do catch them where we know about them, but this was the first time we had seen it for sales.

Question 4: Why aren't all query returns coded to handle nulls?
Answer 4: I don't know; it has never been a priority.

Question 5: Why is it not a priority, as it seems like it's really a weak link in the system?
Answer 5: Agreed. We should add it to the module walk-through checklist.

Here's another example—though this one should not apply to you!

Question 1: Why did you fail the PMI-ACP exam?
Answer 1: I don't know. I guess I didn't answer enough questions on the exam correctly.

Question 2: Why did you not answer enough questions correctly?
Answer 2: Every question has four answers to choose from, and you actually have to know what they are asking about.

Question 3: Why did you not know what they were asking about?
Answer 3: Well, I have been kind of doing agile for a while and thought that would be enough without really studying.

Question 4: What exam areas did you not study for or have experience in?
Answer 4: Well I know Scrum, but the XP, lean, and Kanban stuff was all Greek to me.

Question 5: Why did you not read up on these topics in this book?
Answer 5: Because I underestimated the breadth of the exam, and thought I could bluff my way through.

T&T Fishbone Analysis

A fishbone diagram is a visual tool that often accompanies the Five Whys exercise, since it provides a way to display the root cause analysis of problems. Using this technique, the team identifies factors that are causing or affecting the problem situation and looks for their likely causes.

The process starts with drawing an empty fishbone diagram and writing the problem at the head of the "fish." The next step is to identify the categories of contributing factors, which are also written on the diagram. The team can use categories that are related to the questions in the Five Whys exercise, or they can use a set of commonly used categories, such as:

» People, procedure, policies, place
» Systems, suppliers, skills, surroundings

Another option is to come up with a set of custom categories that make sense for the problem. Then the facilitator of the exercise asks the team members, "What are the [fill in category name] factors contributing to the problem?" The answers are then filled in as the "bones" on the fish.

Figure 7.26: Fishbone Diagram

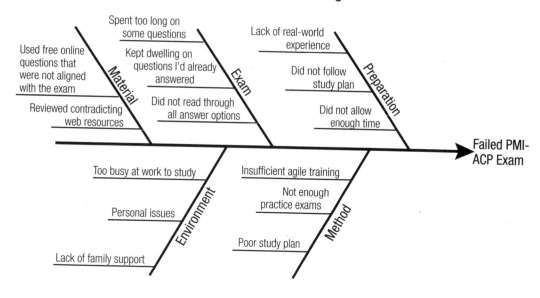

The above fishbone diagram investigates the factors that could contribute to the problem of failing the PMI-ACP exam. (This is, of course, a fictitious example, because it is not going to happen to you, right?)

Step 4: Decide What to Do

The last of the three problem-solving steps involves deciding what to do about the problems we have identified. The activities involved in this step move us from thinking about the iteration we have just completed into thinking about the next iteration—what we will change, and how we will behave differently. These team-based exercises focus on validating our approach and creating measurable goals to track progress and problem resolution. They help build a clear set of actions for solving the problem—we identify the highest-priority action items, create detailed plans for experiments, and set measurable goals to achieve the desired results.

There are several activities we can use to help decide on an action plan, including:

» **Short Subjects**: The team uses an action wheel with categories such as "Keep, Drop, Add" or "Start Doing, Stop Doing, Do More Of, Do Less Of" to identify problem-resolution actions.

» **SMART Goals**: The participants transform their list of action items into goals that are SMART: Specific, Measurable, Attainable, Relevant, and Timely.

» **Circle of Questions**: Each participant asks a question about how to improve one of the issues that has been identified, to be addressed by the next person in the circle.

» **Retrospective Planning Game**: In this exercise, the team plans the tasks required to reach the process improvement goals they have identified for the next iteration.

Let's look at the first two of these techniques (Short Subjects and SMART goals) in more detail.

Short Subjects

This activity helps the team agree on which problem-resolution actions to pursue. The group creates a circle on a flip chart or whiteboard and divides it into categories. For example, they might use:

» What Went Well, Do Differently Next Time
» Keep, Drop, Add
» Start Doing, Stop Doing, Do More Of, Do Less Of

The figure below shows an example of a Short Subjects session that was held for a software project. This session used the categories Start Doing, Stop Doing, Do More Of, and Do Less Of.

Figure 7.27: Short Subjects

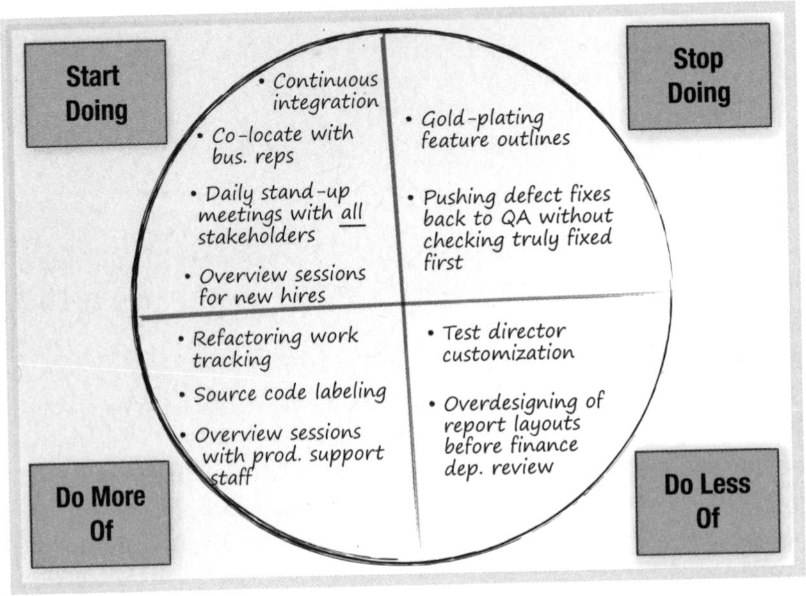

SMART Goals

This activity helps the team create goals that are Specific, Measurable, Attainable, Relevant, and Timely (SMART). The reason for doing this is that goals with these characteristics are more likely to be achieved. To begin the SMART goals exercise, the facilitator lists the SMART characteristics on a flip chart or whiteboard. Then the team is shown the difference between a non-SMART goal, such as *"We need to do more testing,"* and a SMART version of that goal, such as *"Each module must have and pass a unit test, functional test, and system test before iteration end."*

Figure 7.28: SMART Goals

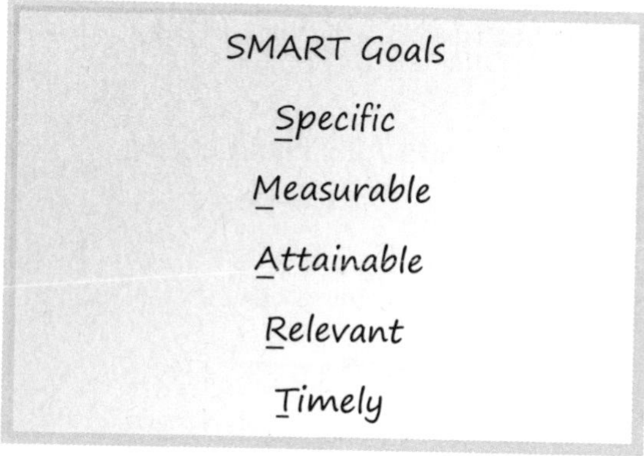

SMART Goals

<u>S</u>pecific

<u>M</u>easurable

<u>A</u>ttainable

<u>R</u>elevant

<u>T</u>imely

Once everyone in the room understands the characteristics of SMART goals, they break out into smaller groups to transform the problem-solving actions they have already identified into SMART goals. In the large group, the participants then review each goal—they discuss whether the goal is indeed SMART and make refinements where necessary. The process of ensuring that the goals have SMART characteristics also confirms everyone's understanding of what will happen and helps the participants form mental models of what will be done to complete the goals and resolve the problems.

Step 5: Close the Retrospective

The final step is closing the retrospective. Here, we have the opportunity to reflect on what happened during the retrospective and express our appreciation to each other. The exercises in this step might summarize what we have decided to keep or change, what we are thankful for, and where we can make the best use of our time going forward. These activities help round out the retrospective and reinforce its value to the project.

There are several team-based activities that can be used to close the retrospective, including:

» **Plus/Delta**: The team records what they want to do more of ("plus") and what they want to change ("delta") in two columns.

» **Helped, Hindered, Hypothesis**: The team members provide feedback about the retrospective itself— what helped, what hindered, and any ideas they came up with ("hypotheses") for improving future retrospectives.

» **Return on Time Invested (ROTI)**: Participants discuss the benefits of retrospectives, and then grade the meeting on a five-point scale to show whether their time was well spent.

» **Appreciations**: Team members have an opportunity to express their appreciation to each other for specific efforts during the iteration.

To illustrate the types of activities that can be undertaken to help close a retrospective, we will examine the first two in more detail—Plus/Delta and Helped, Hindered, Hypothesis.

Plus/Delta

In this exercise, we capture and validate our ideas for what we should do more of (things that are going well) and what we should change (things that are not going well) on a T-diagram on a whiteboard or flip chart, as shown in this example:[9]

Figure 7.29: Plus/Delta

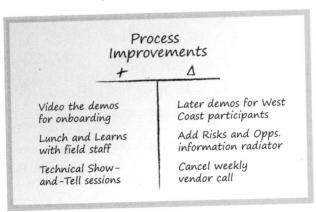

Helped, Hindered, Hypothesis

This exercise helps generate feedback on the retrospective process itself and produces ideas for improvement. To run this session, we first prepare three flip charts, and give them the titles "Helped," "Hindered," and "Hypothesis." We explain to the team that we are looking to improve the retrospective process and would like feedback on what they think helped, what was a hindrance, and any ideas they have (hypotheses) for improving the retrospectives going forward. The team members then write their ideas on sticky notes and post the notes on the appropriate flip chart.

This concludes our discussion of the five steps of the retrospective process. In summary, retrospectives are important agile workshops. They enable the team to take the impediments and problems they faced during the iteration, along with items identified after reflection and observation, and do something to improve the situation while those lessons and actions are still relevant to the project.

K&S Team Self-Assessments

As we've seen, it is standard practice for agile teams to reflect on how well they are doing and look for things they can improve—and another tool that they can use for this purpose is a team self-assessment.

✔ Exam tip

Although the agile term for these activities is "self-assessment," it's important to understand that they focus on evaluating the effectiveness of the team as a whole, not that of the individual team members. To make that clear, we are using the term "team self-assessment" in this discussion. However, if you see this term on the exam, it will simply be called "self-assessment."

Shore's Team Self-Assessment Scoring Model

James Shore offers a self-assessment quiz and scoring graph focused on XP practices that teams can use to gauge their performance.[10] This model measures how teams perform within the following categories:

» Thinking
» Collaborating
» Releasing
» Planning
» Developing

The quiz is completed by answering questions within each category and scoring the answers on a scale of 1 to 100. The following table shows a few of the questions from the Planning category:[11]

Planning Questions	Yes	No	XP Practice
Do nearly all team members understand what they are building, why they're building it, and what stakeholders consider success?	25	0	Vision
Does the team have a plan for achieving success?	4	0	Release Planning
Does the team regularly seek out new information and use it to improve its plan for success?	3	0	Release Planning
Does the team's plan incorporate the expertise of business people as well as programmers, and do nearly all involved agree the plan is achievable?	4	0	The Planning Game

Once the answers to all the questions have been scored, we plot the results on a radar (spider) diagram, as shown here:[12]

Figure 7.30: Team Self-Assessment Scoring Model

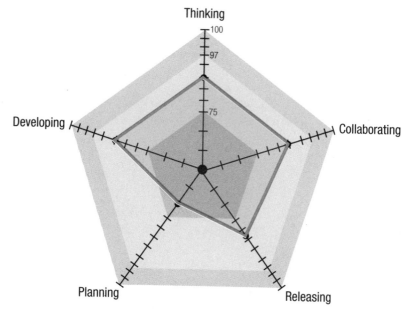

We then gather the team to analyze this chart to identify which areas could use improvement, and decide what actions to take. For example, the types of improvement actions that may be taken include training or adding more working sessions with the business representatives.

Tabaka's Team Self-Assessment Model

Jean Tabaka also offers a model for assessing the attributes of high-performing teams. This model investigates the following areas:[13]

» **Self-organization**: Is the team self-organizing, rather than functioning in a command-and-control, top-down organization?

» **Empowered to make decisions**: Is the team empowered to discuss, evaluate, and make decisions, rather than being dictated to by an outside authority?

» **Belief in vision and success**: Do team members understand the project vision and goals, and do they truly believe that, as a team, they can solve any problem to achieve those goals?

» **Committed team**: Are team members committed to succeed as a team, rather than being committed to individual success at any cost?

» **Trust each other**: Does the team have the confidence to continually work on improving their ability to act without fear, anger, or bullying?

» **Participatory decision making**: Is the team engaged in participatory decision making, rather than submitting to authoritarian decision making or the decisions of others?

» **Consensus-driven**: Are the team decisions consensus-driven, rather than leader-driven? Do team members share their opinions freely and participate in the final decision?

» **Constructive disagreement**: Is the team able to negotiate through a variety of alternatives and impacts surrounding a decision, and craft the one that provides the best outcome?

Figure 7.31: Tabaka's Self-Assessment Model

	High-Performance Teams' Collaboration Criteria	Team Score	Median	Team member assessment									
				1	2	3	4	5	6	7	8	9	10
1	Self-organization *Is the team self-organizing, rather than functioning in a command-and-control, top-down organization?*	5.0	5.0	5	5	5	5						
2	Empowered to make decisions *Is the team empowered to discuss, evaluate, and make decisions, rather than being dictated to by an outside authority?*	4.5	4.5	5	4	5	4						
3	Belief in vision and success *Do team members understand the project vision and goals, and do they truly believe that, as a team, they can solve any problem to achieve those goals?*	4.3	4.0	4	5	4	4						
4	Committed team *Are team members committed to succeed as a team, rather than being committed to individual success at any cost?*	4.8	5.0	5	5	4	5						
5	Trust each other *Does the team have the confidence to continually work on improving their ability to act without fear, anger, or bullying?*	3.5	3.5	4	4	3	3						
6	Participatory decision making *Is the team engaged in participatory decision making, rather than submitting to authoritarian decision making or the decisions of others?*	4.5	4.5	5	4	5	4						
7	Consensus-driven *Are team decisions consensus driven, rather than leader driven? Do team members share their opinions freely and participate in the final decision?*	2.3	2.0	3	2	2	2						
8	Constructive disagreement *Is the team able to negotiate through a variety of alternatives and impacts surrounding a decision, and craft the one that provides the best outcome?*	4.8	5.0	5	5	4	5						
	Total Score	4.2	4.2	4.5	4.3	4.0	4.0						

Legend: 1 = Strongly disagree, 3 = Neutral, 5 = Strongly agree

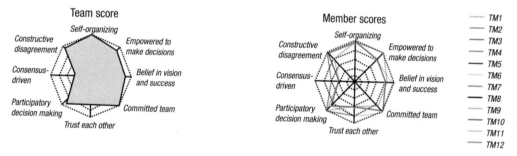

Image copyright © 2012 Edgardo Gonzalez, Projects Recovery Specialists, Ltd., www.prsl.ca

Exam tip

For the exam, you don't need to know the details of these self-assessment models. You might not even see this topic on the exam, since it isn't a key component of agile. However, if the exam does ask about self-assessments, it's likely to focus on how they are used and what their purpose is. You should also understand that these tools are part of the inspect and adapt cycle—they not only help the team improve their product and process, they also help team members build their skills and improve their effectiveness.

K&S **PMI's Code of Ethics and Professional Conduct**

The final topic in the exam content outline that we'll cover is PMI's Code of Ethics and Professional Conduct. While this isn't an agile topic, it is very much aligned with agile's core value of respect for other people. Also, PMI expects every applicant for the PMI-ACP credential to follow this code—you won't be able to complete the exam application process without agreeing to be bound by it. So let's take a look and see what you are getting yourself into.

PMI's Code of Ethics and Professional Conduct is a document that outlines four areas of professional behavior to conduct ourselves by—Responsibility, Respect, Fairness, and Honesty. Each area of the code has a set of aspirational standards and a set of mandatory standards. The aspirational standards describe the ideals PMI is asking practitioners to strive for, or aspire to. The mandatory standards are behaviors PMI expects all practitioners to follow. The structure of the Code is shown below, in a partially expanded mind map of the aspirational components of the Responsibility section.[14]

Figure 7.32: Structure of PMI's Code of Ethics and Professional Conduct

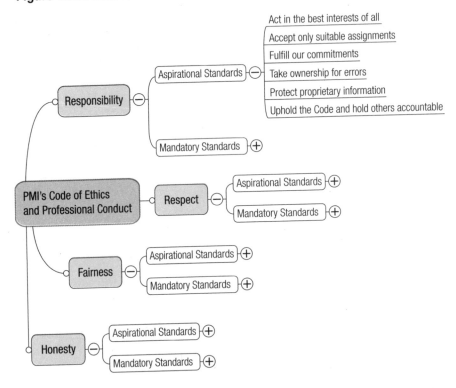

The other aspect of the Code that you should be aware of is the disciplinary and ethics compliance process. If someone believes that you are in violation of the mandatory standards, they are supposed to report you, and if you encounter another PMI member who you believe is in violation of the standards, you are supposed to report them.[15] The PMI Ethics Committee will then investigate any such reports.

Exam tip

PMI provides the full version of the Code of Ethics and Professional Conduct on its website (www.pmi.org). You should read this document and be familiar with it. Although the exam questions won't test the Code directly, you might see situational questions that test it indirectly. The main reason the Code is included in the exam content outline is to make sure you understand what you are agreeing to when you sign up for the exam. We encourage you to read the full version of this code on PMI's website.

Chapter Review

1. Process tailoring would be a good option for:

   A. A new team that has been formed by the merger of two other agile teams
   B. A team that is struggling to get their agile methods to work smoothly
   C. A team that is using agile successfully but would like to improve their methods
   D. A team that wants to start using agile methods

2. A series of project pre-mortems would be most useful on:

   A. A long-term, risky project
   B. A short project with a hard deadline
   C. A proposed project with well-defined technology and requirements
   D. A project that will require process tailoring

3. You hear that another team is using an agile-traditional hybrid approach. What is this most likely to involve?

   A. They are using agile methods to plan the project and traditional methods to track and measure performance.
   B. They are using agile methods for project accounting and traditional methods for the other elements of the project.
   C. They are using traditional methods to estimate the work and agile methods to plan the work.
   D. They are using agile methods to build the product and traditional methods for the procurement workflow.

4. In a retrospective, what technique could your team use to ensure that you really understand a problem you've identified?

   A. Five Whys
   B. Dot voting
   C. Kano analysis
   D. MoSCoW

5. If your total cycle time has remained the same but your value-added time has gone up, what does this mean?

   A. Your nonvalue-added time has increased.
   B. Your value-added time now equals your nonvalue-added time.
   C. The efficiency of your process has decreased.
   D. The efficiency of your process has improved.

6. The best way to improve the efficiency of a process is to:

   A. Do more detailed planning.
   B. Remove waste and bottlenecks.
   C. Minimize common cause variation.
   D. Add more stringent process controls.

7. Your team spends 10 minutes waiting for people to turn up for the daily stand-up meeting, which then lasts 15 minutes. The stand-up meeting is then followed by another 5 minutes of the team discussing the hockey game last night, and yet another 5 minutes brainstorming issues to solve. Calculate the total cycle time of this process, assuming that the hockey discussion is not categorized as a value-added activity.

    A. 15 minutes
    B. 20 minutes
    C. 30 minutes
    D. 35 minutes

8. When generating insights in a retrospective, we ask "why" five times to:

    A. Move beyond our automatic, habitual answers and get to the root cause of an issue.
    B. Get past the denial and resistance that tend to occur in the first three or four answers.
    C. Gather five levels of data around the issue.
    D. Avoid asking "why" six times, since that would be overkill.

9. Which of the following is a form of agile review?

    A. Sprints
    B. Customer demos
    C. Unit testing
    D. Project pre-mortems

10. In agile, the kaizen approach involves:

    A. Continuous, incremental improvements recommended by process experts
    B. Bold, top-down re-engineering initiatives
    C. Small, incremental improvements by the people who are doing the work
    D. Stopping the process as needed to implement essential improvements

11. You have been asked to review an agile team's recently enhanced methodology to assess its effectiveness. The desirable characteristics that you should be looking for include:

    A. A preference for rapid communications, significant process weight, recommendations for larger teams to use lighter methods
    B. A preference for digital communications, not too much process weight, recommendations for larger teams to use simpler methods
    C. A preference for relevant communications, significant process weight, recommendations for larger teams to use heavier methods
    D. A preference for face-to-face communications, not too much process weight, recommendations for larger teams to use heavier methods

12. Continuous improvement is a core benefit of which agile practices?

    A. Pair programming, daily stand-up meetings, WIP limits
    B. User stories, daily stand-up meetings, retrospectives
    C. Pair programming, daily stand-up meetings, retrospectives
    D. Story points, pair programming, WIP limits

13. Which of the following is not one of the questions that we are continually asking in a product feedback loop?

    A. Does it meet the customer needs and expectations?
    B. Does it work in all conditions?
    C. Did we break anything while building this?
    D. Did the project get shipped?

14. The primary reason for doing a project pre-mortem is to:

    A. Gather lessons learned from an iteration.
    B. Summarize the lessons learned at the end of a project.
    C. Identify potential risks so that we can mitigate them.
    D. Accurately forecast the results of the project.

15. On the chart used in the Timeline exercise for gathering data in a retrospective, a horizontal line can be added below the x-axis to represent:

    A. Team velocity over time
    B. Team emotional responses over time
    C. Team risk rankings average
    D. Team hours of work per week

16. Process tailoring is best undertaken on agile projects when:

    A. There are difficulties in implementing agile practices.
    B. Experienced practitioners want to address an issue.
    C. The team needs new processes to keep them engaged.
    D. A boost in team velocity is needed to meet the schedule.

17. Which of the following benefits would your agile team most likely gain from performing a self-assessment?

    A. Improve the team's practices.
    B. Gain insights for individual performance reviews.
    C. Identify personal traits for human resources counseling.
    D. Determine compatibilities for pair programming assignments.

18. When should agile teams collect lessons learned?

    A. At the end of the project
    B. Throughout the project
    C. When projects go well
    D. When projects go poorly

19. Your sponsor is asking about tailoring the company's newly adopted agile methodology. Your advice should be:

    A. Tailoring it will be a good way to learn more about the methodology.
    B. Tailoring it will be a good way to ease into the initial adoption process.
    C. We should tailor it first, then consider adopting it.
    D. We should try it first, then consider tailoring it.

20. Teams gain the most benefits from using an agile approach in:

    A. Projects that are as uncertain as possible

    B. Projects with high levels of technological uncertainty and low levels of complexity in requirements

    C. Projects with medium levels of complexity in both technology and requirements

    D. Projects with low levels of uncertainty in requirements and technology

Answers

1. **Answer**: C

   Explanation: Process tailoring should only be attempted after a team has mastered agile, and the best reason for doing it is to improve upon methods that are basically already working well. Process tailoring would not be a good idea in any of the other scenarios described here.

2. **Answer**: A

   Explanation: Some of the questions on the PMI-ACP exam will be relatively straightforward and can be figured out logically. Here is an example. To answer this question, ask yourself when a series of dedicated workshops for identifying potential risks would be most worthwhile. On a short, time-sensitive project we probably wouldn't have the time for such an effort. On a project with well-defined technology and requirements, the risks might not be significant enough to justify gathering the stakeholders for multiple pre-mortems. And "a project that will require process tailoring" is a distractor, since the decision whether to tailor the team's process is generally up to the team members, not a requirement for the project. Also, that description doesn't really tell us anything about the project concerned, which means that it can't be the BEST answer to the question. This process of elimination leaves a long-term, risky project as the correct answer.

3. **Answer**: D

   Explanation: The options for this question provide four possible scenarios for using an agile-traditional hybrid approach. Three of the options would be unlikely in the real world since they describe an awkward mixture that would be hard to use effectively. Only one of these scenarios would be a common type of real-world hybrid—using an agile approach for incremental development along with a traditional approach for procurement. With a good grasp of agile, this answer should stand out as the most likely approach. However, this question may still be tricky because you have to stop and think through the implications of each option.

4. **Answer**: A

   Explanation: The PMI-ACP exam is unlikely to pose detailed questions about the retrospective process; however, you should have a basic understanding of the four techniques listed in this question. Dot voting is a method for participatory decision making. Kano analysis is a tool that can be helpful for prioritizing product features from the user's perspective. MoSCoW is a commonly used prioritization method. That leaves Five Whys, which is a team exercise used to get to the root cause of a given problem—and therefore the correct answer to this question.

5. **Answer**: D

   Explanation: In value stream mapping, total cycle time is the sum of value-added time and nonvalue-added time, and the efficiency of a process is its value-added time divided by its total cycle time. Based on these formulas, if the value-added time of a process has increased while its total cycle time has remained the same, that means its efficiency has improved.

6. **Answer**: B

   Explanation: Although this question is framed in the terminology of value stream mapping, to identify the correct answer you really just need to understand the importance of removing waste and bottlenecks, which is a key principle of agile/lean approaches. While more detailed planning, reduced common cause variation, and more stringent process controls might also improve the efficiency of a process, they typically aren't the best or most cost-effective ways to do that on an agile project.

7. **Answer**: D

   Explanation: Since the total cycle time is the value-added time + the nonvalue-added time, we do not actually need to determine what falls under the value-added category. Instead, we just need to add up all the times involved: 10 + 15 + 5 + 5 = 35 minutes.

8. **Answer**: A

   Explanation: The goal of the Five Whys exercise is to move beyond our automatic, habitual answers and get to the root cause of an issue. The other options are made-up.

9. **Answer**: B

   Explanation: Agile methods adhere to the scientific method, in which experiments are followed by reviews. From this perspective, sprints, unit testing, and project pre-mortems would all be considered experiments in which the team is doing the work or gathering data. On the other hand, customer demos are a form of review, in which the team gets feedback on the work they completed in the sprint from the customer.

10. **Answer**: C

    Explanation: The kaizen approach focuses on encouraging the team (the people who are doing the work) to frequently initiate and implement small, incremental improvements. This approach is based on a completely different mindset than top-down, management-led re-engineering initiatives, which are typically large, one-time efforts. Although it wasn't mentioned earlier, in manufacturing systems that have a production line, kaizen does stop the process as needed to address issues. However, that isn't a key component of the kaizen approach as applied to knowledge work.

11. **Answer**: D

    Explanation: Desirable methodology characteristics include a preference for face-to-face communications, not too much process weight, and a recommendation that larger teams use heavier methods. The option of significant process weight is generally the opposite of the barely sufficient goal we should be striving for. However, as teams get larger, we will inevitably need to use heavier methodologies to compensate for the reduced face-to-face communications and the increased difficulty of maintaining tacit knowledge.

12. **Answer**: C

    Explanation: Continuous improvement is a core function of the practices of pair programming, daily stand-up meetings, and retrospectives. It has little to do with WIP limits, story points, or user stories.

13. **Answer**: D

    Explanation: The first three options are the questions that are asked in a product feedback loop, at increasingly more granular levels as the project proceeds. The last question is associated with the methodology success criteria for process analysis, not feedback loops.

14. **Answer**: C

    Explanation: The primary reason for doing a project pre-mortem is to identify potential risks to the project so that we can mitigate them. Although it does involve forecasting the results in a way, that is simply the framework used for identifying risks. The other options are incorrect.

15. **Answer**: B

    Explanation: Below the timeline chart, the team can draw a line that tracks their feelings about the iteration or project as it progressed. Tracking how they felt at each point in the process can help them recollect additional data.

16. **Answer**: B

    Explanation: Process tailoring is not to be undertaken just for implementation issues, to entertain the team, or as a scheme to increase velocity (although velocity might improve as a result of a process change, that isn't the "best" reason to do it). The correct answer is that it is best for experienced practitioners to undertake process tailoring when there is an issue to address.

17. **Answer**: A

    Explanation: The most likely benefit of performing a self-assessment is an improvement of the team's practices. While some of the other options might ensue as tangential benefits, they aren't the focus of a self-assessment exercise, which focuses on the team as a whole.

18. **Answer**: B

    Explanation: Lessons learned should be captured throughout the project when the information is still fresh and people remember the most details. This allows the lessons to be used in the remainder of the project.

19. **Answer**: D

    Explanation: Agile methods should be tried as-is first before considering modifications for process tailoring. We needs to first understand how the practices work before we attempt to change them. If we change the method first and then encounter problems, how will we know if the problems are genuine project issues or the result of the changes we made?

20. **Answer**: C

    Explanation: From the perspective of systems thinking, the projects that can gain the most from using an agile approach are those that have medium levels of complexity (i.e., uncertainty) in both technology and requirements.

Conclusion

Congratulations—you have finished this book and survived! Rest assured that you now have had great exposure to all the agile domains, tools and techniques, and knowledge and skills that will be tested on the exam. We have covered some topics more deeply than the exam will ask for and have discussed some ideas that aren't included in the exam content outline. This approach is designed to help you connect the ideas together so you can rely on understanding, rather than memorization, to pass the exam. Such understanding also reinforces the concepts so you can apply what you have learned to your real-world projects. When you know the material in this book, you will be more than ready to ace the exam.

The trouble is, if this is your first time reading through this material, chances are that you do not really know all the information covered in this book yet. I know it is a terrible thing to contemplate right now, since you have only just finished, but you need to go back through the book again. Don't worry—it will be much faster the second time. As you read through the book again, focus on the areas you struggled with. Take an iterative and incremental approach, and soon those four or five topics that you have been dreading will become one or two—and then you'll nail those, too.

A great attribute of agile is that it fits how people think and behave. It is tolerant of mistakes and incorporates feedback and refinement into the process. So practice what you have been learning. Do a retrospective of your studies, make sure you recognize the areas in which you did well and deserve praise, and also create a list of topics to revisit. You can also use the *PM FASTrack®* exam simulation software to help you identify where you still need improvement. Revisit topics, refine your understanding, and retest your knowledge.

And finally, good luck with the exam. If you have genuinely worked through these study materials and have the requisite training and project experience, I am confident you will pass the exam. Also, please share your thoughts and feedback—I refine, revisit, and retest, too. I would love to hear from you and can be reached at mikegriffiths@rmcls.com.

Best regards,

Mike

> Reminder! Purchasing this book gives you access to valuable supplemental study tools for the PMI-ACP exam at shop.rmcls.com/agileprep. Be sure to take advantage of those resources, which include a practice quiz that can help you determine if you are ready to take the exam. Also, be sure to check back from time to time, since we will continue to add more resources to that site. *You will need your copy of this book the first time you access that website.*

Endnotes

Introduction

1. Project Management Institute, "PMI Agile Certified Practitioner (PMI-ACP)®: Examination Content Outline," revised December 2014, http://www.pmi.org/~/media/PDF/Certifications/exam-outline/agile-certified-exam-outline.ashx.
2. Project Management Institute, "Reference Materials for the PMI® Agile Certified Practitioner (PMI-ACP)® Examination, accessed August 18, 2015, http://www.pmi.org/~/media/PDF/Certifications/ACP_Reference_list_v2.ashx.

Chapter 1

1. Mike Griffiths, "Reinventing PM for Knowledge Workers," accessed August 18, 2015, http://www.projectmanagement.com/articles/267966/Reinventing-Project-Management-for-Knowledge-Workers.
2. David Anderson, Sanjiv Augustine, Christopher Avery, Alistair Cockburn, Mike Cohn, Doug DeCarlo, Donna Fitzgerald, et al., "Declaration of Interdependence," accessed August 18, 2015, http://pmdoi.org/.
3. "Manifesto for Agile Software Development," accessed August 18, 2015, http://agilemanifesto.org.
4. "Principles behind the Agile Manifesto," accessed August 18, 2015, http://agilemanifesto.org/principles.html.
5. Jim Johnson, "The Cost of Big Requirements Up Front (BRUF)," keynote presentation at the annual XP (eXtreme Programming) Conference, Alghero, Sardinia, May 2002.
6. "A Brief History of Lean," Lean Enterprise Institute, accessed June 26, 2015, https://www.lean.org/WhatsLean/History.cfm.
7. Mary Poppendieck and Tom Poppendieck, *Lean Software Development: An Agile Toolkit* (Upper Saddle River, NJ: Pearson, 2003), 3.
8. Warren Bennis, *Managing People Is like Herding Cats: Warren Bennis on Leadership* (Provo, UT: Executive Excellence, 1999), 189.
9. James M. Kouzes and Barry Z. Posner, *The Leadership Challenge*, 4th ed. (San Francisco: Wiley, 2007), 33.
10. Jeffrey Pinto, *Project Leadership: from Theory to Practice* (Newtown Square, PA: Project Management Institute, 1998), 88.
11. Kouzes and Posner, *The Leadership Challenge*, 14.
12. Ibid., 28.
13. Matthew May, *The Elegant Solution: Toyota's Formula for Mastering Innovation* (New York: Free Press, 2011).

Chapter 2

1. "Manifesto for Agile Software Development," accessed August 18, 2015, http://agilemanifesto.org.
2. "Principles behind the Agile Manifesto," accessed August 18, 2015, http://agilemanifesto.org/principles.html.
3. Poppendieck and Poppendieck, *Lean Software Development*, 3.
4. "Discount Rate," Investopedia, accessed July 2, 2015, http://www.investopedia.com/terms/d/discountrate.asp.

5. Rita Mulcahy et al., *PMP® Exam Prep*, 8th ed. (Minnetonka, MN: RMC Publications, 2013), 120.

6. Project Management Institute, *A Guide to the Project Management Body of Knowledge (PMBOK® Guide)*, 5th ed. (Newtown Square, PA: Project Management Institute, 2013), 310.

7. Project Management Institute, *A Guide to the Project Management Body of Knowledge (PMBOK® Guide)*, 6th ed. (Newtown Square, PA: Project Management Institute, 2017), 395.

8. Mark Denne and Jane Cleland-Huang, *Software by Numbers: Low-Risk, High-Return Development* (Upper Saddle River, NJ: Prentice Hall, 2003).

9. Donald G. Reinertsen, *Managing the Design Factory* (New York: Simon & Schuster, 1997), 26.

10. Eliyahu M. Goldratt and Robert E. Fox, *The Race* (Great Barrington, MA: North River Press, 1986), 146.

11. Eliyahu M. Goldratt, *Theory of Constraints* (Great Barrington, MA: North River Press, 1990).

12. Jeff Sutherland, "Agile Contracts: Money for Nothing and Change for Free," last modified July 31, 2013, http://jeffsutherland.com/Agile2008MoneyforNothing.pdf.

13. L. Thorup and B. Jensen, "Collaborative Agile Contracts," *Agile 2009 Conference* (2009), 195–200, http://ieeexplore.ieee.org/xpl/freeabs_all.jsp?arnumber=5261083.

14. J. Fewell, "Marriott's Agile Turnaround," *Agile 2009 Conference* (2009), 219–222, http://ieeexplore.ieee.org/xpl/freeabs_all.jsp?arnumber=5261079.

15. The Framework for Integrated Testing (FIT) was developed by Ward Cunningham. For more information about FIT and FitNesse, see http://fitnesse.org and http://fit.c2.com/.

Chapter 3

1. Project Management Institute, *A Guide to the Project Management Body of Knowledge (PMBOK® Guide)*, 5th ed. (Newtown Square, PA: Project Management Institute, 2013), 30.

2. Ibid., 66.

3. James Shore and Shane Warden, *The Art of Agile Development* (Sebastopol, CA: O'Reilly Media, 2008), 156–57.

4. Visio and PowerPoint are trademarks of Microsoft Corporation in the United States and/or other countries.

5. Alistair Cockburn, *Agile Software Development: The Cooperative Game*, 2nd ed. (Upper Saddle River, NJ: Addison-Wesley, 2007), 125.

6. Project is a trademark of Microsoft Corporation in the United States and/or other countries. Primavera is a registered trademark of Oracle and/or its affiliates.

7. Kimiz Dalkir, *Knowledge Management in Theory and Practice*, 2nd ed. (Cambridge, MA: MIT Press, 2011), 169.

8. Robert D. Austin, *Measuring and Managing Performance in Organizations* (Dorset House, 1996), quoted in Mary Poppendieck, "Measure Up," *The Lean Mindset* (blog), January 6, 2003, http://www.leanessays.com/2003_01_01_archive.html.

9. Mary Poppendieck, "Measure Up," *The Lean Mindset* (blog), January 6, 2003, http://www.leanessays.com/2003_01_01_archive.html.

10. Steven L. Yaffee, "Benefits of Collaboration," Ecosystem Management Initiative, School of Natural Resources & Environment, University of Michigan, 2002, http://www.snre.umich.edu/ecomgt/lessons/stages/getting_started/Benefits_of_Collaboration.pdf.

11. Lyssa Adkins, *Coaching Agile Teams: A Companion for ScrumMasters, Agile Coaches, and Project Managers in Transition* (Upper Saddle River, NJ: Addison-Wesley, 2010), 236.

12. Jonah Lehrer, "Groupthink," *The New Yorker*, January 30, 2012, http://www.newyorker.com/magazine/2012/01/30/groupthink.

13. Daniel Goleman, Richard Boyatzis, and Annie McKee, "Primal Leadership: The Hidden Driver of Great Performance," *Harvard Business Review* 79, no. 11 (2001): 42–51.

14. Henry Kimsey-House, Karen Kimsey-House, Phillip Sandahl, and Laura Whitworth, *Co-Active Coaching: Changing Business, Transforming Lives*, 3rd ed. (Boston: Nicholas Brealey, 2011), 33–38.

15. Word is a trademark of Microsoft Corporation in the United States and/or other countries.

16. Speed B. Leas, *Moving Your Church through Conflict* (Herndon, VA: Alban Institute, 1985).

Chapter 4

1. Jon R. Katzenbach and Douglas K. Smith, *The Wisdom of Teams: Creating the High-Performance Organization* (New York: HarperBusiness, 2003), 45.

2. Carl E. Larson and Frank M. J. LaFasto, *Teamwork: What Must Go Right/What Can Go Wrong* (Newbury Park, CA: Sage Publications, 1989).

3. Adkins, *Coaching Agile Teams*, 26.

4. Ronald S. Friedman and Jens Förster, "The Effects of Promotion and Prevention Cues on Creativity," *Journal of Personality and Social Psychology* 81, no. 6 (2001): 1001–1013, http://www.socolab.de/content/files/Jens%20pubs/friedman_foerster2001.pdf.

5. David Rock, "SCARF: A Brain-Based Model for Collaborating with and Influencing Others," *NeuroLeadership Journal*, no. 1 (2008), http://www.scarf360.com/files/SCARF-NeuroleadershipArticle.pdf.

6. Patrick M. Lencioni, *The Five Dysfunctions of a Team: A Leadership Fable* (San Francisco: Jossey-Bass, 2002), 188–89.

7. Cockburn, *Agile Software Development*, 17 and 24.

8. Stuart E. Dreyfus, "The Five-Stage Model of Adult Skill Acquisition," *Bulletin of Science, Technology & Society* 24, no. 3 (2004): 177–181, http://www.bumc.bu.edu/facdev-medicine/files/2012/03/Dreyfus-skill-level.pdf.

9. Bruce Tuckman, "Developmental Sequences in Small Groups," *Psychological Bulletin* 63 (1965): 384–99.

10. "Situational Leadership® II Model" in Ken Blanchard, *Leading at a Higher Level, Revised and Expanded Edition: Blanchard on Leadership and Creating High Performing Organizations* (Upper Saddle River, NJ: FT Press, 2009), 77.

11. Based on Adkins, *Coaching Agile Teams*, 79.

12. Adkins, *Coaching Agile Teams*, 84–90.

13. Cockburn, *Agile Software Development*, 110.

14. Scrum Alliance, "The 2015 State of Scrum Report" (Indianapolis, IN: Scrum Alliance, July 2015), http://www.projectmanagement.com/pdf/StateofScrum2015BenchmarkStudy_FINAL.pdf.

15. Jim Highsmith, *Agile Project Management: Creating Innovative Products*, 2nd ed. (Upper Saddle River, NJ: Addison-Wesley, 2009), 304.

16. Ibid., 303.

17. Jean Tabaka, *Collaboration Explained: Facilitation Skills for Software Project Leaders* (Upper Saddle River, NJ: Addison-Wesley, 2006), 176–79.

18. Excel is a trademark of Microsoft Corporation in the United States and/or other countries.

Chapter 5

1. Alfred Korzybski, "A Non-Aristotelian System and Its Necessity for Rigour in Mathematics and Physics," paper presented at the American Mathematical Society at the meeting of the American Association for the Advancement of Science, New Orleans, December 28, 1931.

2. Project Management Institute, *PMBOK® Guide*.

3. Ibid., 560.

4. Ibid., 6.

5. Bill Wake, "INVEST in Good Stories, and SMART Tasks," accessed August 31, 2015, http://xp123.com/articles/invest-in-good-stories-and-smart-tasks/.

6. Mike Cohn, *User Stories Applied: For Agile Software Development* (Upper Saddle River, NJ: Addison-Wesley, 2004), 87.

7. Laurie Williams, Gabe Brown, Adam Meltzer, and Nachiappan Nagappan, "Scrum + Engineering Practices: Experiences of Three Microsoft Teams," *Proceedings of the 2011 International Symposium on Empirical Software Engineering and Measurement* (Washington, DC: IEEE Computer Society, 2011), 463–471, http://collaboration.csc.ncsu.edu/laurie/Papers/ESEM11_SCRUM_Experience_ CameraReady.pdf.

Chapter 6

1. Cockburn, *Agile Software Development*, 72.

2. Ibid., 91.

3. Ibid., 79–91.

4. W. Edwards Deming, *The New Economics for Industry, Government, Education*, 2nd ed. (Cambridge, MA: MIT Press, 2000), 99.

5. Ibid., 174.

6. Mary Poppendieck and Tom Poppendieck, *Leading Lean Software Development: Results Are Not the Point* (Upper Saddle River, NJ: Addison-Wesley, 2009), 14–15.

7. Project Management Institute, *PMBOK® Guide*, 309.

Chapter 7

1. Paul Arveson, "The Deming Cycle," Balanced Scorecard Institute, accessed August 31, 2015, http:// www.balancedscorecard.org/TheDemingCycle/tabid/112/Default.aspx.

2. David J. Anderson, *Kanban: Successful Evolutionary Change for Your Technology Business* (Sequim, WA: Blue Hole Press, 2010), 17.

3. Scrum Alliance, "The 2015 State of Scrum Report."

4. Cockburn, *Agile Software Development*, 175–79.

5. Ibid., 182.

6. Esther Derby and Diana Larsen, *Agile Retrospectives: Making Good Teams Great* (Dallas, TX: Pragmatic Bookshelf, 2006), 19.

7. Ibid., 43.

8. Ibid., 46.

9. Ibid., 111.

10. Shore and Warden, *The Art of Agile Development*, 67.

11. Ibid., 68.

12. Ibid., 67.

13. Tabaka, *Collaboration Explained*, 22.

14. Project Management Institute, "Project Management Institute Code of Ethics and Professional Conduct," accessed August 31, 2015, http://www.pmi.org/~/media/PDF/Ethics/ap_ pmicodeofethics.ashx, 1–5.

15. Ibid., 2–3.

INDEX

100-point method, 102

A

AC. *See* actual cost
acceptance test–driven development (ATDD), 135–37
active listening, 6, 180–81
actual cost (AC), 96
adaptation (Scrum pillar), 41, 43
adaptive leadership, 7, 213, 215–17
adaptive planning. *See* planning, adaptive
Adjourning. *See* team formation and development stages, Adjourning
Adkins, Lyssa, 9, 172, 206, 210, 221,
affinity estimating, 7, 280–81, 282, 283, 287, 296
agile charters, 6, 154–55
agile contracts, 6, 30, 123–27
    change for free, 125
    customized, 127
    DSDM, 124
    fixed-price, 125
    fixed-price work packages, 126–27
    graduated fixed-price, 126
    money for nothing, 125–26
    time and materials, 125, 126
agile discovery, 7, 253
agile frameworks and terminology, 1, 2, 6, 20, 22, 40–65, 198, 200, 303
agile hybrid models. *See* hybrid models
agile KPIs, 97–8
Agile Manifesto, 22, 27–37, 41, 48, 63, 84, 100, 110, 124, 151, 171, 198, 245, 271, 274, 322
    principles, 6, 31–37
    values, 6, 28–30
agile methodologies, 8, 40–65, 371–74
agile methods and approaches, 6, 8, 40–65, 371–74
agile mindset, 22–27, 71, 98, 120, 133, 245, 253, 377
agile modeling, 6, 158–59
agile planning concepts. *See* planning, concepts of
agile principles. *See* Agile Manifesto, principles
agile process overview, 64–65
agile project accounting principles, 6, 97
agile project chartering. *See* agile charters
agile scaling, 8, 65

agile sizing and estimation. *See* estimating *and* sizing
agile tooling, 6, 110–13
agile triangle, 26–27, 97, 123, 259,
agile value proposition, 84–85, 149–50, 244
agile values. *See* Agile Manifesto, values
agile *vs.* traditional projects
    chartering, 154
    estimating, 309
    planning, 246–48
analysis, value-based. *See* value-based analysis
Anderson, David, 9, 370
anti-patterns. *See* process analysis, methodology anti-patterns
Appreciations (exercise), 403
approved iterations, 7, 390
architectural spikes. *See* spikes, architectural
aspirational standards. *See* PMI's Code of Ethics and Professional Conduct
assessing and incorporating community and stakeholder values, 6, 151
assessing value, 86–100
ATDD. *See* acceptance test–driven development
Austin, Robert, 167

B

backlog. *See* product backlog
backlog, risk-adjusted. *See* risk-adjusted backlog
backlog, sprint. *See* sprint backlog
backlog refinement. *See* product backlog, grooming/refining
Bandwagon effect, 287
Bang-for-the-Buck. *See* collaboration games
Bennis, Warren, 66
big visible chart. *See* information radiators
Blanchard, Ken, 215–16
blank page syndrome, 329
bottlenecks, 59, 115–18, 120–22, 165, 203–5, 333, 339, 377
brainstorming, 6, 173–74, 176, 399
buffers, 124, 276, 301, 304
building agile teams, 7, 202–21
build tools, 132, 229
burn charts, 229–32
burndown charts, 7, 229–31
burnup charts, 7, 231–32
business case, 156
Buy a Feature. *See* collaboration games

C

© 2015 RMC Publications, Inc • 952.846.4484 • info@rmcls.com • www.rmcls.com

problem resolution. *See* problem solving

problems, 29, 32, 35, 37, 52, 71, 73, 74, 98, 99, 111, 115, 123, 128, 131, 133, 150, 158, 171, 173, 203, 208, 224, 225, 258, 293, 304, 310, 322–55, 369, 370, 378, 383–84, 390, 404

  understanding the impact of, 322–24

problem solving, 7, 29, 44, 171, 182, 204, 208, 220, 322, 327, 352–55

  as continuous improvement, 352

  problems that can't be solved, 355

  team engagement in, 352–55

problem-solving steps, 391–92, 397–403

process analysis, 7, 375–77

  methodology anti-patterns, 375–76

  methodology success criteria, 376

  methodology success patterns, 376–77

process cycle efficiency, 379–81

process tailoring, 7, 40, 211, 369–71, 375

processes *vs.* people. *See* people *vs.* processes

procurement. *See* agile contracts

product backlog, 42, 43, 44, 46, 47, 101, 200, 245, 273–76, 383,

  grooming/refining, 7, 44, 47, 253, 274–76

product feedback loops, 7, 388–89

product feedback methods, 389

productivity, 7, 34, 51, 59, 61, 66, 68, 165, 167, 186, 217, 219, 277, 334, 337, 391

product owner (Scrum role), 33, 41, 42, 43, 44, 46, 47, 49, 50, 64, 200, 296, 297, 303, 306, 310, 334, 383, 384, 389, 390

product roadmaps. *See also* story maps, 7, 284, 286–87, 296, 297

programmer (XP role), 49, 50, 51, 52, 130, 405

progressive elaboration, 7, 177, 245, 253–55, 258, 284, 304

project charters. *See* agile charters

project pre-mortems, 7, 383–84

project Tweets, 72, 155

project vision, 42, 43, 64, 70, 156, 161, 257, 296, 406

  communicating, 43, 69, 71, 72, 200, 207

prototypes, 151, 246, 376, 389,

Prune the Product Tree. *See* collaboration games

pull system. *See* Kanban, pull system

PV. *See* planned value

Q

quiet work period, 223, 329

Quiet Writing (exercise), 174, 399

R

ranking. *See* relative prioritization/ranking

rapid application development (RAD), 35

recognition, 35, 69, 216

Red, Green, Refactor (Red, Green, Clean), 134–35, 326

refactoring (XP core practice), 35, 41, 52, 55, 130, 134, 137, 157, 166, 233, 279, 306, 326–27, 370

regulatory compliance. *See* compliance

Reinertsen, Donald, 111

relative prioritization/ranking, 6, 106–7

relative sizing, 7, 276–79

release planning. *See* planning, release

releases, 49, 51, 60, 64, 108, 126, 157, 166, 200, 234, 245, 250, 252, 274, 279, 285, 286, 291, 292, 297, 298, 370

Remember the Future. *See* collaboration games

requirements, 29, 31, 32, 33, 36, 46, 49, 69, 70, 97, 98, 100, 101, 102, 104, 107, 123, 129, 135, 136, 151, 152, 154, 160, 166, 174, 177, 182, 204, 246, 250, 252, 253, 256–58, 261, 265, 266–71, 274, 276, 297, 303, 325, 332, 345, 348, 370, 374–75, 376, 377, 385, 389

requirements decomposition, 266–67

requirements hierarchy, 266, 267,

requirements prioritization model,

requirements reviews, 7, 276

respect (XP core value), 41, 48, 151, 387, 408

Retrospective Planning Game (exercise), 401

retrospective process, 352, 390, 391–404

  step 1 (set the stage), 393–96

  step 2 (gather data), 397–98

  step 3 (generate insights), 399–401

  step 4 (decide what to do), 401–403

  step 5 (close the retrospective), 403–4

retrospectives, 64, 73, 151, 169, 174, 225, 332, 345, 368–69, 390–404

  iteration, 207, 220, 256, 390

  sprint, 42, 46

return on investment (ROI), 22, 55, 86, 87–88, 97, 108, 109, 156, 249, 345
Return on Time Invested (ROTI; exercise), 403
reviews, 7, 22, 43, 56, 62, 65, 128, 129, 198, 200, 220, 245, 276, 325, 330, 352, 366, 368, 385–87, 388, 390
rewards, 69, 153, 167, 208, 328, 330
risk, 36, 52, 84–85, 86, 97, 98–100, 104, 115, 126, 151, 177, 178, 199, 244, 246–49, 274, 279, 293–94, 296, 344–52, 366, 383, 384
risk-adjusted backlog, 7, 100, 257, 344–49
risk-based spikes. *See* spikes, risk-based
risk burndown graphs, 7, 100, 351–52
risk management, 98–100, 322, 344–52
risk probability and impact, 346, 349
risk register, 349
risk severity, 349–51
roadmap, product. *See* product roadmaps
ROI. *See* return on investment
rolling wave planning, 255
Round-Robin (exercise), 174, 399
rules for meetings. *See* facilitation methods

S

safe and open environment, 73–74, 208, 210–11, 327
Sailboat. *See* collaboration games
Satisfaction Histogram (exercise), 397
scaling practices. *See* agile scaling
schedule performance index (SPI), 93, 96
schedule variance (SV), 93, 96
screen designs, 64, 158, 159
Scrum, 8, 22, 40, 41–47, 48, 51, 54, 59, 101, 150, 151, 198, 212, 225, 274, 275, 303, 334, 370, 372, 386
    activities (events, ceremonies), 43–46
    artifacts, 46–47
    pillars, 41
    process, 42
    team roles in, 43, 49, 50, 200
    values, 41, 151
ScrumMaster (Scrum role), 8, 43, 49, 50, 68, 69, 72, 150, 165, 184, 198, 200, 220, 290, 297, 304, 306, 310, 332, 368
scrum of scrums, 44–45, 205
S-curves, 92–94, 96

self-assessments, 404–7
self-directing teams, 207
self-organizing teams, 32, 36, 43, 198, 206–7, 406–7
servant leadership, 6, 43, 68–70, 71, 72, 148, 171, 200, 207
    primary duties, 68–69
set the stage. *See* retrospective process
seven wastes of lean. *See* waste
Shore, James, 157, 405
Short Subjects (exercise), 401, 402
Shu-Ha-Ri model. *See* developmental mastery models, Shu-Ha-Ri
simple design (XP core practice), 41, 49, 52, 166, 370
simple prioritization scheme, 102, 174
simple voting. *See* voting, simple
simplicity (XP core value), 32, 36, 48
simulations. *See* demonstrations
situational leadership. *See* adaptive leadership
situationally specific, 23, 63, 370
sizing, 7, 263, 265, 276–78, 281–84, 287, 296, 297
small releases (XP core practice), 49, 51, 166, 370
SMART goals, 401, 402–3
Smith, Douglas, 202
social media. *See* communication, social media
solving problems. *See* problem solving
source code control system, 132
special cause variation, 338
Speedboat. *See* collaboration games
SPI. *See* schedule performance index
spikes, 49, 293–94
    architectural, 7, 49, 293, 295
    risk-based, 7, 293–94
sprint backlog, 42, 44, 47, 274
sprint goal, 42, 44, 164
sprint planning. *See* planning, sprint
sprint retrospective, 42, 43, 46
sprint review, 41, 42, 43, 46, 150, 385, 386, 390
sprints, 8, 22, 41, 42, 43, 46, 51, 158, 204, 259, 292, 303, 308, 327, 334, 387
stages of team formation and development. *See* team formation and development stages
stakeholder engagement, 6, 148–88, 244
stakeholder management. *See also* stakeholder stewardship, 6, 148,

© 2015 RMC Publications, Inc • 952.846.4484 • info@rmcls.com • www.rmcls.com